Preparing for Integrated Circuit (IC) Design Technical Interviews

IEEE Press
445 Hoes Lane
Piscataway, NJ 08854

Preparing for Integrated Circuit (IC) Design Technical Interviews

50 Questions to Sharpen Your Circuit Intuition

Song Liu

Technical and Product Line Director
Monolithic Power Systems (MPS)
San Jose
California

IEEE Press
WILEY

Published by John Wiley & Sons, Inc., Hoboken, New Jersey.

Library of Congress Cataloging-in-Publication Data:

Names: Liu, Song (Director), author
Title: Preparing for integrated circuit (IC) design technical interviews:
50 questions to sharpen your circuit intuition / Song Liu.
Description: Hoboken, New Jersey: Wiley-IEEE Press, [2026] | Includes
bibliographical references and index.
Identifiers: LCCN 2026020318 | ISBN 9781394431076 hardback | ISBN
9781394431090 adobe pdf | ISBN 9781394431083 epub | ISBN 9781394431106
obook
Subjects: LCSH: Integrated circuits—Design and construction | Electrical
engineering—Vocational guidance
Classification: LCC TK7874. L575 2026
LC record available at https://lccn.loc.gov/2026020318

Cover Design: Wiley
Cover Image: saicle/stock.adobe.com

To: Nancy

Contents

About the Author

Song Liu is a Technical and Product Line Director at Monolithic Power Systems (MPS) in San Jose, California. He was born in Hohhot, Inner Mongolia, China, and earned his M.S.E.E. from the University of Idaho in 1999. With a background in both control systems and microelectronics, Liu has dedicated his career to advancing analog and mixed-signal integrated circuit design over the last 25 years. Specializing in data converters and analog front ends, he has led the technical development of a wide range of circuits for applications such as wireline and wireless communications, industrial automation, and consumer electronics.

Preface

Each year, many new Electrical Engineering graduates interview for their first circuit design jobs, and experienced design engineers seek better career opportunities. Having worked in the circuit design industry for over 25 years and interviewed hundreds of candidates along the way, two questions that I was frequently asked were: "What are the key things that interviewers look for in a technical interview?" and "How do I prepare for a technical interview for a circuit design position?"

Before offering my thoughts on these questions, I would like to share an observation from a quick question that I often asked candidates. The question was: "What is the input capacitance of a CMOS source follower?" Many new graduates would try to draw the small-signal diagram and derive the input capacitance. Under the time pressure of a technical interview, quite a few got lost in the process. Some experienced engineers, however, believed that it is roughly the gate-to-source capacitance of the MOSFET.

The purpose of the question is in fact quite simple: to test candidates' knowledge on the transfer function of a source follower and the concept of equivalent input capacitance. Since the output of a source follower tracks its input with a gain close to one within the circuit's bandwidth, the equivalent input capacitance of the gate-to-source capacitor of the MOSFET is minimal, because the voltage across it does not change much. The effective input capacitance is mainly the gate-to-drain capacitance, which is not significant when the transistor is in saturation.

This example reveals one of the most important capabilities that many interviewers value and seek from candidates: circuit intuition, or, as many say, circuit sense. Circuit sense is the ability to grasp the essential nature of circuits and identify the root cause of a circuit problem quickly. It is beneficial to remember all equations, technical terms, and circuit topologies, and to have extensive project experiences, but without circuit intuition, one cannot apply the knowledge effectively, or even correctly, to solve real-world circuit design problems. This means that preparation for a successful technical interview should go beyond simply memorizing circuit equations and topologies or a list of past projects.

How can one build circuit sense? It is important to realize that circuit sense is not magic or a sudden flash of brilliance. On the contrary, it is grounded in a comprehensive understanding of device mathematics and circuit laws. How could one truly understand the origin of the second-order harmonic distortion in a CMOS amplifier without being able to tell the basic square-law equations for MOSFET? Revisit every device equation and circuit law that you have learned and ask yourself: "Do I really understand what this equation or theorem states?" Imagine and simulate in your brain the mechanism behind a device equation and circuit law. Draw the small-signal diagram and analyze every important circuit type in detail to understand the dominant factors of circuit behavior and the secondary effects. The process can be somewhat tedious, but it is similar to professional musicians' daily practice on scales and arpeggios: tedious exercises that enable them to play virtuoso pieces in concerts.

Following this path, we will gradually discover many commonalities among different circuits. For example, by studying the inverting amplifier in Figure P1 below, we conclude that the negative-feedback loop (voltage-sampling, shunt-mixing) lowers the input resistance looking into the virtual ground (A) of the op-amp by a factor roughly equal to the amplifier's open-loop gain.

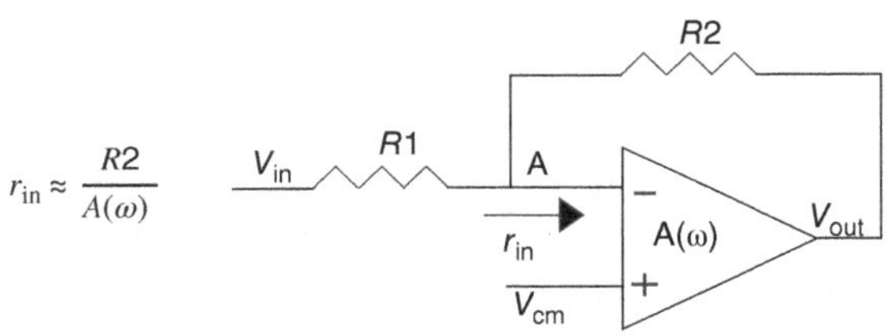

Figure P1

This conclusion is important: it also applies to other closed-loop circuits with the same feedback type. For example, the input impedance of a transimpedance amplifier (Figure P2(a)) and that of the second stage of a two-stage op-amp compensated by a Miller capacitor (Figure P2(b)) are both lowered by the open-loop gain of the amplifier. Note that in the latter case, lowered input impedance means a larger equivalent input capacitance, which is another way to understand the Miller effect. At this point, circuit intuition begins to emerge.

The purpose of this book is to help readers establish and enhance circuit intuition through detailed analyses of a collection of frequently asked questions in technical interviews. This is NOT a typical question-and-answer book. Instead, for most questions in the book, we provide a two-part analysis: an intuitive analysis

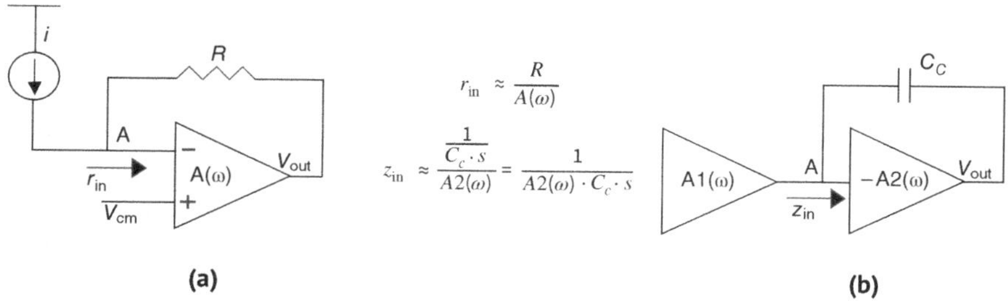

Figure P2

and a quantitative analysis. In the intuitive analysis, we focus on using circuit intuition to analyze the problem and answer the question in a manner similar to how candidates would respond in a technical interview. Detailed mathematical derivations and proofs of the intuition, or at least of the nontrivial parts, are provided in the quantitative section. Although real interviews often do not allow time for thorough quantitative analysis, doing this exercise at least once for every type of circuit under study is essential to laying the foundation for building circuit intuition. Each question concludes with a short answer section, offering a summary of the answers for quick reference.

The interview questions are carefully chosen to cover the most essential circuit design concepts as well as fundamental circuit types. The questions are grouped into six chapters. Chapter 1 studies a critical step in circuit analysis: impedance analysis. Chapter 2 focuses on amplifiers in both closed-loop and open-loop configurations, followed by Chapter 3 on bias and reference circuits. Chapter 4 addresses discrete-time analog circuits, in particular switched-capacitor circuits, comparators, and switched-inductor circuits. Digital circuits, from an analog perspective, are discussed in Chapter 5, covering topics from basic digital building blocks, including inverters and D flip-flops, to complex systems such as digital phase-frequency detectors commonly used in phase-locked loops (PLLs). The last chapter, Chapter 6, examines an often-ignored part of interview preparation: circuit layout.

Throughout the book, we identify commonalities and connections among various types of circuits as well as different analytical approaches: a digital inverter behaves like an analog amplifier near the middle of its input range, and phase margin can be used to judge whether an inverter-based ring oscillator will oscillate at all; a degeneration resistor in a common-source amplifier plays the same role as in a matched current mirror, reducing the circuit's sensitivity to transistor parameter variations through negative feedback. Finding similarities and relationships among circuits is one of the best ways to develop intuition.

For the convenience of readers preparing for interviews, the book also offers four appendices. Appendix 1 summarizes important device equations (MOSFET, bipolar, diode, and passive components). Appendix 2 covers fundamental circuit theorems (Kirchhoff's circuit laws, Thévenin's and Norton's theorem, the superposition theorem, and Miller's theorem). For readers without a control system background, an introduction to circuit analysis in the complex frequency domain (s-domain) is provided in Appendix 3. Appendix 4 is for readers who lack, or wish to brush up on, some fundamental circuit analysis knowledge and skills. It illustrates the most basic circuit analytical methods through examples on fundamental circuit types and concepts, such as voltage and current dividers, RC circuits, basic amplifiers, op-amp based closed-loop circuits, types of negative feedback in amplifiers, circuit noise, and basic digital circuits.

Securing a circuit design career begins with a technical interview, but it is strong circuit sense and a passion for innovation that fuel a lifelong journey of excitement and fulfillment. It would be my true delight if this book both aids readers in their interview preparation and rekindles their curiosity about the circuits that they know.

Cupertino, California, 2026 *Song Liu*

Acknowledgments

I am profoundly grateful to Professor R. Jacob Baker for introducing me to the intuitive circuit analysis approach, a methodology that has been invaluable throughout my 25 years of IC design practice. His foundational guidance and mentorship were essential to my career and the development of this work. I would also like to extend sincere gratitude to the editorial staff at Wiley and the team of anonymous peer reviewers for their thorough and insightful feedback. Their expertise and critical reading were instrumental in refining the technical accuracy and clarity of this text.

Unless otherwise noted, we employ the convention that the case of the first letter of the symbol indicates the signal type: lowercase for small-signal (AC) quantities (e.g., v_{in}, v_{Test}, v_{gs}, i_2, and r_o), and uppercase for large-signal (DC or DC+AC) quantities (e.g., V_{in}, V_{out}, V_{GS}, I_D, and $R3$). This convention is observed consistently throughout this book.

1

Impedance

Impedance analysis is one of the most important aspects of circuit design and analysis. Impedance can be resistive, reactive (i.e., caused by capacitors or inductors), or a combination of both. It can be a deliberately designed property or an unwanted parasitic effect. No matter which way impedance presents itself, it plays a critical role in determining almost all aspects of the circuit characteristics: from the gain and bandwidth of an amplifier to the settling time and noise of a sampling network. It is fair to say that if you can get the impedances in a circuit correctly, you are at least halfway toward a successful circuit analysis.

The analysis of impedance might not always be straightforward, however. For example, impedance can change over frequency, and it can be altered by feedback loops. Impedance can even change signs under certain conditions. A good understanding of circuit impedance is a must for mastering circuit design at both the IC and board levels. This is certainly critical knowledge that interviewers may wish to address during a technical interview.

In this chapter, we present a collection of six questions to test readers' knowledge of circuit impedance. Question 1 studies the impact of negative feedback on the input and output resistance of a transimpedance amplifier. Question 2 examines the input capacitance of a source follower. Question 3 investigates the input capacitance of a differential amplifier under different input modes (differential versus common-mode), followed by Questions 4 and 5 on impedance in cascode circuit structures. The last question (Question 6) explores impedance in a positive-feedback circuit.

Question 1

A current source is connected to a closed-loop amplifier implemented with a CMOS op-amp, as shown in Figure Q1a. The open-loop gain (magnitude) of the op-amp is $A(\omega)$ and the output resistance of the op-amp is r_o. What is the input resistance of the amplifier (excluding the current source), and what is its output resistance?

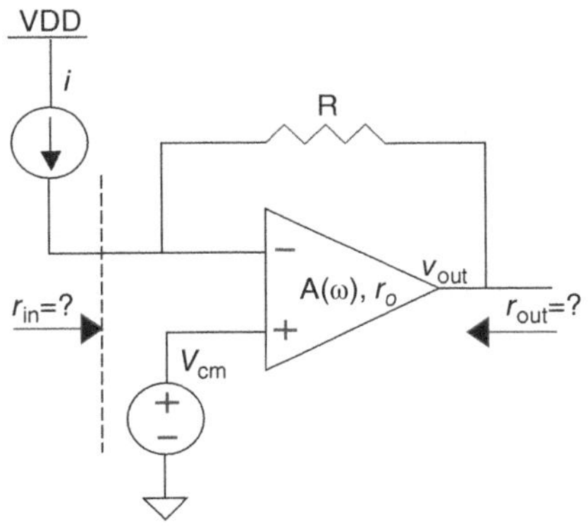

Figure Q1a

Intuitive Analysis: The circuit is known as a transimpedance amplifier (TIA) and converts an input (AC) current i to an output (AC) voltage v_{out}. It is not difficult to conclude that the DC transfer function of the amplifier is:

$$v_{out} = -R \cdot i$$

The negative sign in the equation states the fact that when the input current increases, the output voltage will drop due to the inverting amplifier configuration.

Typical incorrect answers to this question include $r_{in} = R$ and $r_{out} = R||r_o$ ("||" stands for "in parallel with," a designation we will maintain for the remainder of the book), assuming the input resistance of the CMOS op-amp is very large. It overlooks a key fact in this topology: the presence of the negative feedback through the resistor R. This negative feedback is *voltage-sampling, shunt-mixing* (or simply *voltage-shunt*; see *Appendix 4, Section 5, Types of negative feedback in amplifier circuits*), and it reduces both the input and output resistances of the amplifier by a factor approximately equal to the loop gain. This yields:

$$r_{in} \approx \frac{R}{A(\omega)}; \; r_{out} \approx \frac{r_o}{A(\omega)}$$

Quantitative Analysis: For a quantitative derivation of the r_{in} and r_{out}, we need a small-signal model of the amplifier, which is illustrated in Figure Q1b.

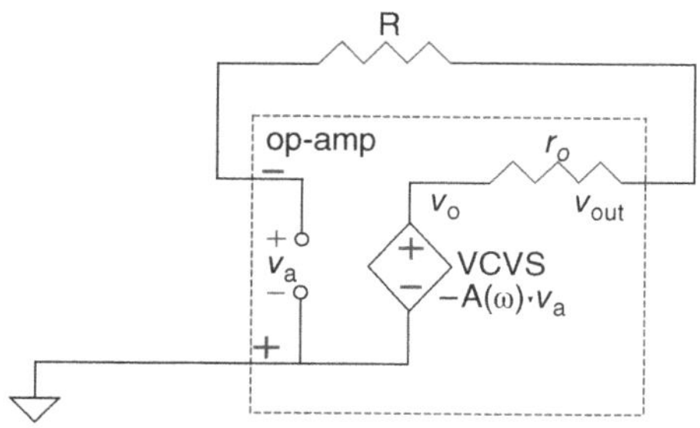

Figure Q1b

In our small-signal model, the op-amp is represented by a voltage-controlled voltage source (VCVS) with output resistance of r_o. Alternatively, we can also use voltage-controlled current source (VCCS) to model the op-amp. The gain of the VCVS is negative due to the inverting amplifier configuration. Note that, in small-signal models for input or output resistance calculation, independent current sources are treated as open circuits and independent voltage sources as short circuits. Controlled sources, such as a VCVS, need to remain in the circuit (see *Appendix 2, Section 2, Thévenin's theorem and Norton's theorem*).

To analyze the input or output resistance, we can insert a test voltage source at the input or output and calculate the resulting current. In this example, we can insert a test voltage source, v_{Test1}, at the input of the amplifier (Figure Q1c), and calculate the input current, i_{Test1}. The input resistance is:

$$r_{in} = \frac{v_{Test1}}{i_{Test1}}$$

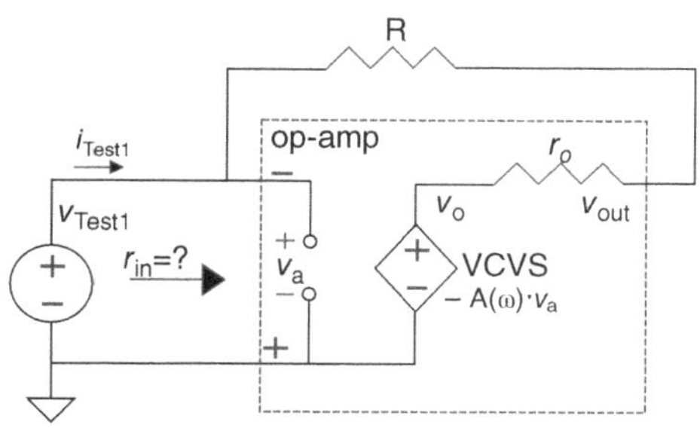

Figure Q1c

Similarly, we can insert a voltage source v_{Test2} at the output of the amplifier (Figure Q1d) and the output resistance can be calculated by:

$$r_{\text{out}} = \frac{v_{\text{Test2}}}{i_{\text{Test2}}}$$

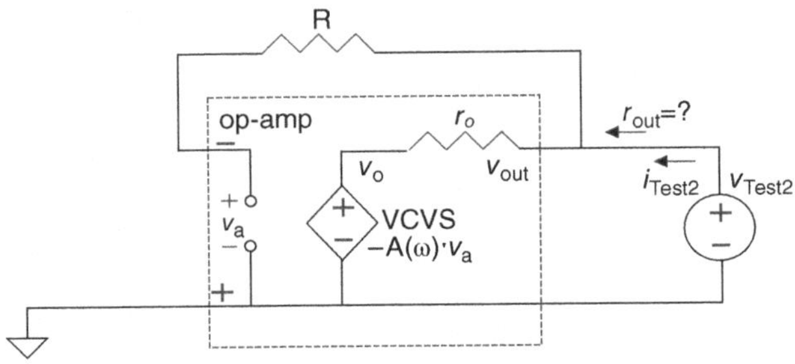

Figure Q1d

Let us start with the input resistance. With the input voltage of the op-amp, v_a, being set to v_{Test1} by the test voltage source, the output voltage, v_o, of the VCVS is:

$$v_o = -A(\omega) \cdot v_{\text{Test1}}$$

Because there is no current flowing into the op-amp, the current i_{Test1} can be calculated as:

$$i_{\text{Test1}} = \frac{v_{\text{Test1}} - v_o}{R + r_o} = \frac{v_{\text{Test1}} - [-A(\omega) \cdot v_{\text{Test1}}]}{R + r_o}$$

And the input resistance, r_{in}, is:

$$r_{\text{in}} = \frac{v_{\text{Test1}}}{i_{\text{Test1}}} = \frac{R + r_o}{1 + A(\omega)}$$

In practice, $A(\omega) \gg 1$ and $r_o \ll R$. This yields the same answers as our intuitive analysis:

$$r_{\text{in}} \approx \frac{R}{A(\omega)}$$

To calculate the output resistance, we insert a test voltage source, v_{Test2}, at the amplifier output and determine the current flowing into the amplifier, i_{Test2}. As the input of the op-amp is open circuit (the input current source is treated as open circuit in the output resistance calculation), there is no current flowing through the feedback resistor, R. This fact leads to the following equations:

$$v_a = v_{\text{Test2}}$$

$$i_{\text{Test2}} = \frac{v_{\text{Test2}} - \left[-A(\omega) \cdot v_{\text{Test2}}\right]}{r_o}$$

And therefore, the output resistance is:

$$r_{\text{out}} = \frac{v_{\text{Test2}}}{i_{\text{Test2}}} = \frac{r_o}{1 + A(\omega)} \approx \frac{r_o}{A(\omega)}$$

Typical mistakes in small-signal analysis include assuming that the voltage at the inverting input of the op-amp is "virtual ground" or zero, when using test voltage sources at the input or output of the amplifier. The fact that a test voltage source is connected to and forces the voltage of either input or output of the op-amp means the feedback is taken out of action.

Short Answer to Question 1

What is the input resistance of the amplifier (excluding the current source), and what is its output resistance?

Both input and output resistance are reduced by the negative feedback of the TIA.

$$r_{\text{in}} \approx \frac{R}{A(\omega)};\; r_{\text{out}} \approx \frac{r_o}{A(\omega)}$$

Question 2

A CMOS voltage source follower circuit is shown in Figure Q2a. M1 is an NMOS transistor and I_1 is an ideal current source. V_{in} is the input and V_{out} is the output. What is the input capacitance of this circuit?

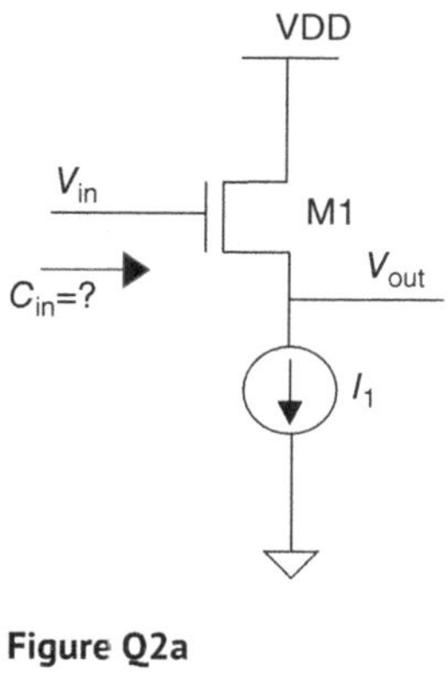

Figure Q2a

Intuitive Analysis: CMOS transistors have parasitic capacitances between their terminals, including C_{gs} (gate-to-source), C_{gd} (gate-to-drain), C_{db} (drain-to-bulk), and C_{sb} (source-to-bulk). The most relevant ones that would affect the input capacitance are the gate capacitances C_{gs} and C_{gd}. We add these two parasitic capacitances in Figure Q2b.

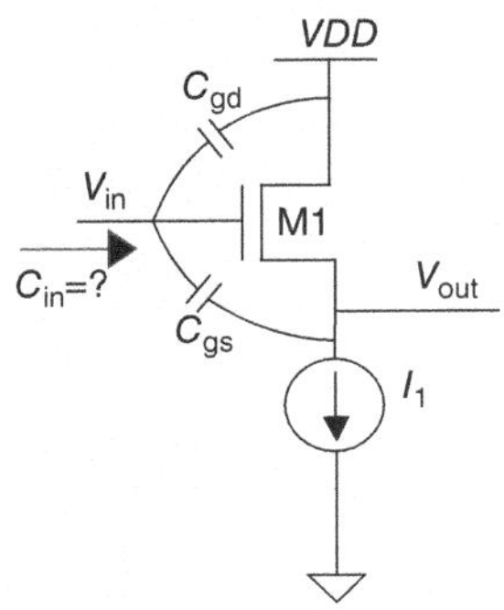

Figure Q2b

Typical incorrect answer: the input capacitance C_{in} is taken to be equal to the gate-to-source capacitance C_{gs}. This mistake arises from a misconception regarding a fundamental characteristic of capacitors. In the basic capacitor relation $\Delta Q = C \cdot \Delta V$, ΔV is the voltage change across the two terminals of the capacitor. In other words, if the voltage across a capacitor is not changing, we are not charging (or discharging) the capacitor. This is the scenario in this case, as within the bandwidth of the source follower, the V_{out} tracks V_{in} and there is no charging or discharging of the capacitance C_{gs}. This means the input does not see C_{gs} as its load. The input must charge and discharge C_{gd}, however, because one terminal of C_{gd} is at the fixed potential of VDD. The input capacitance is therefore C_{gd} within the bandwidth of the source follower. Outside the bandwidth of the source follower, C_{gs} starts to contribute because the output voltage is no longer well regulated.

Another way to understand the effective input capacitance of C_{gs} is to apply Miller's theorem (see *Appendix 2, Section 4, Miller's theorem*). C_{gs} connects between the input and output of the source follower (a gain stage). The equivalent input capacitance is:

$$C_{in} = C_{gs} \cdot (1 - A_v)$$

where A_v is the gain of the source follower.

Within the bandwidth of the source follower, $A_v \approx 1$ and $C_{in} \approx 0$.

Quantitative Analysis: It turns out that the small-signal analysis on even the "ideal" source followers in Figure Q2b can be a little more involved than that of a common-source amplifier, but doing such an analysis is worthwhile as we will gain some key insights into the source follower. Figure Q2c is the small-signal model with a voltage source as the input to the source follower. M1 is modeled as a VCCS with transconductance g_m and output impedance r_o. To facilitate the circuit analysis, the scheme in Figure Q2c is rearranged as shown in Figure Q2d.

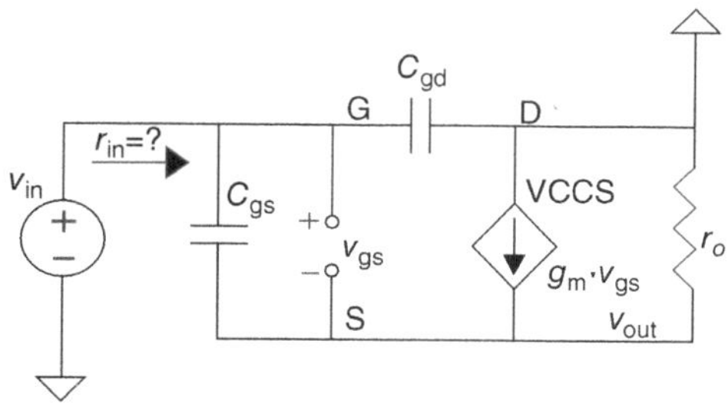

Figure Q2c

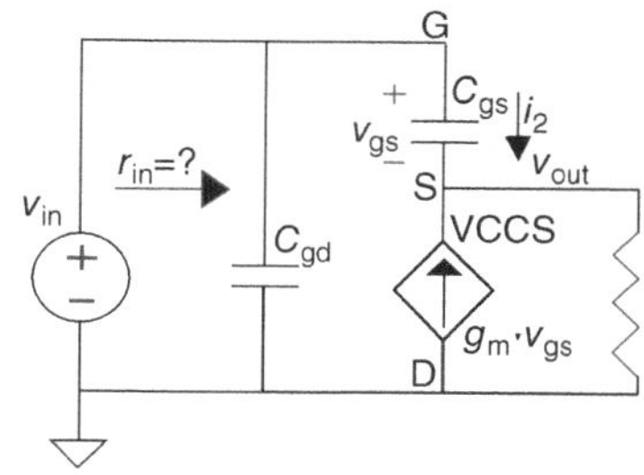

Figure Q2d

Let us first derive the transfer function (the gain) of the source follower. The current flowing through C_{gs}, i_2, can be written as:

$$i_2 = -g_m \cdot (v_{in} - v_{out}) + \frac{v_{out}}{r_o}$$

Keep in mind that the impedance of a capacitor C, denoted as Z_c, is $1/(C \cdot s)$, where s is the complex frequency (see *Appendix 3, Section 1, Laplace transform and transfer functions*). The output voltage v_{out} can be written as:

$$v_{out} = v_{in} - v_{gs} = v_{in} - i_2 \cdot \frac{1}{C_{gs} \cdot s}$$

The transfer function is:

$$H(s) = \frac{v_{out}}{v_{in}} = \frac{1 + \frac{g_m}{C_{gs} \cdot s}}{1 + \frac{g_m}{C_{gs} \cdot s} + \frac{1}{r_o \cdot C_{gs} \cdot s}} = \frac{C_{gs} \cdot r_o \cdot s + g_m \cdot r_o}{C_{gs} \cdot r_o \cdot s + g_m \cdot r_o + 1}$$

When $g_m \cdot r_o \gg 1$, the gain approaches one. It is important to realize that the gain attenuation is due to the negative feedback, or *degeneration*, by the current source I_1. From the transfer function, we also see that both a pole and a zero exist in the source follower, and they are very close to each other. One could wonder if this makes the –3 dB bandwidth of the source follower unlimited. In reality we have not considered other capacitances of the source follower, for example, the source-to-bulk capacitance (C_{sb}), and the load capacitance in the above derivation. These capacitances will be put in parallel with r_o of the transistor M1, making the pole move to a lower frequency and split from the zero. In addition, series resistance present on the gate of M1 causes an additional pole, further limiting the bandwidth of the source follower. We will dive in a little further in another source follower question: Question 20, *Frequency response of a source follower.*

Now back to the input capacitance. The current flowing through the C_{gs} can also be written as:

$$i_2 = \frac{v_{in} - v_{out}}{\frac{1}{C_{gs} \cdot s}} = (v_{in} - v_{out}) \cdot C_{gs} \cdot s = v_{in}\left[1 - H(s)\right] \cdot C_{gs} \cdot s$$

Substituting the transfer function $H(s)$ into the expression, we obtain:

$$i_2 = v_{in} \cdot \frac{C_{gs} \cdot s}{C_{gs} \cdot r_o \cdot s + g_m \cdot r_o + 1}$$

The impedance looking into the C_{gs} branch, Z_{Cgs}, is:

$$Z_{Cgs} = \frac{v_{in}}{i_2} = r_o + \frac{1}{\frac{C_{gs}}{g_m \cdot r_o + 1} \cdot s}$$

The capacitance component looking into the C_{gs} branch is:

$$C_{Cgs} = \frac{C_{gs}}{g_m \cdot r_o + 1} \approx 0; \text{ if } g_m \cdot r_o \gg 1$$

It is attenuated by the loop gain $g_m \cdot r_o$. The input capacitance is:

$$C_{in} \approx C_{gd}$$

The result verifies the conclusion in our intuitive analysis of the input capacitance.

Short Answer to Question 2

What is the input capacitance of this circuit?

Due to Miller effect, the gate-to-source capacitance C_{gs} of the transistor is negligible within the bandwidth of the source follower. The equivalent input capacitance is:

$$C_{in} \approx C_{gd}$$

Question 3

A differential amplifier is shown in Figure Q3a. What is the input capacitance for differential input signals and for common-mode input signals?

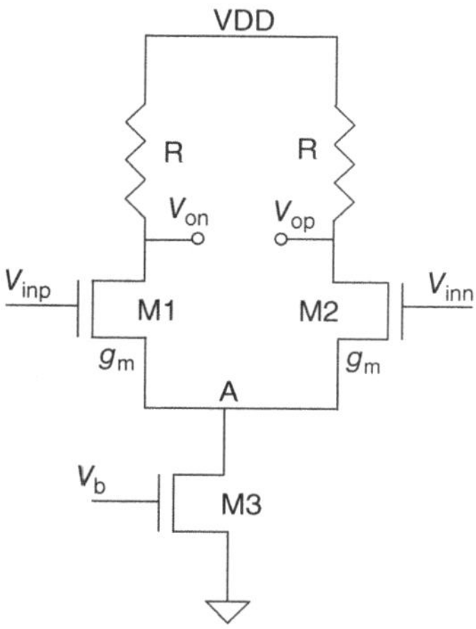

Figure Q3a

Intuitive Analysis: The major capacitances that affect inputs are the parasitic capacitances of the transistors M1 and M2, namely C_{gs1}, C_{gd1}, and C_{gs2}, C_{gd2}, respectively, as shown in Figure Q3b. The key to answering this question is to understand that a differential pair behaves differently for differential input signals and common-mode input signals.

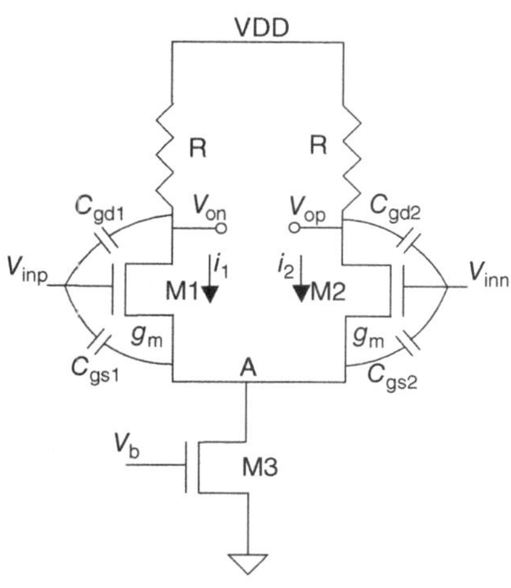

Figure Q3b

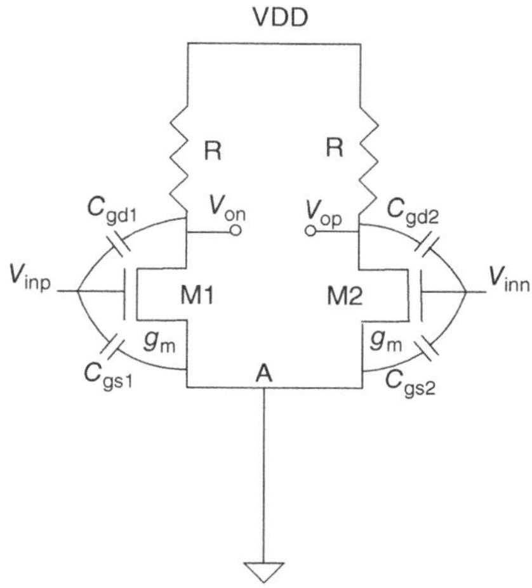

Figure Q3c

For differential input signals, the drain of M3 (node A) is at AC ground. This is because the AC current i_1 caused by v_{inp} cancels the AC current i_2 caused by v_{inn}; v_{inp} and v_{inn} are opposite in differential mode, and the transconductance g_m of M1 and M2 are matched. The result is zero AC current flowing through M3, and node A becomes an AC ground. The differential pair can now be simplified as two common-source amplifiers as shown in Figure Q3c. It is clear that C_{gs1} and C_{gd2} are part of the capacitive load by V_{inp} and V_{inn}. What about C_{gd1} and C_{gd2}? Note that these gate-to-drain capacitances are not connected to a fixed potential as in the source follower example (Question 2). They are instead connected between the gate (input) and drain (output) of a common-source amplifier – similar to the Miller compensation capacitor in a two-stage op-amp. The capacitive load of C_{gd1} and C_{gd2} on V_{inp} and V_{inn} increases by a factor equal to the gain of the amplifier due to *Miller effect*. The effective input capacitance from each side of the differential inputs is:

$$C_{in}(\text{diff}) = C_{gs1}\left(\text{or } C_{gs2}\right) + g_m \cdot R \cdot C_{gd1}(\text{or } C_{gd2})$$

For CMOS transistors in saturation, $C_{gs} \gg C_{gd}$, so we often ignore C_{gd} when estimating the input capacitance. For common-source amplifiers, keep in mind that the contribution of C_{gd} to the input capacitance could be significant due to the Miller effect.

For common-mode input signals, node A is no longer AC ground. This is similar to two source followers sharing the same current-source load, except that node A is not used as an output (they are, in fact, two common-source amplifiers sharing the same degeneration). Source degeneration reduces the common-mode gain of the amplifier, thereby enhancing its common-mode rejection ratio (CMRR – defined as the absolute value of the ratio between an amplifier's differential-mode voltage gain and its common-mode voltage gain). Recall from the results of Question 2 that within the bandwidth of the source followers, C_{gs} does not load the inputs because the gate and drain voltages of M1 (or M2) track each other. Also note that the Miller effect on C_{gd} is minimal because of the source degeneration. The effective input capacitance for the common-mode input signals therefore is:

$$C_{in}(\text{com}) \approx C_{gd1}(\text{or } C_{gd2})$$

Quantitative Analysis: The quantitative analysis is left as an exercise for interested readers.

Short Answer to Question 3

What is the input capacitance for differential input signals and for common-mode input signals?

The tail current source (M3) has no effect on the amplifier for differential input signals, but it serves as source degeneration for common-mode input signals.

$$C_{\text{in}}(\text{diff}) = C_{\text{gs1}}\left(\text{or } C_{\text{gs2}}\right) + g_m \cdot R \cdot C_{\text{gd1}}(\text{or } C_{\text{gd2}})$$

$$C_{\text{in}}(\text{com}) \approx C_{\text{gd1}}(\text{or } C_{\text{gd2}})$$

Question 4

Two NMOS transistors M1 and M2 are connected in series with their gate tied together as shown in Figure Q4a. What is the output resistance looking into the drain of M1?

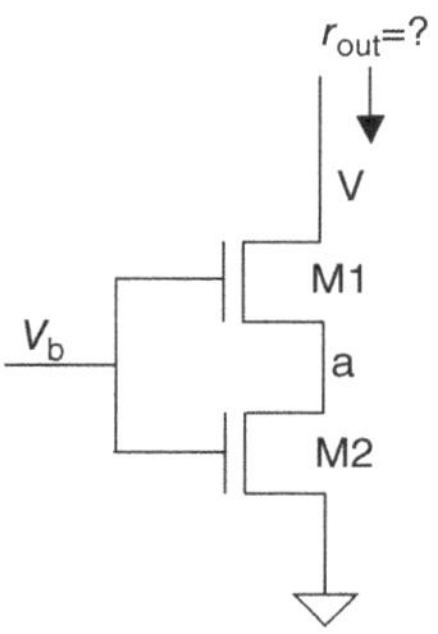

Figure Q4a

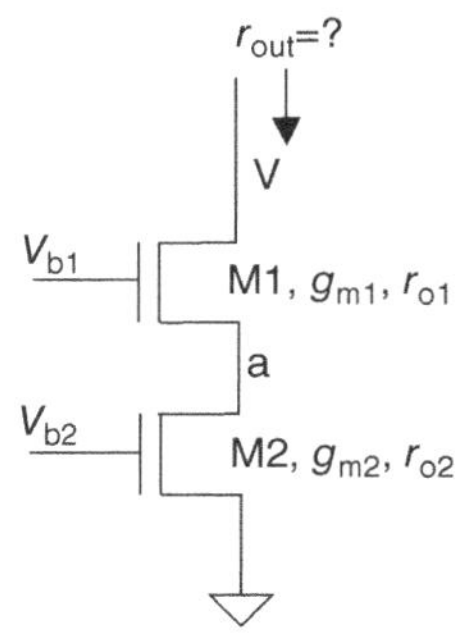

Figure Q4b

Intuitive Analysis: If the gates of the two transistors are biased separately, as depicted in Figure Q4b, so that both transistors operate in the saturation region, the circuit becomes a classical *cascode* topology: the current source M2 serves as source degeneration, and the resistance looking into the cascode structure is:

$$r_{\text{out}} = r_{o1} + r_{o2} + r_{o2} \cdot r_{o1} \cdot g_{m1} \approx r_{o2} \cdot r_{o1} \cdot g_{m1}$$

Now the question is: can we keep both transistors M1 and M2 in saturation if their gates are tied together? To determine this, we need to examine the saturation condition for each transistor. To keep M2 in saturation, the following condition must be met:

$$V_b < V_a + V_{\text{th}}$$

To keep M1 turned on, V_b must satisfy:

$$V_b > V_a + V_{\text{th}}$$

Assume both transistors have the same threshold voltage V_{th}. The two inequalities show that once M1 is turned on, M2 is no longer in saturation. This means M1 and M2 cannot both be in the saturation region. However, we can choose an appropriate V_b to keep M1 in saturation while M2 operates in the linear region. Condition for keeping M1 in saturation is:

$$V_a + V_{\text{th}} < V_b < V + V_{\text{th}}$$

The rest of the analysis is straightforward: the topology in Figure Q4a turns out to be a cascode structure with a degeneration transistor (M2) in the linear region. The on-resistance of a transistor in the linear region, r_l, is approximately equal to:

$$r_l \approx \frac{1}{\beta \cdot \left(V_{\text{gs}} - V_{\text{th}}\right)}$$

where β is the device transconductance parameter of the transistor, which is defined as $Kp \cdot W/L$. The resistance in linear region (resembles a linear resistor) is significantly less than that in saturation region (resembles a current source). The output resistance, r_{out}, in Figure Q4a is:

$$r_{\text{out}} = r_{o1} + r_{l2} + r_{l2} \cdot r_{o1} \cdot g_{m1} \approx r_{l2} \cdot r_{o1} \cdot g_{m1}$$

The output resistance in Figure Q4a is less than a classical cascode topology (Figure Q4b) where both transistors are in saturation, but is higher than a single transistor, thanks to the degeneration.

Quantitative Analysis: The key part of the quantitative analysis is on derivation of the output resistance of a cascode topology. Figure Q4c is the small-signal model for the classical cascode topology shown in Figure Q4b. Same as in Question 1, we use a test voltage source, v_{Test}, to facilitate the calculation of the output resistance.

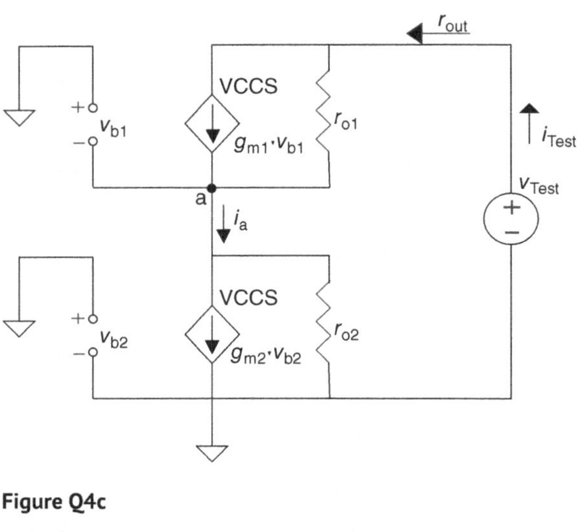

Figure Q4c

The current flowing through M1, i_a, is:

$$i_a = -g_{m1} \cdot v_a + \frac{v_{\text{Test}} - v_a}{r_{o1}}$$

It also equals the current flowing through M2. Note that there is no current in the VCCS of M2 (g_{m2}) as $v_{b2} = 0$.

$$i_a = i_{\text{Test}} = \frac{v_a}{r_{o2}}$$

The two equations yield:

$$v_a = \frac{r_{o2} \cdot v_{\text{Test}}}{r_{o1} + r_{o2} + r_{o1} \cdot r_{o2} \cdot g_{m1}}$$

And

$$i_{\text{Test}} = \frac{v_a}{r_{o2}} = \frac{v_{\text{Test}}}{r_{o1} + r_{o2} + r_{o1} \cdot r_{o2} \cdot g_{m1}}$$

The r_{out} is solved as:

$$r_{\text{out}} = \frac{v_{\text{Test}}}{i_{\text{Test}}} = r_{o1} + r_{o2} + r_{o1} \cdot r_{o2} \cdot g_{m1} \approx r_{o1} \cdot r_{o2} \cdot g_{m1}$$

Short Answer to Question 4

What is the output resistance looking into the drain of M1?

When the gates of two cascoded transistors are tied together, only one of the transistors can be in the saturation region, and the other one operates in the triode region. The output resistance in this condition is:

$$r_{\text{out}} \approx r_{l2} \cdot r_{o1} \cdot g_{m1} = \frac{1}{\beta_2 \cdot \left(V_{\text{gs2}} - V_{\text{th}}\right)} \cdot r_{o1} \cdot g_{m1}$$

Question 5

A fully differential amplifier is shown in Figure Q5a. What is the input capacitance for differential input signals?

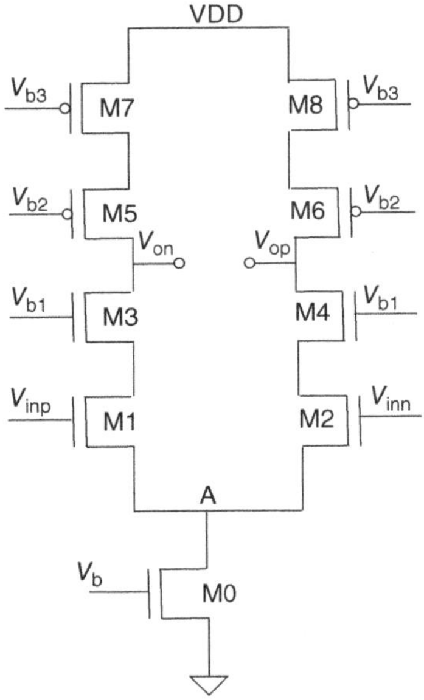

Figure Q5a

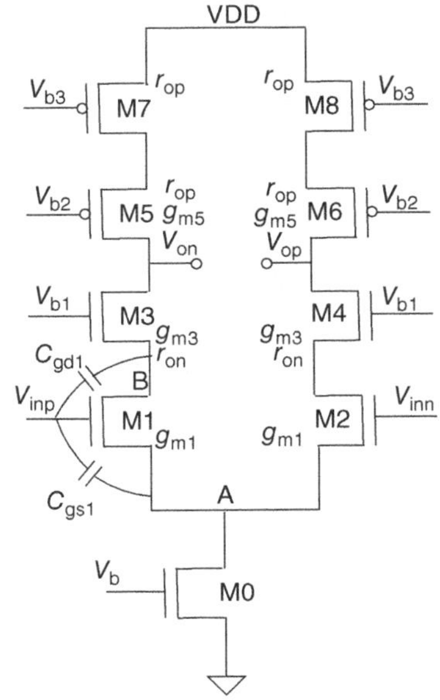

Figure Q5b

Intuitive Analysis: This is a classical telescopic amplifier, known for its high gain in a single stage. This is because the output impedances at nodes V_{op} and V_{on} are enhanced by the cascode structure of both the NMOS and PMOS loads. We have seen in Question 3 that the voltage gain of the common-source amplifier could trigger the Miller effect and potentially cause a large input capacitance. Will the high gain in the telescopic amplifier cause this problem here?

To simplify the analysis, we assume all NMOS transistors have an output impedance of r_{on}, and all PMOS transistors have an output impedance of r_{op}, as shown in Figure Q5b. We also assume the gate-to-source and gate-to-drain capacitances of the input transistor M1 to be C_{gs1} and C_{gd1}, respectively. The transconductances (g_m) of M1, M3 and M5 are denoted g_{m1}, g_{m3}, and g_{m5}, respectively. Due to the fully symmetrical nature of the topology, we can focus on one side, for example the left side of the circuit, for the analysis.

The key to determining the input capacitance is the gain from V_{inp} to the drain of the input transistor M1, namely node B in Figure Q5b. This is related to the impedance looking into the source of transistor M3. The impedance looking into the source of a transistor in saturation, r_s, is low and is approximately equal to the reciprocal of the transistor transconductance g_m.

$$r_s \approx \frac{1}{g_m}$$

The voltage gain from the input to node B, A_{iB}, therefore, is:

$$A_{\mathrm{iB}} \approx \frac{g_{m1}}{g_{m3}}$$

Note that this gain is significantly lower than the voltage gain of the amplifier itself (A), which approximates:

$$A \approx g_{\mathrm{m1}} \cdot \left(g_{\mathrm{m3}} \cdot r_{\mathrm{on}}^2 \parallel g_{m5} \cdot r_{\mathrm{op}}^2\right)$$

Recall from Question 3 that for the differential input signals, the drain of the tail transistor (node A in Figure Q5b) is AC ground. The input capacitance for differential input signals therefore is:

$$C_{\mathrm{in}} = C_{\mathrm{gs1}} + \frac{g_{m1}}{g_{m3}} \cdot C_{\mathrm{gd1}}$$

The gate-to-source capacitance of the input transistors could dominate the input capacitance in a telescopic amplifier, as the impact of the Miller effect on the input capacitance is reduced thanks to the cascode structure.

Quantitative Analysis: We will verify, by the quantitative analysis, that the impedance looking into the source of transistor M3 in the telescopic amplifier in Figure Q5a is approximately $1/g_{m3}$. Figure Q5c is the small-signal model of the left side of the amplifier. For the top portion (M5 and M7) of the circuit, we model it using its output impedance $\left(g_{m5} \cdot r_{\mathrm{op}}^2\right)$ to simplify the derivation.

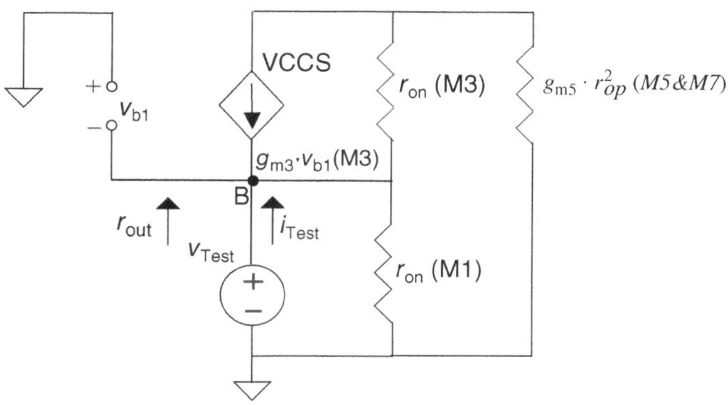

Figure Q5c

Again, we use a test voltage source v_{Test} as the stimulus to calculate the impedance looking into the node B (source of M3). As the output impedance of M1 is in parallel with the rest of the circuit, we take it out for now and consider it at the final calculation. Using Kirchhoff's current law, we obtain:

$$g_{m3} \cdot v_{\text{Test}} + \frac{v_{\text{Test}} - i_{\text{Test}} \cdot g_{m5} \cdot r_{\text{op}}^2}{r_{\text{on}}} = i_{\text{Test}}$$

As such, the impedance looking into the source of M3, not counting the current branch of M1, is:

$$r'_{\text{out}} = \frac{v_{\text{Test}}}{i_{\text{Test}}} = \frac{r_{\text{on}} + g_{m5} \cdot r_{\text{op}}^2}{1 + g_{m3} \cdot r_{\text{on}}} \approx \frac{g_{m5} \cdot r_{\text{op}}^2}{g_{m3} \cdot r_{\text{on}}}$$

The final impedance is:

$$r_{\text{out}} \approx r_{\text{on}} \parallel \frac{g_{m5} \cdot r_{\text{op}}^2}{g_{m3} \cdot r_{\text{on}}}$$

The impedance looking into the source of M3 is significantly smaller than the output impedance of the amplifier, but nevertheless substantially larger than $1/g_{m3}$ as estimated in the intuitive analysis. If we assume $g_{m3} = g_{m5}$ and $r_{\text{on}} = r_{\text{op}} = r_o$, then $r_{\text{out}} \approx r_o/2$. Although this discrepancy might not change the fact that the gate-to-source capacitance of M1 is the major contributor of the input capacitance, we do run the risk of underestimating the impact of C_{gd} on the total input capacitance.

Thus, the impedance looking into the source of a transistor can be significantly higher than $1/g_m$ if the device is loaded by a high impedance at the drain. This example shows that although it is often convenient to use a "rule of thumb" in intuitive analysis, one needs to understand the conditions or constraints of such a guideline. A quick small-signal analysis can often refresh our memories.

Telescopic amplifiers are widely used in high-gain and high-bandwidth applications. They are simple in topology and efficient in power consumption. The major drawback is less voltage headroom due to their large number of stacked transistors. Folded cascode amplifiers find more applications in low-voltage designs where voltage headroom is a concern, with the trade-off of lowered power efficiency.

Figure Q5d shows the topology of a folded cascode amplifier. Interested readers can do Question 5 again on Figure Q5d and verify the input capacitance.

Please note that for fully differential amplifiers with high gains, such as topologies in Question 5, common-mode feedback control (CMFB) is needed. We did not include the CMFB circuit here for the sake of simplicity to focus on the input capacitance analysis. We will visit the topics of CMFB in *Chapter 2, Amplifiers – Open-loop and Closed-loop*.

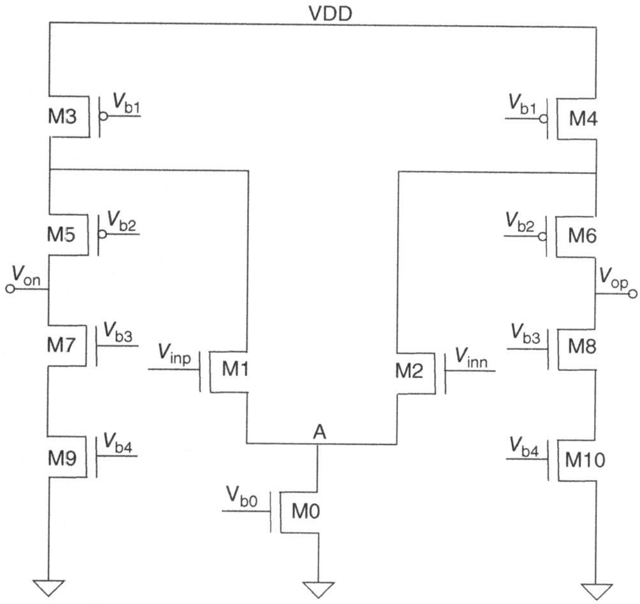

Figure Q5d

Short Answer to Question 5

What is the input capacitance for differential input signals?

Cascode load effectively reduces the gain across the C_{gd} of the input transistors. A quick estimation of the input capacitance is:

$$C_{in} \approx C_{gs1} + \frac{g_{m1}}{g_{m3}} \cdot C_{gd1}$$

Note that in fact the impedance looking into the source of M3 is significantly higher than $1/g_{m3}$ due to the high impedance at the drain of M3.

Question 6

A fully differential amplifier is shown in Figure Q6a. What is the impedance at node V_{on} for differential input signals, and what is the gain of the amplifier?

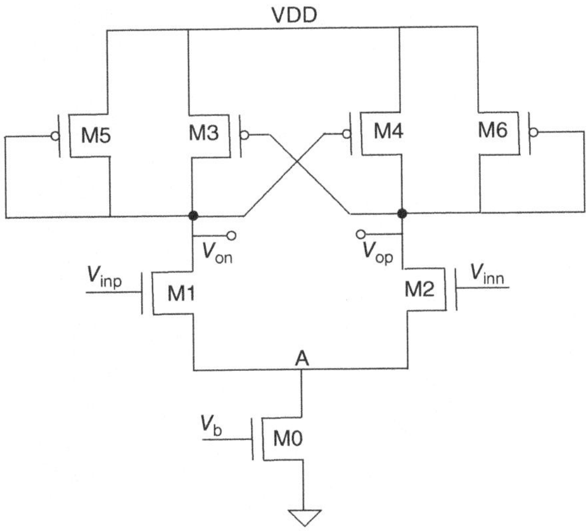

Figure Q6a

Intuitive Analysis: A key feature of this topology is that it contains a pair of cross-coupled transistors, M3 and M4. It is relatively straightforward to figure out that this cross-coupled pair implements positive feedback. Without the cross-coupled pair, the load of the amplifier is simply the diode-connected (drain tied to gate) transistors M5 and M6. If you do not already know the impedance looking into the drain of a diode-connected transistor, a small-signal analysis shows that the impedance is approximately $1/g_m$. This is a very useful rule of thumb to remember in circuit analysis.

The question is: what is the impedance looking into the drains of this cross-coupled transistor pair? Unlike negative feedback, positive feedback is less common in amplifier circuits. However, positive feedback can provide electrical characteristics that negative feedback cannot, for example, negative impedance. We can think of M3 and M4 as special diode-connected transistors in which the gate of each transistor is not connected to its own drain but to the drain of the paired transistor. In a fully differential amplifier, the two outputs are differential–possessing equal magnitude but opposite phase. Consequently, we can intuitively see that the drain of a cross-coupled pair presents an impedance of $-1/g_m$, effectively acting as a negative impedance.

As such, both a positive impedance ($1/g_{m5}$) and a negative impedance ($-1/g_{m3}$) are present at the output node (V_{on}). Because M3 and M5 are in parallel, the resulting impedance at node V_{on} is $1/(g_{m5}-g_{m3})$. To keep the impedance positive so that the circuit does not latch up, the transconductance of M5 must be larger than that of M3, i.e., $g_{m5} > g_{m3}$. The gain of the amplifier is $g_{m1}/(g_{m5} - g_{m3})$. A large positive output impedance, and therefore a large gain can be achieved by making g_{m5} slightly larger than g_{m3} in value.

It is clear that the purpose of the positive feedback is to create a local negative impedance, for the enhancement of the gain of the amplifier. In practical applications, we do not wish to make g_{m5} and g_{m3} too close to each other to avoid large gain sensitivity to process, supply voltage, and temperature variations (also known as PVT variations). This topology is widely used as pre-amplifiers for comparators where large gain is needed but the precision of the gain is not critical. A variant of the topology is shown in Figure Q6b. Current sources M7 and M8 divert a portion of the tail current away from transistors M3–M6, providing more design flexibility.

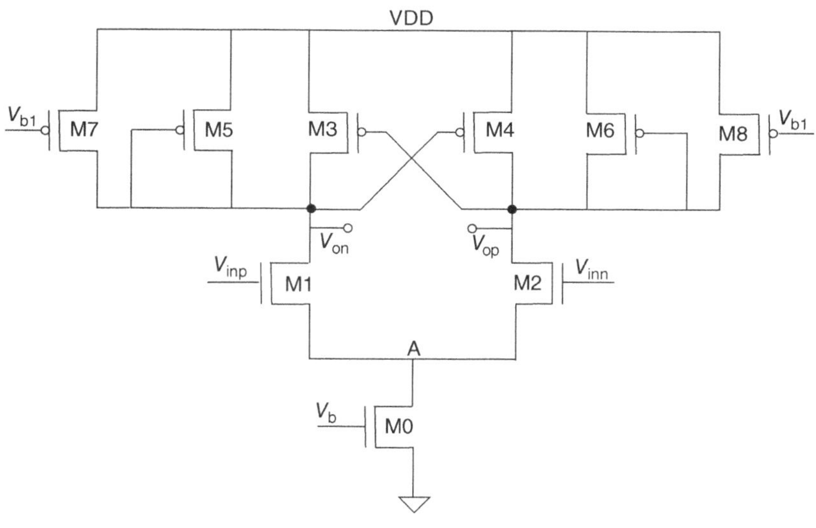

Figure Q6b

Quantitative Analysis: Let us verify quantitatively that the impedance of node V_{on} (and node V_{op}) in Figure Q6a matches our intuitive guess. The small-signal model of the topology in Figure Q6a is shown in Figure Q6c. As the circuit is symmetrical between the two differential inputs, we only draw the left side (V_{inp}) of the topology in our small-signal model. A test voltage source v_{Test} is connected to the node V_{on}. Notice that for the analysis of the impedance on node V_{on}, the input V_{inp} and V_{inn} are considered AC ground. Importantly, the (gate) control voltage for the VCCS of M3 is v_{op}, and $v_{\text{op}} = -v_{\text{on}}$.

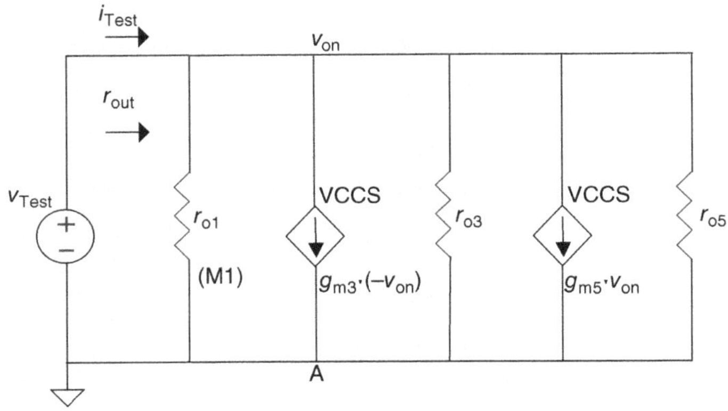

Figure Q6c

The test current i_{Test} is the sum of the branch currents of M1, M3, and M5:

$$i_{Test} = \frac{v_{Test}}{r_{o1}} + \frac{v_{Test}}{r_{o3}} + \frac{v_{Test}}{r_{o5}} - g_{m3} \cdot v_{Test} + g_{m5} \cdot v_{Test}$$

The impedance of node V_{on}, r_{out}, therefore is:

$$r_{out} = \frac{1}{\frac{1}{r_{o1}} + \frac{1}{r_{o3}} + \frac{1}{r_{o5}} + g_{m5} - g_{m3}} \approx \frac{1}{g_{m5} - g_{m3}}$$

The gain of the amplifier for differential input signals is:

$$A = \frac{g_{m1}}{\frac{1}{r_{o1}} + \frac{1}{r_{o3}} + \frac{1}{r_{o5}} + g_{m5} - g_{m3}} \approx \frac{g_{m1}}{g_{m5} - g_{m3}}$$

Short Answer to Question 6

What is the impedance at node V_{on} for differential input signals, and what is the gain of the amplifier?

Cross-coupled load transistors form positive-feedback loop which results in negative impedance. The impedance at node V_{on} and the gain of the amplifier are:

$$r_{out} \approx \frac{1}{g_{m5} - g_{m3}}$$

$$A \approx \frac{g_{m1}}{g_{m5} - g_{m3}}$$

General References

1 Allen PE and Holberg DR. *CMOS Analog Circuit Design*. 3rd ed. New York (NY): Oxford University Press; 2012.
2 Baker RJ. *CMOS: Circuit Design, Layout, and Simulation*. 3rd ed. Hoboken (NJ): Wiley/IEEE Press; 2010.
3 Boylestad RL. *Introductory Circuit Analysis*. 13th ed. [place unknown]: Pearson; 2015.
4 Carusone TC, Johns DA and Martin KW. *Analog Integrated Circuit Design*. 2nd ed. Hoboken (NJ): Wiley; 2012.
5 Gray PR, Hurst PJ, Lewis SH and Meyer RG. *Analysis and Design of Analog Integrated Circuits*. 6th ed. Hoboken (NJ): Wiley; 2024.
6 Razavi B. *Design of Analog CMOS Integrated Circuits*. 2nd ed. New York (NY): McGraw-Hill Education; 2017.
7 Sansen WMC. *Analog Design Essentials*. New York (NY): Springer; 2006.

2

Amplifiers – Open-loop and Closed-loop

Amplifiers are among the most widely used components in analog circuits, and their appearance in a technical interview for circuit design is highly likely. The challenge in the analysis and design of amplifiers is their complexity.

By "complexity," we mean a lot. Amplifiers have many types and configurations, and designing an amplifier involves tradeoffs among many factors: DC gain, gain-bandwidth product, offset, slew rate, input and output impedance, distortion, noise, supply rejection, stability, and more. Amplifiers are used in both open-loop and closed-loop applications, in either continuous-time or discrete-time format, each with its own challenges. To many, amplifiers are a lifetime topic of study.

We have carefully chosen fourteen questions in an effort to cover some of the most important aspects of amplifier analysis and design. The first three questions, Questions 7 to 9, are about amplifier stability and compensation. Questions 10 and 11 examine the input-referred noise and slew-rate limitation of amplifiers, respectively. The next three questions, Questions 12 to 14, look into the amplifier distortion and methods to reduce its effects. Question 13, in particular, concerns a bipolar common-emitter amplifier, offering readers some exercises on BJTs. Question 15 deals with the common-mode feedback control of a fully differential amplifier. Some closed-loop op-amp applications are studied in Questions 16 and 17, including the impact of the feedback factor. Question 18 explores the power supply rejection ratio of a closed-loop amplifier, followed by Question 19 on power optimization of multistage open-loop amplifiers. In the last question of the chapter, Question 20, a special type of amplifier, namely common-drain amplifier, or source follower, is studied.

All questions in Chapter 2 deal with amplifiers in continuous-time applications. Amplifiers in discrete-time circuits are discussed in Chapter 4, *Switched-capacitor Circuits and Comparators*.

Question 7

A two-stage op-amp is shown in Figure Q7a. Assume the output resistances at the output of the first stage (node B) and second stage (V_{out}) are r_{o1} and r_{o2}, respectively. The total capacitances at the output of the first stage and second stage are C_1 and C_2, respectively. Ignore the parasitic capacitance at node C. The transconductances of transistors M1 and M5 are g_{m1} and g_{m5}, respectively. The op-amp will be used in a unity-gain buffer. To maintain a phase margin of around 45°, what would be the design constraint on r_{o1}, r_{o2}, C_1, C_2, g_{m1}, and g_{m5}? Assume the first stage dominates the time constant.

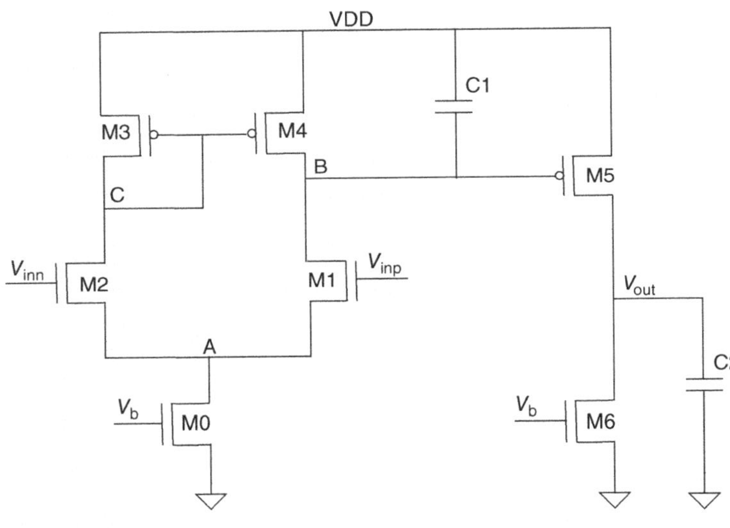

Figure Q7a

Intuitive and Quantitative Analysis: A two-stage op-amp without (Miller) compensation? Can the topology even be stable in closed-loop applications? The question might appear somewhat naïve at first glance, but the process of analyzing this problem helps us gain a better understanding of some of the fundamental concepts of the two-stage amplifier compensation.

First, we need to figure out what makes the (open-loop) phase margin around 45°. As there are only two poles in this amplifier, the phase margin depends on the location of the two poles and the DC gain. If we draw a Bode plot of a two-pole system, as shown in Figure Q7b, it is not difficult to conclude that if the frequency of nondominant pole (Pole #2) is in the neighborhood of the unity-gain frequency (f_u) of the amplifier, the phase margin is about 45°. Note that this is assuming that the two poles are sufficiently separated from each other and the op-amp approximates a first-order system up to the unity-gain frequency.

The frequency of the dominant pole, p_1, is associated with the time constant of the first stage:

$$fp_1 = \frac{1}{2 \cdot \pi \cdot r_{o1} \cdot C_1}$$

Similarly, the frequency of the nondominant pole, p_2, is determined by the time constant of the second stage:

$$fp_2 = \frac{1}{2 \cdot \pi \cdot r_{o2} \cdot C_2}$$

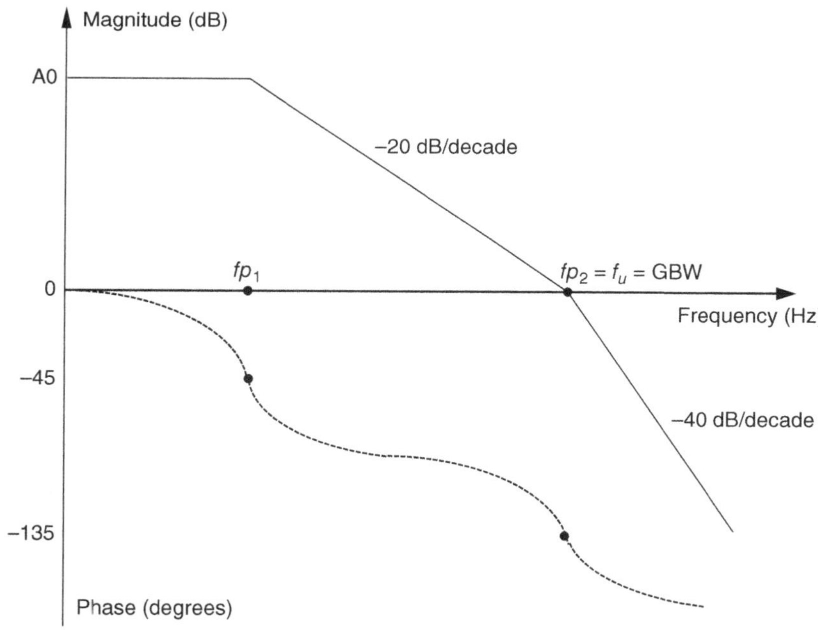

Figure Q7b

The unity-gain frequency f_u equals approximately the product of the op-amp DC gain $A(0)$ and the -3 dB cutoff frequency fp_1, also known as the *gain-bandwidth product* (GBW), as long as up to f_u the op-amp behaves like a single-pole system:

$$f_u \approx \text{GBW} = A(0) \cdot fp_1 = \frac{g_{m1} \cdot r_{o1} \cdot g_{m5} \cdot r_{o2}}{2 \cdot \pi \cdot r_{o1} \cdot C_1}$$

Set $fp_2 = \text{GBW}$ yields:

$$C_1 = g_{m1} \cdot g_{m5} \cdot (r_{o2})^2 \cdot C_2$$

Note that $g_{m1} \cdot g_{m5} \cdot (r_{o2})^2$ is on the order of the DC gain of the op-amp if the second stage gain is comparable to that of the first stage. If the DC gain of the op-amp is 60 dB, this means that to maintain 45° of phase margin C_1 must be about a thousand times larger than C_2! Keep in mind that C_2 is the load capacitance which could be large by itself in some applications. If a higher phase margin is needed, we need to push the pole of the second stage to an even higher frequency than the unity-gain frequency. For example, to make a phase margin around 60°, the frequency of the nondominant pole needs to be twice the unity-gain frequency, assuming the op-amp approximates a second-order system (see *Appendix 3, Section 4, Second-order systems*).

This is how Miller compensation can be helpful to practical two-stage op-amps if they are used in closed-loop applications. Some people might ask: "how about an OTA (operational transconductance amplifier) in which the internal pole is nondominant and the load capacitance of the output stage compensates the amplifier?" Notice that an OTA's high output impedance limits its applications to capacitive load only. An OTA has its niche but cannot replace a "classical" two-stage op-amp in many applications.

Amplifier compensation is one of the most asked topics in technical interviews. We will continue to study this topic with the next two questions.

Short Answer to Question 7

To maintain a phase margin of around 45°, what would be the design constraint on r_{o1}, r_{o2}, C_1, C_2, g_{m1}, and g_{m5}?

To maintain a phase margin of 45°, the (first) nondominant pole needs to be placed around the unity-gain frequency of the op-amp. The design constraint for this to happen is:

$$C_1 = g_{m1} \cdot g_{m5} \cdot (r_{o2})^2 \cdot C_2$$

Question 8

Use the same two-stage op-amp from Question 7 (Figure Q7a), but this time a compensation capacitor, C_3, is connected between the gate and drain of M5 as shown in Figure Q8a. What are the frequencies of the two poles of the op-amp in the new circuit diagram?

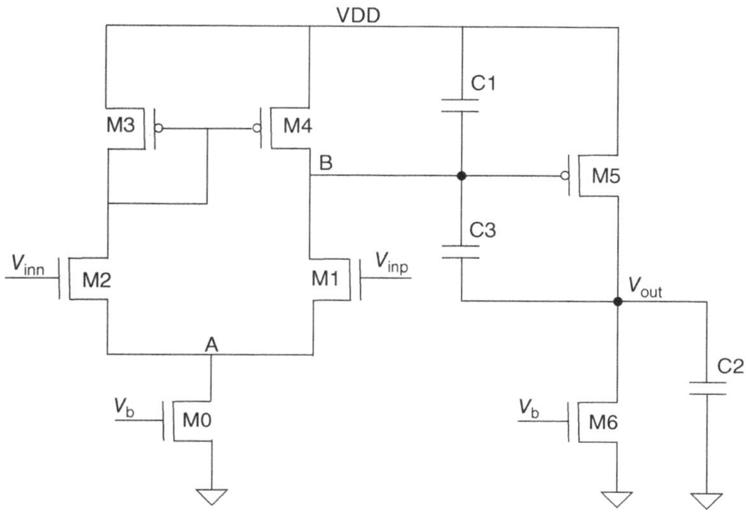

Figure Q8a

Intuitive Analysis: This is a classical op-amp (Miller) compensation question. To analyze the impact of C_3 on the time constant of the first stage, we can apply Miller's theorem to replace C_3 with an equivalent capacitance, C_m. As illustrated in Figure Q8b, C_m is connected between the output of the first stage (node B) and VDD (*see Appendix 2, Section 4, Miller's theorem* for further details).

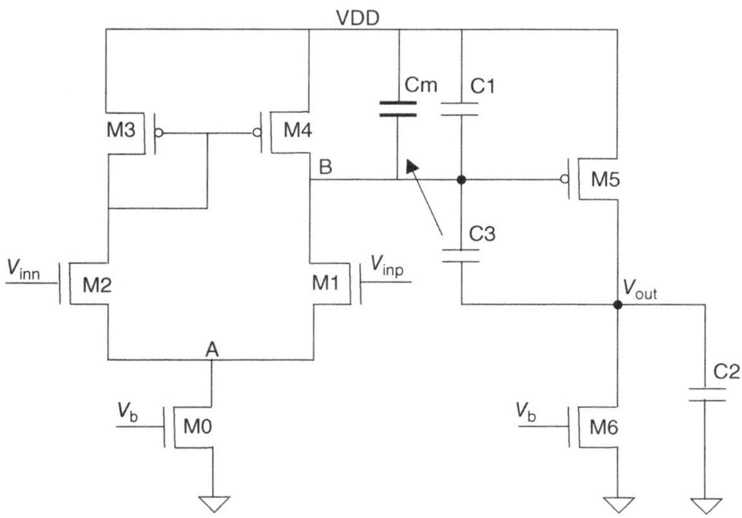

Figure Q8b

The value of C_m is given by the following equation:

$$C_m = C_3 \cdot (1 - A_v) \approx g_{m5} \cdot r_{o2} \cdot C_3$$

where A_v is the voltage gain of the second stage. Adopting the same notation as in Question 7, r_{o1} and r_{o2} represent the output resistances of the first and second stages, respectively, while g_{m1} and g_{m5} denote transconductance of M1 and M5, respectively.

As C_m is boosted by the gain of the second stage, it can significantly increase the total capacitance at the output of the first stage so that we do not need to have a large C_1 for closed-loop stability. It is obvious that the pole (p_1) of the first stage moves to a lower frequency:

$$fp_1 \approx \frac{1}{2 \cdot \pi \cdot r_{o1} \cdot g_{m5} \cdot r_{o2} \cdot C_3}; \text{ assuming } g_{m5} \cdot r_{o2} \cdot C_3 \gg C_1$$

Now, what happens to the pole of the second stage, p_2, after C_3 is connected? A typical incorrect answer states that the frequency of p_2 does not change much because the equivalent Miller capacitance at the output node (V_{out}) is not magnified by the second stage gain.

This misconception is the same as the one mentioned in Question 1. We should not ignore the fact that by connecting the gate and drain of M5, C_3 also establishes a negative-feedback path at the second stage. Similar to Question 1, this negative feedback is of *voltage-sampling, shunt-mixing* type and lowers both the input and output impedance of the second stage. An easy way to perceive this is that at higher frequencies around Pole 2, the impedance of C_3 becomes small and M5 can be treated as a diode-connected transistor with an impedance of $1/g_{m5}$ at its drain. The output impedance at the second stage is therefore reduced to $1/g_{m5}$ from r_{o2} (the original output impedance at the second stage), assuming $1/g_{m5} \ll r_{o2}$.

As such, the frequency of the nondominant pole, p_2, becomes:

$$fp_2 \approx \frac{g_{m5}}{2 \cdot \pi \cdot C_2} \gg \frac{1}{2 \cdot \pi \cdot C_2 \cdot r_{o2}}$$

It is evident that the pole of the second stage moves to a significantly higher frequency due to negative feedback, while the pole of the first stage moves to a significantly lower frequency due to Miller effect. This is sometimes referred to as "*pole splitting*."

The DC gain, $A(0)$, of the op-amp is:

$$A(0) = g_{m1} \cdot r_{o1} \cdot g_{m5} \cdot r_{o2}$$

The gain-bandwidth product (GBW) of the op-amp is:

$$\text{GBW} \approx A(0) \cdot fp_1 = \frac{g_{m1} \cdot r_{o1} \cdot g_{m5} \cdot r_{o2}}{2 \cdot \pi \cdot r_{o1} \cdot g_{m5} \cdot r_{o2} \cdot C_3} = \frac{g_{m1}}{2 \cdot \pi \cdot C_3}$$

Quantitative Analysis: Interested readers can conduct a complete small-signal analysis on this two-stage op-amp. Instead of working on a small-signal model of the whole amplifier, we will focus around the second stage with the compensation (Miller) capacitance. Figure Q8c illustrates the topology for this investigation. Note that we use a voltage source v_1 and output impedance r_{o1} to represent the first stage of the original amplifier. To simplify the analysis, C_1 is ignored as the Miller capacitance dominates the output node of the first stage. The corresponding small-signal model is shown in Figure Q8d.

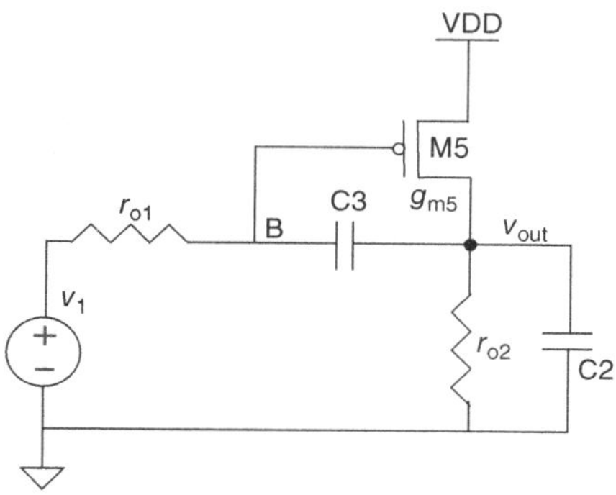

Figure Q8c

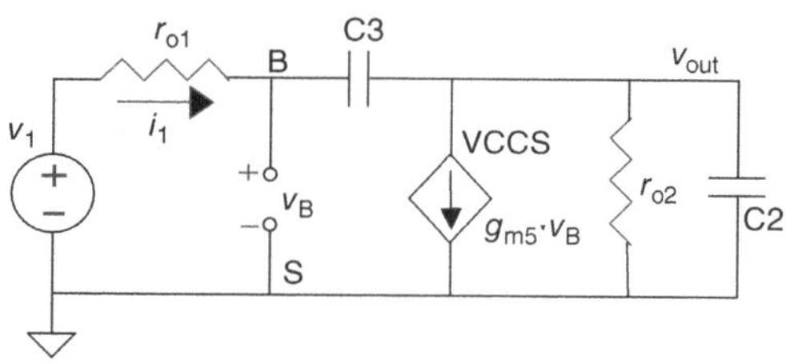

Figure Q8d

Using KVL and KCL (Kirchhoff voltage and current law), we have the following system of equations for the small-signal model in Figure Q8d:

$$
\begin{cases}
i_1 = g_{m5} \cdot v_B + \dfrac{v_{\text{out}}}{z_2} \\
v_B = v_1 - i_1 \cdot r_{o1} \\
v_1 = i_1 \cdot (r_{o1} + z_3) + v_{\text{out}} \\
z_2 = r_{o2} \parallel \dfrac{1}{C_2 \cdot s} = \dfrac{r_{o2}}{r_{o2} \cdot C_2 \cdot s + 1} \\
z_3 = \dfrac{1}{C_3 \cdot s}
\end{cases}
$$

Note that we use the *impedance in the s-domain* for capacitances. The system of equations yields,

$$H_2(s) = \frac{v_{out}}{v_1} = \frac{1 - g_{m5} \cdot z_3}{\frac{r_{o1} + z_3}{z_2} + g_{m5} \cdot r_{o1} + 1} \approx \frac{1 - g_{m5} \cdot z_3}{\frac{r_{o1} + z_3}{z_2} + g_{m5} \cdot r_{o1}}; \; g_{m5} \cdot r_{o1} \gg 1$$

Replacing z_2 and z_3 with their s-domain impedance, the transfer function becomes:

$$H_2(s) = \frac{r_{o2} \cdot (C_3 \cdot s - g_{m5})}{r_{o1} \cdot r_{o2} \cdot C_2 \cdot C_3 \cdot s^2 + r_{o1} \cdot C_3 \cdot s + r_{o2} \cdot C_2 \cdot s + r_{o1} \cdot r_{o2} \cdot C_3 \cdot g_{m5} \cdot s + 1}$$

At relatively low frequencies, the term $r_{o1} \cdot r_{o2} \cdot C_3 \cdot g_{m5} \cdot s$ dominates the denominator, $H(s)$ can be simplified to the following (ignoring the high-frequency non-dominant pole):

$$H_2(s) = \frac{v_{out}}{v_1} = \frac{r_{o2} \cdot (C_3 \cdot s - g_{m5})}{r_{o1} \cdot r_{o2} \cdot C_3 \cdot g_{m5} \cdot s + 1}$$

The open-loop transfer function of the op-amp is:

$$H(s) = \frac{v_{out}}{v_1} \cdot \frac{v_1}{v_{in}} = \frac{-r_{o2} \cdot (C_3 \cdot s - g_{m5}) \cdot g_{m1} \cdot r_{o1}}{r_{o1} \cdot r_{o2} \cdot C_3 \cdot g_{m5} \cdot s + 1} = \frac{r_{o2} \cdot (g_{m5} - C_3 \cdot s) \cdot g_{m1} \cdot r_{o1}}{r_{o1} \cdot r_{o2} \cdot C_3 \cdot g_{m5} \cdot s + 1}$$

Now we can see mathematically the two consequences due to the compensation capacitance C_3:

First, the system has a dominant pole at $-1/(2 \cdot \pi \cdot r_{o1} \cdot r_{o2} \cdot g_{m5} \cdot C_3)$, this verifies the Miller effect associated with C_3: the loading effect of C_3 is amplified by the second stage gain of $g_{m5} \cdot r_{o2}$. As a sanity check, at DC ($s = 0$), the open-loop DC gain is: $g_{m1} \cdot g_{m5} \cdot r_{o1} \cdot r_{o2}$.

Second, the system now also has a zero at g_{m5}/C_3. It is a right-hand-plane zero (RHP zero) because it is located at the right side of the complex frequency plane (s-plane), or in other words, the real part of the zero is a positive number. RHP zero causes an increase of magnitude at higher frequencies, same as left-hand-plane zero (LHP zero), but with a phase lag instead of a phase lead (see *Appendix 3, Section 5, S-domain and frequency domain*). This phase lag could degrade the phase margin of the two-stage op-amp. A well-known method to cancel this phase lagging effect of C_3 is to put a nulling resistor in series with C_3 and transform the RHP zero into a LHP zero, which contributes a phase lead at higher frequency instead (the op-amp in Question 11 is an example of Miller compensation with nulling resistors). For high-speed amplifier design, indirect compensation is a good candidate for compensating the amplifier and maintaining a high bandwidth/fast settling. Unlike direct Miller compensation, which connects a capacitor directly between the input and output of a gain stage, indirect compensation feeds the compensation current back through an internal low-impedance node such as a cascode node (the op-amp in Question 10 and Question 15 are examples of indirect compensation). As there are many important aspects of amplifiers besides stability that we would like to cover in this chapter, we are not going to dive into these compensation schemes. Interested readers are encouraged to explore and analyze these approaches as compensation is nonetheless one of the most important design concerns for closed-loop amplifiers.

If we do not ignore the s^2 term in the denominator of the op-amp transfer function, we can derive and verify the second high-frequency pole, or the non-dominated pole. The details of this process are left for the interested reader to explore.

Short Answer to Question 8

What is the frequency of the two poles of the op-amp in the new circuit diagram?

Miller capacitor (C_3) lowers the frequency of the dominant pole of the first stage. The negative feedback through C_3 shortens the time constant of the second stage. The pole splitting improves the phase margin of the op-amp.

The frequency of the two poles of the op-amp:

$$fp_1 \approx \frac{1}{2 \cdot \pi \cdot r_{o1} \cdot g_{m5} \cdot r_{o2} \cdot C_3}$$

$$fp_2 \approx \frac{g_{m5}}{2 \cdot \pi \cdot C_2}$$

Question 9

An op-amp is used as a unity-gain buffer, and its open-loop frequency response is shown in Figure Q9a. The phase lag reaches 180° at the frequency f_1 (f_1 is less than the unity-gain frequency, f_c). Between the frequency f_1 and f_2 ($f_1 < f_2 < f_c$), the phase lag is beyond 180°. At frequency f_2, the phase lag returns to the 180° mark, remaining below this value across the unity-gain frequency (f_c). Question: is the op-amp stable when used as a unity-gain buffer?

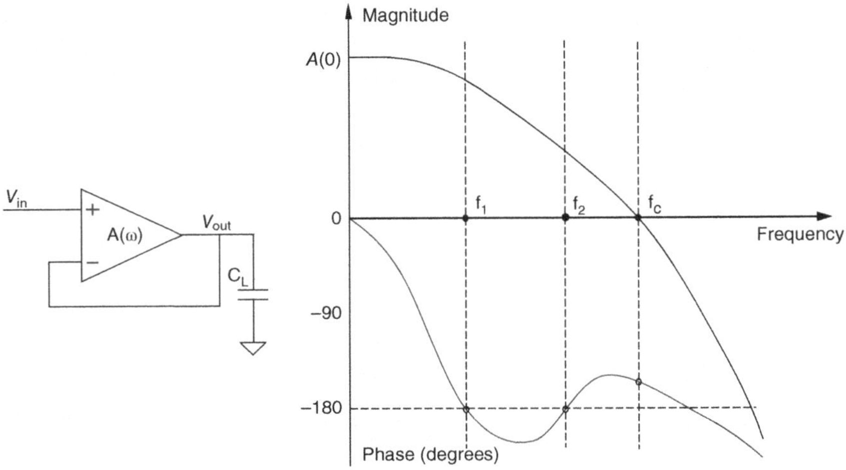

Figure Q9a

Intuitive and Quantitative Analysis: This has been a controversial question not only in interviews, but also in daily design work. The most common wrong belief is that the op-amp would not be stable in a unity-gain feedback loop, and this is based on flawed reasoning that some people find hard to argue with – if we apply a sinusoidal signal at frequency f_1, it is amplified by the op-amp with a gain larger than 1 and with a phase lag of 180°. This amplified signal is then fed back to the input with another 180° shift due to negative feedback, or in other words, with the same phase (360°) as the input signal. This in fact forms positive feedback with a return ratio larger than 1, and the loop will oscillate.

To some, this reasoning is hard to argue with, partially because it is essentially the same reasoning process that they understand the stability criterion: if at the unity-gain frequency, the phase lag is already 180°, the feedback signal would be the same as input signal for both magnitude and phase, which would trigger a sustainable oscillation.

Maybe the best way to explain the flaw of this reasoning is to examine how a feedback system works visually through its block diagram. Figure Q9b illustrates the flawed reasoning process described above. The input sinusoidal signal, v_{in}, with the frequency f_1 drives the input of the amplifier directly, and the output of the amplifier, v_{out}, is larger than vin in magnitude (open-loop gain > 1 at the frequency of f_1) and lags the v_{in} by 180°. With the unity-gain negative feedback, the v_{out} is fed back and added to V_{in} with a 360° phase shift, the input signal to the op-amp now becomes larger than $2 \cdot v_{in}$. And so on.

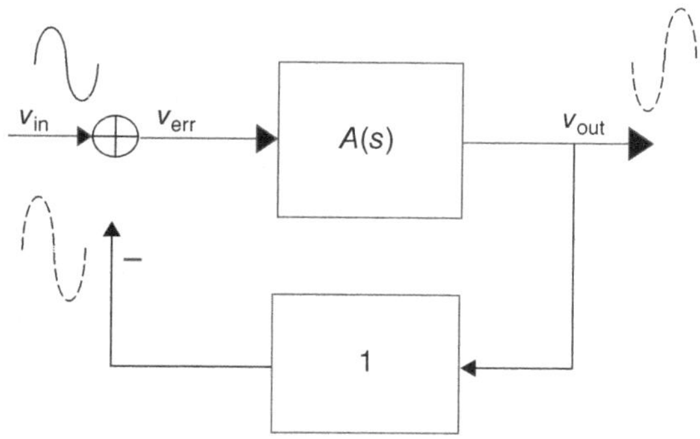

Figure Q9b

We should have already found the flaw. In a closed-loop feedback system, the v_{in} would first be deducted by the feedback signal, which is v_{out} in a unity-gain feedback, and the difference, called the error signal, v_{err}, is the input to the op-amp. That is:

$$v_{out} \neq v_{in} \cdot A(s)$$

$$v_{out} = (v_{in} - v_{out}) \cdot A(s) = v_{err} \cdot A(s)$$

So unfortunately, this reasoning is mistaken at the very beginning. It is important to recognize that while phase margin is an open-loop metric used to assess stability, it characterizes a system that ultimately operates in a closed-loop configuration.

What happens in reality? The v_{in} sees a closed-loop system, which has a transfer function of:

$$H(s) = \frac{v_{out}}{v_{in}} = \frac{A(s)}{1 + A(s)}$$

Note that the sign of $A(s)$ is positive in Figure Q9b because the amplifier's inverting property is already incorporated into the negative feedback at the input mixing point. Compared to the open-loop transfer function (or the magnitude of the op-amp gain), $A(s)$, the closed-loop gain is less than 1, and the system responds to the input with less phase lag and extended bandwidth. This is, of course, to assume that the closed-loop system is stable – it is a little tricky because the condition of the stability is under discussion here, and the closed-loop system may or may not be stable (a *petitio principii* situation – also known as *circular reasoning*). However, if the closed-loop system is in fact not stable, the v_{in} still needs to mix with the feedback signal from the output before reaching the input of the op-amp, and the reasoning under investigation is flawed either way.

One might ask: why not disregard the external input signal and analyze the loop in isolation? As shown in Figure Q9c, if, at certain instant of time, there exists a "trigger" sinusoidal signal at the input of the op-amp with the frequency f_1, now this signal is directly amplified by $A(s)$ and the output will be a larger sinusoidal signal with the same frequency and 180° phase shift, fed back with 360° phase shift (due to the negative feedback) and enhancing the assumed initial sinusoidal signal at the input of the op-amp.

This argument originates from a misconception about the circuit frequency response. It is important to realize that the frequency response is a steady-state response, meaning that the output's magnitude and phase relationship to the input frequency is determined when the output reaches the steady state. If, in the time domain, we input a sinusoidal signal with a certain frequency to the op-amp, the output will experience an initial transient process governed by the circuit time constant until it reaches a steady state and responds with a sinusoidal signal with the same frequency (under the condition that we have continuously input the sinusoidal signal during this process, which also cannot be not true in the above argument), as shown in Figure Q9d. Again, the assumption of the above argument is flawed.

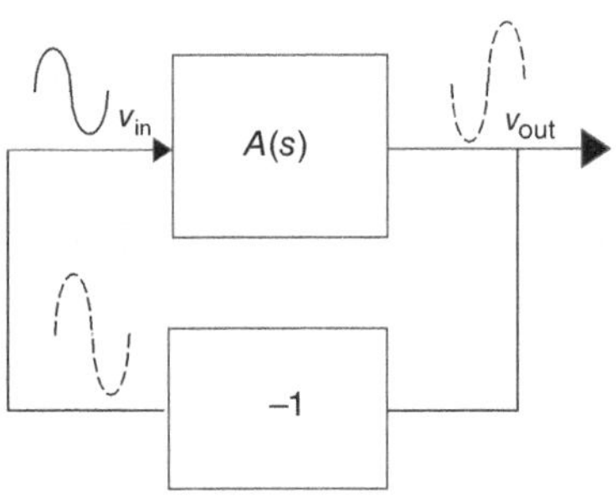

Figure Q9c

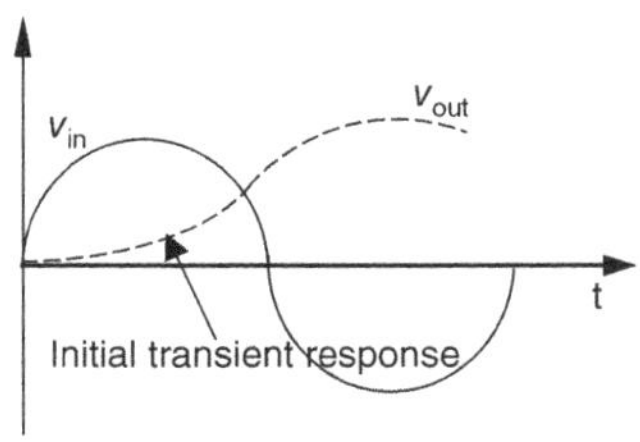

Figure Q9d

The stability of closed-loop negative feedback systems is mathematically studied in the *feedback control theory*, and the derivation of stability criterion is an involved procedure and beyond the scope of this book. There are several stability criteria that exist. The most widely used one in engineering is the *Nyquist stability criterion*, which determines the stability of a closed-loop system by analyzing Nyquist plot of the open-loop system. A Nyquist plot is drawn in a Cartesian complex plane and might not look familiar to some circuit design students and engineers who mostly work with the Bode plot. An equivalent version of Nyquist stability criterion using open-loop Bode plot is therefore presented here.

The Nyquist stability criterion: For a linear time-invariant system, assume the number of closed-loop poles in the right-hand plane (RHP) is Z, and the number of open-loop RHP poles is P. Define N_+ as the number of times that the phase frequency response crosses the $-180°$ line in the positive direction (the direction in which the phase increases) for the frequency range where gain is greater than 1 (0 dB). Similarly, define N_- as the number of times that the phase frequency response crosses the $-180°$ line in the negative direction (the direction in which the phase decreases) for the frequency range where the gain is greater than 1 (0 dB). Let N be the difference between N_+ and N_-, i.e.,

$$N = N_+ - N_-$$

Nyquist stability criterion states that Z, P, and N are related by the following expression:

$$Z = P - 2 \cdot N$$

The closed-loop system is stable if and only if Z equals zero:

$$Z = P - 2 \cdot N = 0$$

Note that if a pole of a system is in the right-hand plane (i.e., a pole with positive real part), the system is not stable. In most applications, the op-amp is stable in open-loop so that $P = 0$. A stable closed-loop system requires $Z = 0$, and the stability criterion reduces to:

$$N = N_+ - N_- = 0; \text{ or } N_+ = N_-$$

This means a closed-loop system is stable if, within the frequency range at which the gain is greater than 1 (0 dB), the number of times the open-loop phase response crosses the $-180°$ line in the positive direction equals the number of times it crosses in the negative direction.

Using the Nyquist stability criterion on the open-loop frequency response in Figure Q9a, we conclude that at the frequency range where gain is greater than 0 dB (i.e., $f < f_c$), there is one negative crossing of $-180°$ line at f_1 and one positive crossing of $-180°$ line at f_2 in the phase frequency response (as shown in Figure Q9e), and therefore the op-amp is stable when used as a unity-gain buffer.

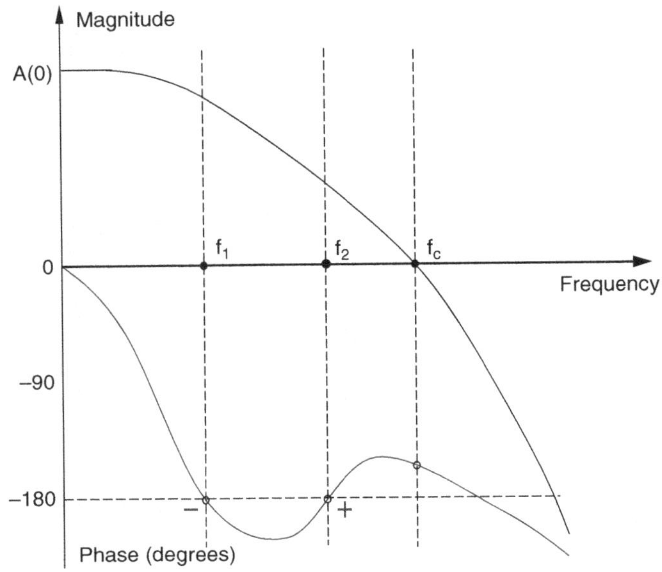

Figure Q9e

Just a reminder: the condition that the op-amp is used in a unity-gain buffer is important here so that we can use the open-loop frequency response of the op-amp alone to judge the closed-loop stability. If the feedback factor is not unity, then we need to consider the frequency response of the complete open-loop circuit including the feedback path to apply the Nyquist stability criterion. We will study feedback factors of closed-loop amplifiers in Question 17.

Interested readers are encouraged to use the Nyquist stability criterion to reconfirm the stability status of simpler cases where the phase response either crosses $-180°$ line just once or never before the gain drops to 0 dB. As a matter of fact, the popular phase margin method for judging stability is a subset of Nyquist stability criterion. The complete rule set should be applied to determine the closed-loop stability in more complicated frequency response situations such as the one in this problem.

In some applications, an op-amp with the frequency response shown in Figure Q9a is considered *conditionally stable*. This is because, as shown in Figure Q9f, a change in the op-amp gain – consequently its gain-bandwidth product – can lead to instability. This occurs whether the gain is decreased $(N = N_+ - N_- = 0 - 1 = -1)$ or increased $(N = N_+ - N_- = 1 - 2 = -1)$. Because op-amp gain can decrease under large-signal conditions, systems using conditionally stable op-amps may exhibit signal-dependent stability. It should be noted that while Figure Q9f shows a constant phase response for simplicity, real-world phase responses typically shift in response to the factors that affect gain.

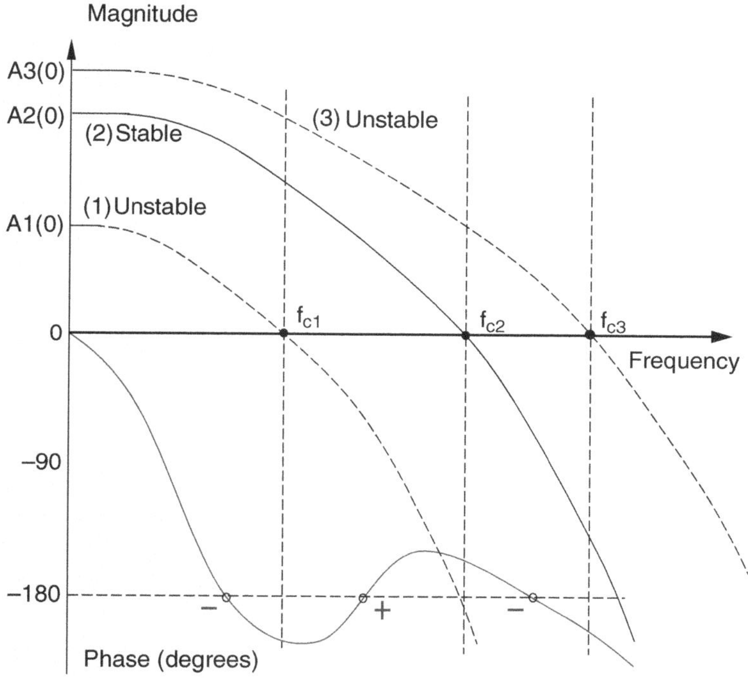

Figure Q9f

Short Answer to Question 9

Is the op-amp stable when used as a unity-gain buffer?

The phase response crossing the −180° line before the gain reaches 0 dB does not necessarily indicate instability. The complete rule set of Nyquist stability criterion should be applied to determine the closed-loop stability in this situation: The op-amp with a frequency response in Figure Q9a is stable when used as a unity-gain buffer.

Question 10

The schematic of a fully differential op-amp is shown in Figure Q10a. Please list the devices that have a significant contribution to the input-referred noise and explain your choices. Only the devices' intrinsic noise is of concern here; ignore any noise coupled from the power supplies.

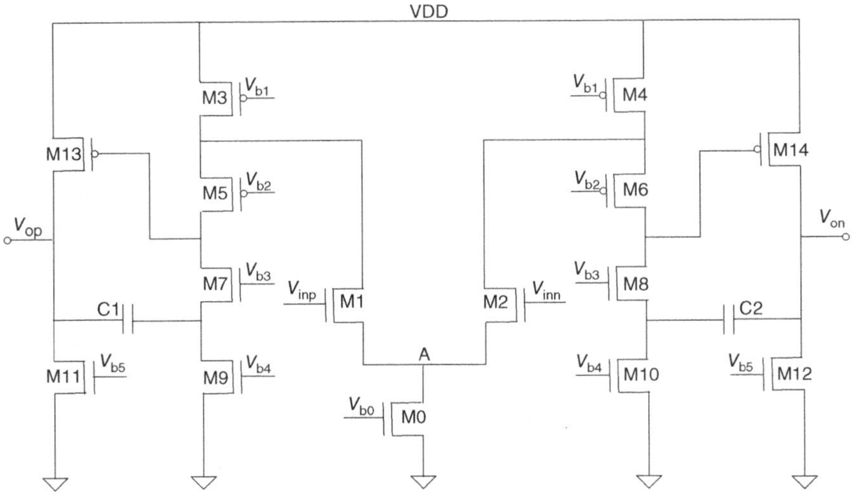

Figure Q10a

Intuitive Analysis: Noise in circuits is inherently a complex topic. Before we move into the analysis, it is beneficial to refresh some basic concepts about circuit noise. First of all, the input-referred noise is to attribute the impact of all noise sources in an amplifier to an equivalent noise source at the input of the amplifier. In the voltage amplifier example here, this means all the noise currents in the amplifier need to be referred to an equivalent noise voltage source at the input of the amplifier, by dividing the total current noise (RMS) with the transconductance, g_m, of the input transistor, as illustrated in Figure Q10b.

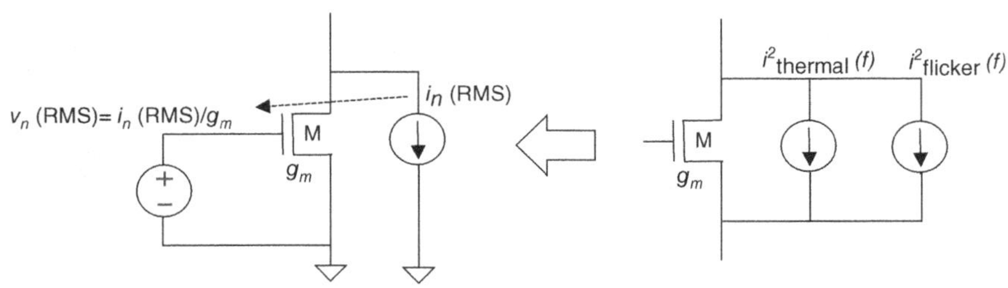

Figure Q10b-c

The two major intrinsic noise sources in CMOS transistors are thermal noise and flicker noise. As shown in Figure Q10c, the noise model of a MOSFET includes two current-noise sources representing these components.

Thermal noise depends on the equivalent resistance and the temperature of the device. The power spectral density (PSD, see *Appendix 4, Section 6, Circuit noise fundamentals*) of the thermal current noise in a resistor is:

$$i^2_{\text{thermal}}(f) = \frac{4 \cdot K \cdot T}{R}$$

where K is Boltzmann's constant and T is absolute temperature (in K). The PSD of a resistor's thermal noise is independent of frequency, or "white," meaning that the noise power is evenly spread over the frequency domain.

For a CMOS transistor, the PSD of the thermal current noise depends on its channel resistance:

$$i^2_{\text{thermal}}(f) = \frac{4 \cdot K \cdot T}{\frac{3}{2} \cdot \frac{1}{g_m}} = \frac{8 \cdot K \cdot T \cdot g_m}{3}$$

where g_m is the transconductance of the transistor.

The model of flicker noise is more complicated and depends on several factors including the process characteristics of the devices. One approximate expression for the flicker-noise current PSD in CMOS transistors is:

$$i^2_{\text{flicker}}(f) = \frac{\text{KF} \cdot I_D^{\text{AF}}}{(C'_{\text{ox}})^2 \cdot L \cdot W \cdot f}$$

where KF is the flicker-noise coefficient, I_D is the drain current of the transistor, AF is flicker-noise exponent (generally close to 1), C'_{ox} is gate capacitance (C_{ox}) per area, and W and L are gate width and length, respectively.

Unlike thermal noise, flicker noise has greater power at low frequencies. Specifically, its noise power density is approximately inversely proportional to the frequency, so flicker noise is also referred to as *1/f noise*.

The PSD of the total noise of a CMOS transistor is:

$$i^2_{n\_\text{total}}(f) = \frac{8 \cdot K \cdot T \cdot g_m}{3} + \frac{\text{KF} \cdot I_D^{\text{AF}}}{(C'_{\text{ox}})^2 \cdot L \cdot W \cdot f}$$

To obtain the RMS value of a transistor's total noise current, we integrate the PSD over the frequency range (bandwidth) of interest.

$$i_{n\_\text{total}}(\text{RMS}) = \sqrt{\int_{f_{\min}}^{f_{\max}} \left[\frac{8 \cdot K \cdot T \cdot g_m}{3} + \frac{\text{KF} \cdot I_D^{\text{AF}}}{(C'_{\text{ox}})^2 \cdot L \cdot W \cdot f}\right] \cdot df}$$

where $f_{\max}$ and $f_{\min}$ are the highest and lowest frequency of interest.

Now let us discuss the contributions of the transistors in this two-stage amplifier to the input-referred voltage noise. Because the amplifier is fully symmetrical, we only need to analyze one side of the circuit. Figure Q10d shows the right side of the circuit with current-noise sources added to all major transistors. Note that we do not attach a noise current source to the tail device M0, because noise current in M0 appears as a common-mode noise to the amplifier and does not affect differential signals (assuming well-matched differential halves). For simplicity, we combine the thermal and flicker noise of each device into a single equivalent current-noise source and focus on each device's contribution to the input-referred noise voltage.

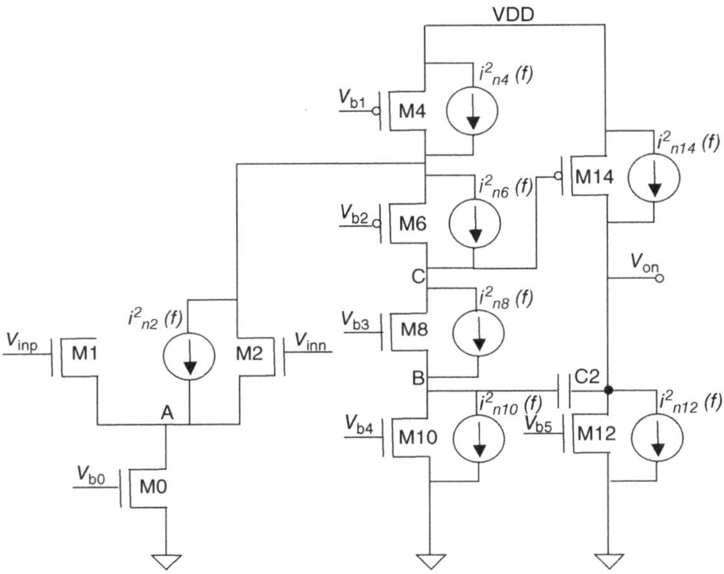

Figure Q10d

In determining which devices are the most significant contributors to the input-referred noise, let us start by excluding the least significant ones. The current noise of the two transistors in the second stage, M12 and M14, can be referred to a voltage noise source, v_{nc}, at the input of the second stage (node C):

$$v_{nc}^2(f) = \frac{i_{n14}^2(f) + i_{n12}^2(f)}{(g_{m14})^2}$$

The voltage noise v_{nc} can then be referred back to the amplifier input, v_{nin}, by the voltage gain of the first stage:

$$v_{nin}^2(f) = \frac{v_{nc}^2(f)}{(g_{m2} \cdot r_{o1})^2} = \frac{I_{n14}^2(f) + I_{n12}^2(f)}{(g_{m14} \cdot g_{m2} \cdot r_{o1})^2}$$

where r_{o1} is the output impedance of the first stage.

The noise generated by M12 and M14 in the second stage is attenuated by the (typically large) voltage gain of the first stage and therefore contributes only weakly to the input-referred voltage noise.

In the first stage, the input transistor M2 and the folded-branch transistors M4, M6, M8, and M10 all generate current noise. At first glance, it may appear that all of these noise currents directly contribute to the total noise current at the drain of M2 and therefore should all be significant contributors to the input-referred noise voltage. However, this is a common misconception.

The key point is that transistors M6 and M8 each have a degeneration device (M4 and M10, respectively) cascoded with them. Unlike M4 and M10, which do not have source degeneration, the noise currents in M6 and M8 are strongly suppressed by this degeneration. An intuitive way to understand this is to imagine that the noise current of M8 momentarily increases the current flowing into the drain of M10. Because the drain of M10 sees a large impedance (r_{o10}), its drain voltage rises, which reduces the gate-to-source voltage of M8 (V_{GS8}). This, in turn, reduces the drain current in M8 and therefore the current flowing into the drain of M10. This negative feedback suppresses changes in the drain current of M8, including those caused by noise.

It is now clear that the major contributors to the input-referred noise voltage of this two-stage amplifier are the input transistors M1, M2 and the folded-branch cascode transistors M3, M4, M9, and M10. The input-referred noise due to these devices can be calculated as:

$$v_{\text{nin}}^2(f) = \frac{i_{n1}^2(f) + i_{n2}^2(f) + i_{n3}^2(f) + i_{n4}^2(f) + i_{n9}^2(f) + i_{n10}^2(f)}{(g_{m1})^2}$$

Input-referred noise in RMS is:

$$v_{\text{nin}}(\text{RMS}) = \frac{\sqrt{\int_{f_{\min}}^{f_{\max}} \left[i_{n1}^2(f) + i_{n2}^2(f) + i_{n3}^2(f) + i_{n4}^2(f) + i_{n9}^2(f) + i_{n10}^2(f)\right] \cdot df}}{g_{m1}}$$

The limits $f_{\min}$ and $f_{\max}$ are determined by the equivalent noise bandwidth, which is related to the gain-bandwidth product of the amplifier.

Quantitative Analysis: Most of the above noise analysis is straightforward, except for the cascode current sources in the folded branches. Let us use the small-signal model to verify the claim that we have made in the intuitive analysis: noise of transistors M6 and M8 is reduced by their degeneration. We use cascode transistors M8 and M10 as an example in the following quantitative analysis. Figure Q10e shows the small-signal model of the circuit around transistors

M8 and M10. Note that we model the impedance of cascode PMOS transistors M6, M4 and M2 as a single resistance, rop, which can be obtained by the following equation:

$$r_{\text{op}} = r_{o4} \parallel r_{o2} + r_{o6} + g_{m6} \cdot (r_{o4} \parallel r_{o2}) \cdot r_{o6} \approx g_{m6} \cdot (r_{o4} \parallel r_{o2}) \cdot r_{o6}$$

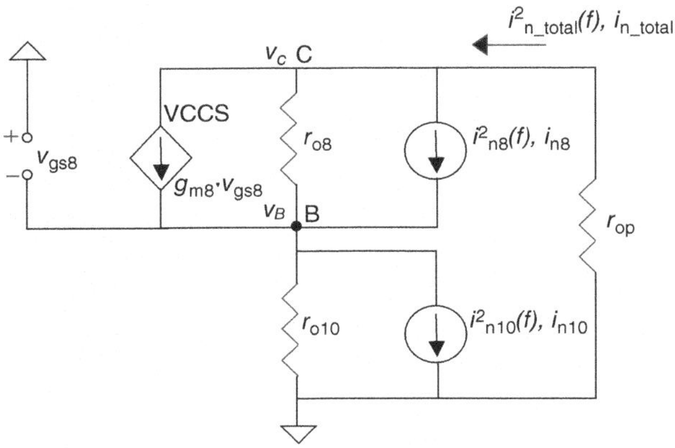

Figure Q10e

Because there are two current-noise sources (M8 and M10), we use superposition and analyze one noise source at a time. For the time-domain noise currents generated by M8 and M10, we use the notations i_{n8} and i_{n10}, respectively, and we denote the total noise current as i_{n_total}.

We first consider the noise of M10 and assume M8 is noise-free. Applying Kirchhoff's circuit laws yields the following system of equations:

$$\begin{cases} \dfrac{v_B}{r_{o10}} + i_{n10} = -g_{m8} \cdot v_B + \dfrac{v_C - v_B}{r_{o8}} \\ \dfrac{v_B}{r_{o10}} + i_{n10} = -\dfrac{v_C}{r_{\text{op}}} \\ \quad i_{n\_\text{total}} = -\dfrac{v_C}{r_{\text{op}}} \end{cases}$$

Solving the equations for v_B, we have:

$$v_B = \frac{(r_{o8} + r_{\text{op}}) \cdot i_{n10}}{\dfrac{r_{o8}}{r_{o10}} + r_{o8} \cdot g_{m8} + \dfrac{r_{\text{op}}}{r_{o10}} + 1}$$

For a rough estimation, let us assume that the transconductances of all NMOS and PMOS devices are equal to g_m, and the output resistances of all NMOS and PMOS devices are equal to ro.

$$v_B \approx \frac{1}{3} \cdot r_o \cdot i_{n10}$$

The total current noise i_{n_total} is:

$$i_{n\_total} = \frac{v_B}{r_{o10}} + i_{n10} = \frac{v_B}{r_o} + i_{n10} \approx \frac{4}{3} \cdot i_{n10}$$

The current noise generated by M10 directly contributes to the total current noise in the folded branch, and therefore the current noise of input transistor M2.

Now let us look at noise generated by M8 and assume M10 is noiseless. Again, by applying KCL and KVL, we have the following system of equations:

$$\begin{cases} \dfrac{v_B}{r_{o10}} = -\dfrac{v_C}{r_{op}} \\ \dfrac{v_B}{r_{o10}} = -g_{m8} \cdot v_B + \dfrac{v_C - v_B}{r_{o8}} + i_{n8} \\ i_{n\_total} = \dfrac{v_B}{r_{o10}} \end{cases}$$

Solving the system of equations for v_B and i_{n_total}, we obtain:

$$v_B = \frac{r_{o8} \cdot i_{n8}}{\frac{r_{o8}}{r_{o10}} + r_{o8} \cdot g_{m8} + \frac{r_{op}}{r_{o10}} + 1} \approx \frac{i_{n8}}{\frac{3}{2} \cdot g_{m8}}$$

$$i_{n\_total} = \frac{v_B}{r_{o10}} \approx \frac{i_{n8}}{\frac{3}{2} \cdot g_{m8} \cdot r_{o10}}$$

It is now evident that the noise generated by M8 is attenuated by degeneration (negative feedback) by a factor of $g_{m8} \cdot r_{o10}$ and therefore becomes an insignificant contributor to the input-referred noise.

Short Answer to Question 10

Please list the devices that have a significant contribution to the input-referred noise and explain your choices.

The most significant contributors to the input-referred noise are:
Input transistors: M1 and M2
Cascode transistors: M3, M4, M9, and M10

Transistors M5, M6, M7, and M8 are not significant noise contributors because of their source degenerations. Transistors M11, M12, M13, and M14 are also not significant noise contributors because their noise is attenuated by the gain of the first stage.

Question 11

The schematic of a two-stage op-amp is shown in Figure Q11a. What is the gain-bandwidth product (GBW) of the op-amp? Assuming the op-amp is used as a unity-gain amplifier (buffer) and the second stage is the limiting factor in terms of the output slew rate, how to design the bias current I_{b2}, so that the amplifier output would not be limited by slew rate?

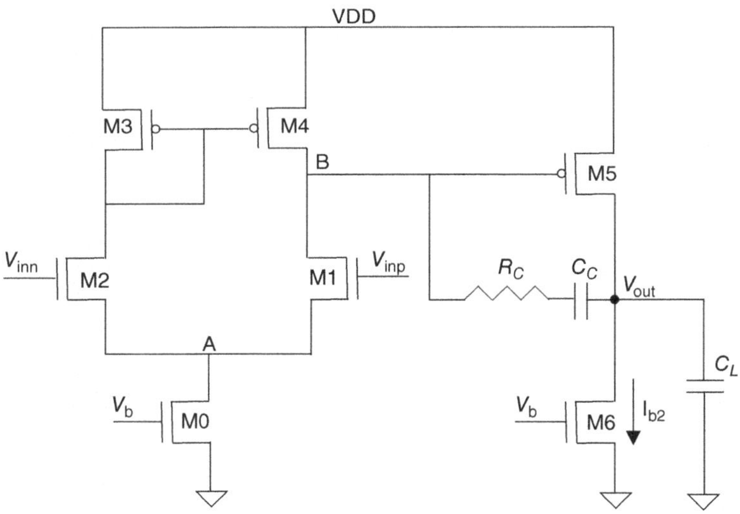

Figure Q11a

Intuitive and Quantitative Analysis: The first part of the question is relatively straightforward. We know from the analysis of Question 8 that, due to the Miller capacitance C_C, the frequency of the dominant pole of the two-stage op-amp is at approximately:

$$fp_1 \approx \frac{1}{2 \cdot \pi \cdot r_{o1} \cdot g_{m5} \cdot r_{o2} \cdot C_C}$$

where r_{o1} is the output resistance of the first stage, g_{m5} is the transconductance of the input transistor of the second stage (M5), and r_{o2} is the output resistance of the second stage.

For a well-compensated op-amp with a phase margin of 60° or higher, the frequency of the second (nondominant) pole is well above the gain-bandwidth product (GBW) of the op-amp (see *Appendix 3, Section 4, Second-order systems*). As a result, the slope of the magnitude frequency response is approximately the same as that of a single-pole system (first-order system), or −20 dB/decade, up to around the GBW.

The GBW is approximately:

$$\text{GBW} \approx A(0) \cdot fp_1 = \frac{g_{m1} \cdot g_{m5} \cdot r_{o1} \cdot r_{o2}}{2 \cdot \pi \cdot r_{o1} \cdot g_{m5} \cdot r_{o2} \cdot C_C} = \frac{g_{m1}}{2 \cdot \pi \cdot C_C}$$

where g_{m1} is the transconductance of the first-stage input transistor M1.

Therefore, the GBW is the ratio of the input stage transconductance to the value of the compensation capacitor.

For the second part of the question, we need to realize that when the op-amp is used as a unity-gain buffer, the closed-loop −3 dB bandwidth is the GBW of the op-amp. The question now becomes what bias current must be provided at the second stage to avoid slewing when the closed-loop amplifier has a −3 dB bandwidth equal to the GBW of the op-amp.

To answer the question, one must understand the mechanisms behind slew rate limiting in the amplifier. An amplifier output slews when it cannot supply sufficient current to charge or discharge the load capacitance at the rate allowed by its (small-signal) bandwidth. This often happens when the output experiences a fast transition with a large voltage swing and the output current is limited by a bias current (current source) of the amplifier. For the op-amp in Figure Q11a, the worst-case condition occurs during a full-scale negative step transition (V_{FS}), as shown in Figure Q11b.

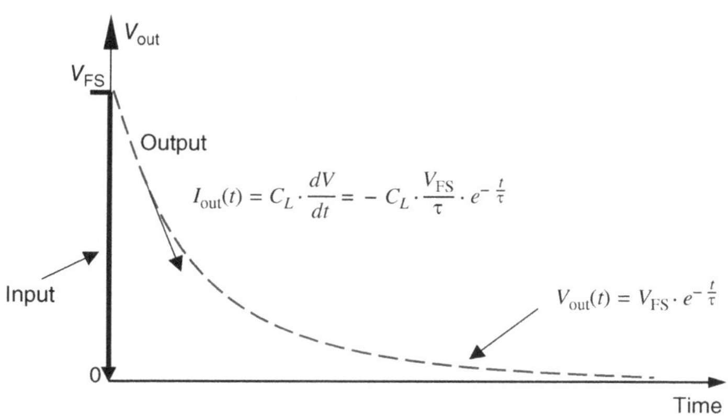

Figure Q11b

Utilizing a single-pole approximation for the closed-loop amplifier, with a time constant of τ_{cl}, the output response to a full-scale negative step input is:

$$V_{\text{out}}(t) = V_{\text{FS}} \cdot e^{-\frac{t}{\tau_{\text{cl}}}}$$

And the current needed to support this step response is:

$$I_{\text{out}}(t) = C_L \cdot \frac{dV_{\text{out}}(t)}{dt} = -C_L \cdot \frac{V_{\text{FS}}}{\tau_{\text{cl}}} \cdot e^{-\frac{t}{\tau_{\text{cl}}}}$$

where C_L is the load capacitance of the amplifier.

In this scenario, transistor M6 provides the discharge current (I_{b2}) required to pull the output voltage down across the load capacitance C_L.

The maximum current happens at start of the transition $(t = 0)$, which is:

$$I_{\text{out}}(0) = C \cdot \frac{V_{\text{FS}}}{\tau_{\text{cl}}} = 2 \cdot \pi \cdot f_{\text{cl}-3\text{dB}} \cdot C_L \cdot V_{\text{FS}} = 2 \cdot \pi \cdot \text{GBW} \cdot C_L \cdot V_{\text{FS}}$$

where $f_{\text{cl}-3\text{dB}}$ is the -3 dB frequency of the closed-loop amplifier, which is equal to the GBW of the op-amp in unity-gain feedback. Note that the relationship between $f_{-3\text{dB}}$ and τ for a first-order system is (see *Appendix 3, Section 2, First-order systems*):

$$\tau = \frac{1}{\omega_{-3\text{dB}}} = \frac{1}{2 \cdot \pi \cdot f_{-3\text{dB}}}$$

For a slew-free settling, the minimum discharge current needed is:

$$I_{\text{out}}(\text{min}) = I_{\text{out}}(0) = 2 \cdot \pi \cdot \text{GBW} \cdot C_L \cdot V_{\text{FS}}$$

Note that only a portion of the bias current I_{b2} is available to discharge the load capacitance. The rest needs to support the current in M5. With a fast negative step input, the current flowing through M5 decreases significantly. The bias current (I_{b2}) required is approximately:

$$I_{b2} > I_{\text{out}}(\text{min}) = 2 \cdot \pi \cdot \text{GBW} \cdot C_L \cdot V_{\text{FS}}$$

Evidently, the required output-stage bias current is directly proportional to the output full-scale voltage, the load capacitance, and the amplifier bandwidth. Note that I_{b2} is used to discharge the load capacitance. To charge the load capacitance (i.e., in a positive step response), transistor M5 needs to provide the current. Because M5 does not function as a constant current source – as its source-to-gate voltage varies to accommodate charging demands – it typically does not serve as the limiting factor for the slew rate.

We have assumed that the second stage is the limiting factor for slew rate in this question. In many cases this is a reasonable assumption as the output stage often needs to drive large load capacitances with a large voltage swing. But this assumption is not always true and sometimes internal nodes can limit the slew rate. For example, if the bias current of the first stage is too weak, the output of the first stage may experience slewing when driving a large Miller capacitance, even though the voltage swing is attenuated by the second stage gain.

In many practical applications, meeting the bias current criterion for slew-free transient response could result in large power consumptions. In fact, many amplifiers allow some slewing at large swings as long as the distortion caused by slewing is within specifications.

Note that the bias current requirement derived in the preceding discussion is driven solely by slew-rate considerations. Besides the slew-rate concern, the values of bias currents depend on many other design targets, for example, the DC gain, GBW, phase margin, distortion, noise, input common-mode range, power consumption budget, etc. The design of an op-amp is often a trade-off of many design considerations.

Short Answer to Question 11

What is the gain-bandwidth product (GBW) of the op-amp? How to design the bias current I_{b2}, so that the amplifier output would not be limited by slew rate?

The GBW of the op-amp is:

$$\text{GBW} \approx A(0) \cdot fp_1 = \frac{g_{m1}}{2 \cdot \pi \cdot C_C}$$

The bias current required for a slew-free settling depends on the voltage swing, load capacitance, and the closed-loop bandwidth of the op-amp:

$$I_{b2} > 2 \cdot \pi \cdot \text{GBW} \cdot C_L \cdot V_{\text{FS}}$$

Question 12

Identify and explain the mechanisms responsible for second-order harmonic distortion in the single-ended CMOS common-source amplifier shown in Figure Q12a. How to reduce this distortion? Assuming the resistor R is perfectly linear, and device M1 is a long-channel NMOS transistor. What about third-order harmonic distortion?

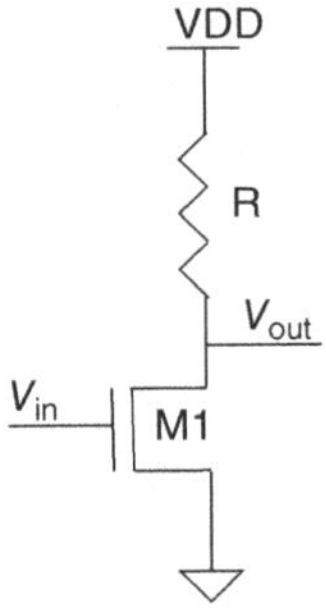

Figure Q12a

Intuitive and Quantitative Analysis: Nonlinearity in circuits generates harmonic distortions, such as second-order (HD2) and third-order (HD3) harmonics. In this circuit the NMOS transistor M1 is the source of nonlinearity. Even though we often treat a CMOS transistor as a linear device, i.e., voltage-controlled current source (VCCS), and use its transconductance (g_m) to calculate the gain of the amplifier, this is just an approximation for small signals. Recall that for long-channel NMOS transistors operating in the saturation region, the *I-V* transfer characteristic equation (large-signal model) is:

$$I_D = \frac{K_p \cdot W}{2 \cdot L} \cdot (V_{GS} - V_{TH})^2 [1 + \lambda \cdot (V_{DS} - V_{DSsat})]; V_{DSsat} = V_{GS} - V_{TH}$$

where K_p is the transconductance parameter, V_{TH} is the threshold voltage, and λ is the channel-length modulation parameter of the transistor.

The *I-V* equation indicates that the drain current relates to the square of the V_{GS}, which is the source for the second-order harmonic distortion. As the channel-length modulation parameter (λ) is typically a small number far less than 1 (for example, 0.01), for a long-channel CMOS transistor, we assume $\lambda = 0$ in the following analysis.

The small-signal transconductance (g_m) of the NMOS transistor is:

$$g_m = \frac{dI_D}{dV_{GS}} = \frac{K_p \cdot W}{L} \cdot (V_{GS} - V_{TH})$$

Notice that, in the single-ended common-source amplifier in Figure Q12a, V_{GS} equals V_{in}. If the AC input signal is small, we can assume V_{GS} is a constant number, and therefore the transconductance (g_m) of the transistor is constant as well. For large input signals, however, the g_m changes significantly, causing harmonic distortions in the circuit.

Let us assume the input signal (large-signal) to the common-source amplifier in Figure Q12a is:

$$V_{IN} = V_{GS} = V_{GSDC} + A \cdot \sin(\omega \cdot t)$$

where V_{GSDC} is the DC bias voltage on the gate of M1, A is the amplitude of the input AC signal, and ω is the angular frequency of the input signal.

The output voltage (large-signal), V_{OUT}, is:

$$V_{OUT} = \text{VDD} - I_D \cdot R = \text{VDD} - \frac{K_p \cdot W \cdot R}{2 \cdot L} \cdot [V_{GSDC} + A \cdot \sin(\omega \cdot t) - V_{TH}]^2$$

Simplify the equation, and ignore the DC terms, we obtain the expression for the AC output voltage, v_{out}:

$$\begin{aligned} v_{out} = &- \frac{K_p \cdot W \cdot R}{L} \cdot (V_{GSDC} - V_{TH}) \cdot A \cdot \sin(\omega \cdot t) \\ &+ \frac{K_p \cdot W \cdot R}{4 \cdot L} \cdot A^2 \cdot \cos(2 \cdot \omega \cdot t) \end{aligned}$$

Hint: Some basic trigonometry is needed in this type of analysis and *half-angle formula* is useful here:

$$\sin^2\left(\frac{\alpha}{2}\right) = \frac{1 - \cos(\alpha)}{2}$$

The first term of v_{out} is what we expect from the amplifier, i.e., the linear output with a gain of $g_m \cdot R$. The second term is not what we desire, but the second-order harmonic term with the frequency equal to twice the input signal frequency.

The magnitude of the second-order harmonic distortion, HD_2 in dBc, is then:

$$\begin{aligned} HD_2 &= 20 \cdot \log_{10} \frac{\text{Amplitude of the output second-order harmonic } (V)}{\text{Amplitude of the output fundamental } (V)} \\ &= 20 \cdot \log_{10} \frac{A}{4 \cdot (V_{GSDC} - V_{TH})} \end{aligned}$$

Note that magnitude of the harmonic distortion is relative to the fundamental and measured in dBc – c stands for carrier (the fundamental frequency of the input signal).

The HD_2, given by the above equation, is relatively large with a single-ended CMOS common-source amplifier. This is due to the square-law nature of the *I-V* characteristic of CMOS transistors. Increasing the overdrive voltage $(V_{GSDC} - V_{TH})$, or decreasing the input signal amplitude, can reduce the second-order harmonic distortion to some degree but not dramatically.

Importantly, even-order harmonic distortions, such as HD_2, are cancelled at the outputs in a fully differential amplifier (both inputs and outputs are differential). This is evident in that if the inputs flip the polarity the second-order harmonic term does not change sign as it is squared. In fact, HD_2 and other even-order harmonic distortions often do not dominate the total harmonic distortion (THD) in fully differential circuit topologies.

What about the third-order harmonic distortion (HD_3)? If we use the complete long-channel CMOS I-V transfer characteristic equation and do not ignore the term associated with the channel-length modulation, we will find the third-order harmonic term at the output of the amplifier with a sinusoidal input. We omit the derivation here for the sake of conciseness. The HD_3 is significantly smaller than the HD_2 in a simple single-ended CMOS common-source amplifier when the transistor is in saturation region. Please note that the differential topology, which is effective in reducing even-order harmonic distortion, is not effective in reducing third-order (and other odd-order) harmonic distortion – the odd-order harmonic terms change sign if the inputs flip the polarity.

Short Answer to Question 12

Identify and explain the mechanisms responsible for second-order harmonic distortion in the single-ended CMOS common-source amplifier shown in Figure Q12a. How to reduce this distortion? What about third-order harmonic distortion?

Harmonic distortion is a large-signal behavior. The second-order harmonic distortion (HD_2) is caused by the square-law nature of the I-V characteristic of CMOS transistors. Fully differential circuit topologies are effective in reducing the second-order (and other even-order) harmonic distortions.

One of the causes of the third-order harmonic distortion is the CMOS I-V square-law characteristic coupled with channel-length modulation. The differential topology is not effective in reducing third-order (and other odd-order) harmonic distortion. Negative feedback is often used to suppress both even-order and odd-order harmonic distortions.

Question 13

A single-ended bipolar amplifier is shown in Figure Q13a. Assuming the output impedance of Q_1 is r_{o1}, and $r_{o1} \gg R$ or R_E. What is the voltage gain of this amplifier? What are the benefits of having R_E in this topology?

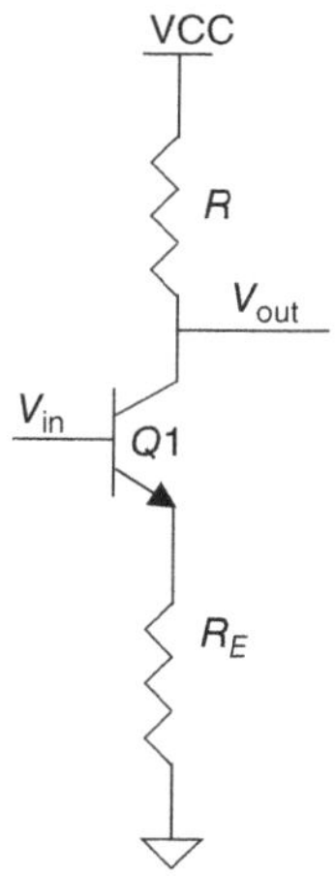

Figure Q13a

Intuitive Analysis: Questions on bipolar transistors (BJTs) are not as popular as those on their CMOS counterparts, primarily because CMOS technology dominates today's industry due to its high density and low power. Bipolar devices, however, offer higher speed, better noise performance, and higher current driving capability, and are still widely used in areas such as RF amplifiers, audio amplifiers, high-current switching, and mixed-signal BiCMOS circuits. If you are to work in one of these fields, the chances of encountering discussions on bipolar circuits in interviews are high.

The methodology of analyzing bipolar circuits is the same as that of CMOS. In fact, both bipolar and CMOS transistors are considered VCCS and they share a lot of common features in circuits. If readers need to refresh fundamental bipolar transistor equations and small-signal models, please refer to *Appendix 1, Section 2: Bipolar transistor (BJT) equations* for more information.

Let us get back to the circuit in Figure Q13a. This is a common-emitter amplifier with emitter degeneration resistor R_E. Degeneration implements negative feedback and lowers the gain of the amplifier by a factor approximately equal to its loop gain. The feedback here is current-sampling, series-mixing (see *Appendix 4, Section 5, Types of negative feedback in amplifier circuits*) – R_E converts the output current I_E into a voltage and feeds it back in series with the input voltage V_{in}. The loop gain with degeneration is:

$$G_{loop} = g_{m1} \cdot R_E$$

where g_{m1} is the transconductance of $Q1$.

As r_{o1} (output impedance of Q_1) $\gg R$, the AC impedance at the output is approximately the resistance of R. The voltage gain without degeneration is:

$$\text{Gain} = -g_{m1} \cdot R$$

With degeneration, the closed-loop gain reduces to:

$$\text{Gain (degenerated)} \approx \frac{-g_{m1} \cdot R}{g_{m1} \cdot R_E} = -\frac{R}{R_E}$$

As we expected, the negative feedback lowers the gain of the amplifier. What are the benefits of having the degeneration-resistance R_E? Note that with emitter degeneration, the voltage gain of the amplifier is independent of the transconductance (g_{m1}) of the input transistor Q_1, and only depends on the resistance ratio between R and R_E. This means the voltage gain is more stable and not sensitive to variations of the transistor's transconductance. Transconductance of transistors can vary due to many factors, including changes of the DC bias current, process parameters, junction temperature, and base to emitter voltage. Harmonic distortions, including both odd-order and even-order, are also suppressed as the gain is not sensitive to the input dependent transconductance with the degeneration. This is an example of how negative feedback can help to reduce circuit nonlinearity and harmonic distortions.

Quantitative Analysis: Now let us verify our conclusion of the voltage gain through the quantitative analysis. Figure Q13b is the small-signal model for the circuit topology in Figure Q13a. Note that the NPN transistor Q_1 is modeled as an input resistance (r_π), and a VCCS (g_{m1}). As r_{o1} (output impedance of Q_1) $\gg R$ or R_E, we ignore the AC current in r_{o1}.

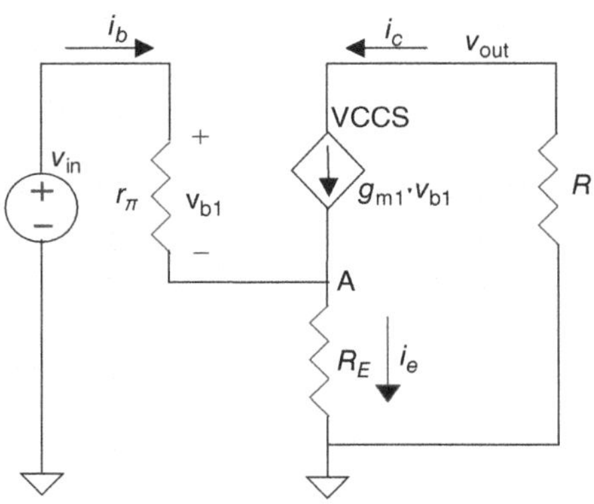

Figure Q13b

Applying KCL and KVL, we obtain the following system of equations:

$$\begin{cases} v_{\text{out}} = -i_c \cdot R \\ i_e = i_c + i_b \\ i_e = \dfrac{v_A}{R_E} \\ i_c = g_{m1} \cdot (v_{\text{in}} - v_A) \\ i_b = \dfrac{i_c}{\beta} \end{cases}$$

which yields:

$$\text{Gain} = \frac{v_{\text{out}}}{v_{\text{in}}} \approx \frac{-g_{m1} \cdot R}{1 + g_{m1} \cdot R_E} \approx -\frac{R}{R_E}; \text{if } \beta \gg 1, g_{m1} \cdot R_E \gg 1$$

Another effect of the emitter degeneration is the increased input resistance of the amplifier (thanks to the series-mixing feedback), which is beneficial to bipolar amplifiers due to the relatively low input resistance (r_π) of bipolar transistors. This is one of the major differences between a bipolar transistor and a MOSFET. To drive a bipolar amplifier, a low-output impedance and current driving capability are needed. Emitter followers (common-collector amplifiers) are often used to serve as a buffer between bipolar amplifier stages because of this reason.

Short Answer to Question 13

What is the voltage gain of this amplifier? What are the benefits of having R_E in this topology?

Emitter degeneration resistor R_E forms negative feedback, reducing the voltage gain of the amplifier:

$$\text{Voltage Gain} \approx \frac{-g_{m1} \cdot R}{1 + g_{m1} \cdot R_E} \approx -\frac{R}{R_E}; \text{ if } \beta \gg 1, \ g_{m1} \cdot R_E \gg 1$$

R_E lowers the circuit's sensitivity to the characteristics of the bipolar transistor. The gain is more stable by relying on the ratio between two resistors. Harmonic distortions (both even-order and odd-order) are reduced as well.

Question 14

A CMOS amplifier is shown in Figure Q14a. What is the maximum and minimum input common-mode voltage to keep all transistors in the saturation region so that the amplifier does not exhibit large distortion?

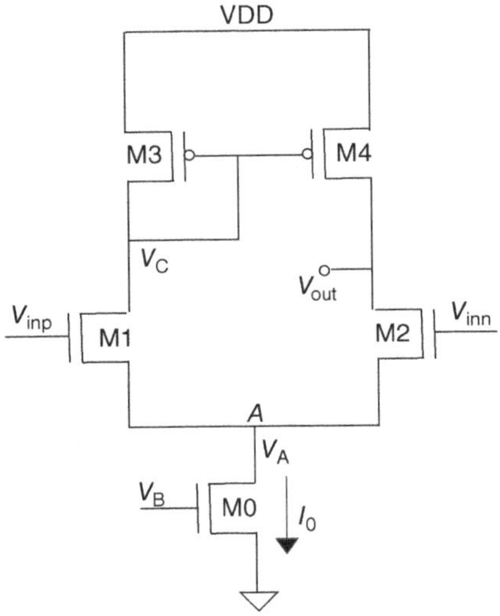

Figure Q14a

Intuitive and Quantitative Analysis: As we have discussed in Question 3, for differential inputs to a differential pair, the drain of the tail transistor (node A in Figure Q14a) is at AC ground. For common-mode inputs, however, node A is no longer at AC ground. The tail transistor, M0, serves as a source degeneration to both input transistors M1 and M2. Note that when $V_{inp} = V_{inn} = V_{cm}$, the current flowing through M1 and M2 is half of the tail current I_0 $(I_0/2)$. Another observation is that the drain voltage of M1 (V_C) is kept one source-to-gate voltage below VDD due to the diode-connected M3. When V_{inp} and V_{inn} are equal and all paired devices are perfectly matched, V_{out} is at the same voltage as V_C. That is, $V_{out} = V_C = VDD - V_{SG3}$.

We first investigate the minimum input common-mode voltage required to keep all transistors in saturation region. When the input common-mode voltage drops, voltage of node A drops as well – under common-mode input conditions, we can view M1, M2 and M0 as two source followers sharing the same source degeneration current source. Eventually, this would push M0 out of saturation first. The condition to keep M0 in saturation is:

$$V_A \geq V_B - V_{THN}$$

For the tail device M0, as long as it is in saturation, the *I-V* relationship is:

$$V_B - V_{THN} = \sqrt{\frac{2 \cdot L_0 \cdot I_0}{K_{pn} \cdot W_0}}$$

Moreover, V_A is one gate-to-source voltage below V_{inp} (or V_{inn}):

$$V_A = V_{\text{inp}} - V_{\text{GS1}} = V_{\text{inp}} - \sqrt{\frac{L_1 \cdot I_0}{K_{pn} \cdot W_1}} - V_{\text{THN}}$$

where V_{THN} is the threshold voltage of NMOS transistors (the body effects of M1 and M2 are ignored), K_{pn} is the transconductance parameter of NMOS transistors, W_0 and L_0 are the width and length of the transistor M0, and W_1 and L_1 are the width and length of the transistor M1.

Therefore, the minimum input common-mode voltage to keep all transistors in saturation is:

$$V_{\text{cm}}(\text{min}) = \sqrt{\frac{2 \cdot L_0 \cdot I_0}{K_{pn} \cdot W_0}} + \sqrt{\frac{L_1 \cdot I_0}{K_{pn} \cdot W_1}} + V_{\text{THN}}$$

When V_{inp} and V_{inn} increase, M1 and M2 will eventually leave the saturation region. This is because the voltages at the drains of M3 and M4 are maintained at VDD − V_{SG3}. The condition to keep M1 in saturation is:

$$V_C - V_A \geq V_{\text{inp}} - V_A - V_{\text{THN}}$$

Note that:

$$V_C = \text{VDD} - V_{\text{SG3}} = \text{VDD} - \sqrt{\frac{L_3 \cdot I_0}{K_{\text{pp}} \cdot W_3}} - V_{\text{THP}}$$

The maximum input common-mode voltage to keep all transistors in saturation is therefore:

$$V_{\text{cm}}(\text{max}) = \text{VDD} - \sqrt{\frac{L_3 \cdot I_0}{K_{pp} \cdot W_3}} - V_{\text{THP}} + V_{\text{THN}}$$

where V_{THP} is the threshold voltage of PMOS transistors, K_{pp} is the transconductance parameter of PMOS transistors, W_3 and L_3 are the width and length of the transistor M3.

Once the input common-mode voltage exceeds $V_{\text{cm}}(\text{max})$, transistors M1 and M2 are out of saturation and the gain of the amplifier drops significantly. When the input common-mode voltage drops below $V_{\text{cm}}(\text{min})$, the tail device will be out of saturation, the bias current drops dramatically, and the gain of the amplifier is affected. In reality, transistors do not "suddenly" go out of saturation at a certain point – it is rather a gradual process. This means the common-mode voltages within, but close to, the maximum and minimum limits cause gain variations. More precisely, as long as the input common-mode voltage changes, the bias current will change due to channel-length modulations, and so does the gain of the amplifier. This is one of the reasons why in applications where minimum distortion and high linearity are desired, (closed-loop) inverting amplifiers are more popular than (closed-loop) non-inverting amplifiers. In the former configuration, the op-amp input common-mode voltage is constant, while in the latter one, the op-amp input common-mode voltage is determined by the input and changes substantially during large input swings.

Common-mode voltage variations and devices operating outside the saturation region constitute another major source of amplifier nonlinearity besides the transistor square-law *I-V* characteristic in saturation region (as discussed in Question 12).

Short Answer to Question 14

What is the maximum and minimum input common-mode voltage to keep all transistors in the saturation region so that the amplifier does not exhibit large distortion?

The maximum input common-mode voltage to keep all transistors in saturation is:

$$V_{\mathrm{cm}}(\mathrm{max}) = \mathrm{VDD} - \sqrt{\frac{L_3 \cdot I_0}{K_{pp} \cdot W_3}} - V_{\mathrm{THP}} + V_{\mathrm{THN}}$$

The minimum input common-mode voltage to keep all transistors in saturation is:

$$V_{\mathrm{cm}}(\mathrm{min}) = \sqrt{\frac{2 \cdot L_0 \cdot I_0}{K_{pn} \cdot W_0}} + \sqrt{\frac{L_1 \cdot I_0}{K_{pn} \cdot W_1}} + V_{\mathrm{THN}}$$

Question 15

A fully differential two-stage folded cascode amplifier is shown in Figure Q15a. Also shown in Figure Q15a is a common-mode feedback (CMFB) control loop to stabilize the common-mode output voltage of the amplifier. Vcm is the reference (target) common-mode voltage. First, what is the DC gain of the CMFB loop? Second, how does the phase margin of the CMFB loop compare to that of the signal path? Finally, name some ideas to improve the stability of the CMFB loop.

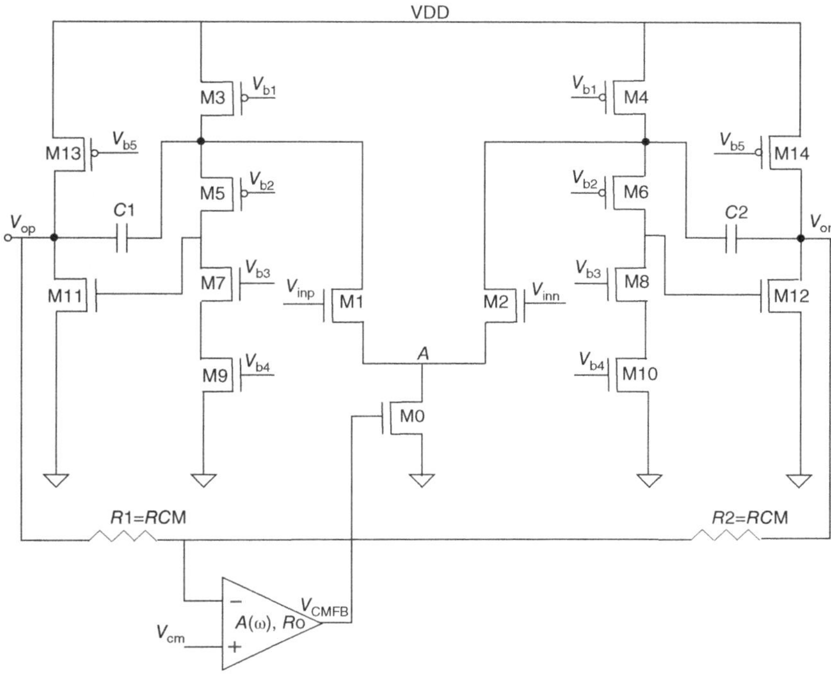

Figure Q15a

Intuitive Analysis: As is typical for fully differential amplifiers, the analysis can be performed on a single half-circuit. We redraw the left half of the amplifier in Figure Q15b.

First, we need to analyze the DC loop gain of the CMFB. As shown in Figure Q15b, the loop of the CMFB includes a voltage divider (R_1 and R_2) to obtain the output common-mode voltage by averaging the two differential outputs; an error amplifier $A(\omega)$ for amplifying the difference (error) between the reference (target) common-mode voltage (V_{cm}) and the actual output common-mode voltage; the first-stage tail device, M0, which converts this error voltage into an error current; the input transistor M1 and all the cascode current-source devices at the folded branch through which the error current flows to the output node of the first stage and is converted back into a voltage; and finally the second-stage voltage amplifier formed by M11 and M13.

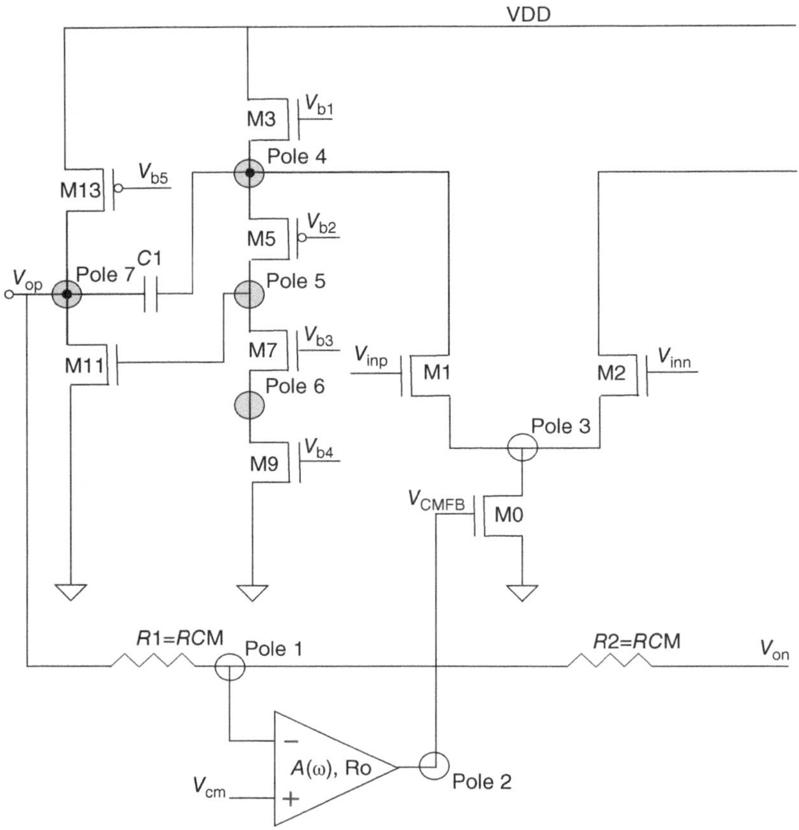

Figure Q15b

The DC gain of the CMFB loop is:

$$G(0) = A(0) \cdot \frac{1}{2} \cdot g_{m0} \cdot r_{o1} \cdot g_{m11} \cdot r_{o2}$$

where $A(0)$ is the DC voltage gain of the error amplifier $A(\omega)$, g_{m0} is the transconductance of transistor M0, g_{m11} is the transconductance of transistor M11, and r_{o1} and r_{o2} are the output impedance of the first stage and second stage, respectively. The coefficient of 1/2 is due to the fact that each branch of the input differential pair gets half of the tail current change.

Recall the DC voltage gain of the signal path of the amplifier, $G_{\text{sig}}(0)$ is:

$$G_{\text{sig}}(0) = g_{m1} \cdot r_{o1} \cdot g_{m11} \cdot r_{o2}$$

where g_{m1} is the transconductance of the input transistor M1.

The loop gain of the CMFB is comparable to, and could potentially be larger than, the signal gain of the amplifier due to the gain of its error amplifier.

Another important observation is that the signal path and CMFB path are largely overlapped. The CMFB includes the first and second gain stage of the amplifier, same as the signal path. This implies that the compensation for the amplifier also helps the stability of the CMFB loop. The CMFB loop, however, has additional stages: the voltage divider and the error amplifier. This turns out to be an adversary to the stability of the CMFB. This is evident if we list the dominant and nondominant poles for the signal path and the CMFB path (Figure Q15b):

Signal path: Pole 4, Pole 5 (dominant), Pole 6, Pole 7.

CMFB path: Pole 1, Pole 2, Pole 3, Pole 4, Pole 5 (dominant), Pole 6, Pole 7.

The three additional nondominant poles (Pole 1, Pole 2, and Pole 3) and possible larger loop gain make the phase margin of the CMFB loop potentially lower than that of the amplifier signal path. Compensating the CMFB loop is often a more challenging task than compensating the signal path.

How to improve the stability of the CMFB loop? One thing to realize is that in a fully differential amplifier, the common-mode voltage must be stable, though it does NOT require high-precision regulation. In other words, the output common-mode voltage must be maintained within a specific range, such as VDD/2 $\pm$ 50 mV, but high precision is not critical, for example, a tolerance of a few mV is not required. This means that we can lower the DC gain of the CMFB loop which helps stability. The error amplifier does not need to have a high gain – usually a single-stage low gain amplifier is used. Another effective way to lower the CMFB loop gain is to lower the effective transconductance of the M0. This can be done by splitting the tail device (M0) into two parts – one with constant bias voltage, and the other one is controlled by the output of the error amplifier, as shown in Figure Q15c.

We need to also try to alleviate the impact of the nondominant poles. For example, as shown in Figure Q15c, capacitances C_3 and C_4, can be put in parallel with the divider resistors R_1 and R_2, to compensate for the phase delay caused by the two resistances.

In some cases, compensating the CMFB loop is so challenging that designers might choose to break the CMFB loop into two local compensation loops and compensate each stage separately. This often turns out to be effective due to the reduction of the loop gain and number of poles in each loop.

Last but not least, fully differential amplifiers find many applications in switched-capacitor circuits (analog signal processing and analog-to-digital converters). In these applications, a switched-capacitor circuit, which fulfills the functions of the output voltage averaging and level shifting, is utilized to replace the continuous-time resistive voltage divider and error amplifier. Due to the low gain and fast operation of the switched-capacitor network, compensation is usually found to be easier than the continuous-time solution. We will discuss

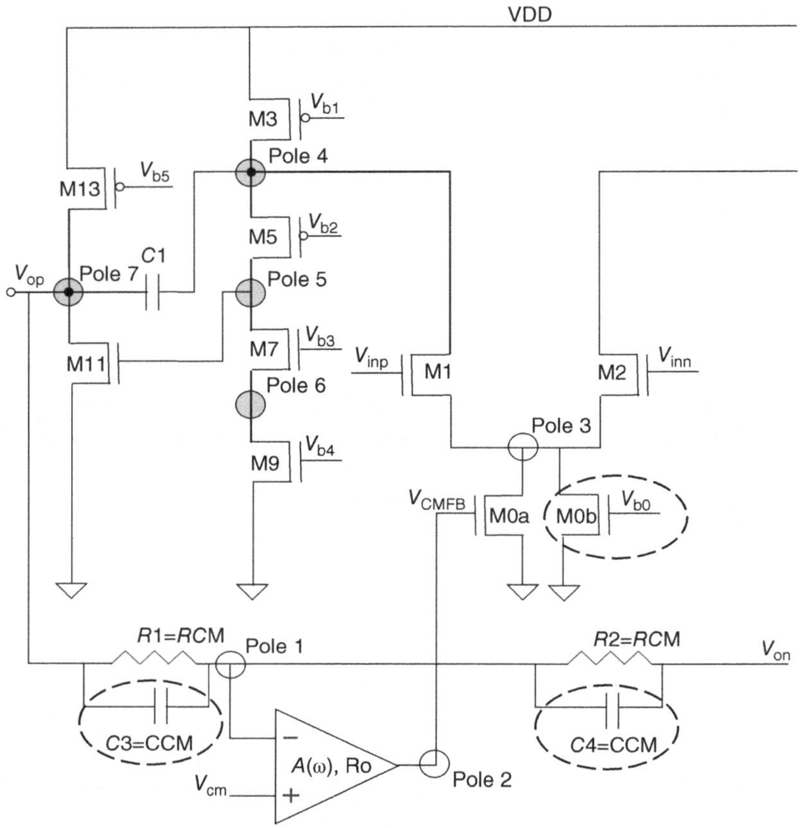

Figure Q15c

an example of the switched-capacitor CMFB in Question 32, *Switched-capacitor common-mode feedback (CMFB) circuits.*

Short Answer to Question 15

What is the DC gain of the CMFB loop? How does the phase margin of the CMFB loop compare to that of the signal path? Name some ideas to improve the stability of the CMFB loop.

The DC gain of the CMFB loop is:

$$G(0) = A(0) \cdot \frac{1}{2} \cdot g_{m0} \cdot r_{o1} \cdot g_{m11} \cdot r_{o2}$$

The additional nondominant poles associated with the CMFB and the gain of the error amplifier make the phase margin of the CMFB loop potentially lower than that of the amplifier signal path. Compensating the CMFB loop is often a more challenging task than compensating the signal path.

Methods to improve the stability of the CMFB loop include lowering the gain of the error amplifier, controlling the common-mode voltage of each amplifier stage separately, and adding shunt capacitors in the output resistor divider. Switched-capacitor based CMFB circuits can also be used in discrete analog circuits (analog-to-digital converters, switched-capacitor filters, and so on).

Question 16

Figure Q16a and Figure Q16b illustrate two op-amp based closed-loop amplifier topologies. Assume both op-amps are ideal. Please describe the transfer function, i.e., the relationship between the input and output of each of the topologies and draw the corresponding Bode plot (magnitude frequency response only).

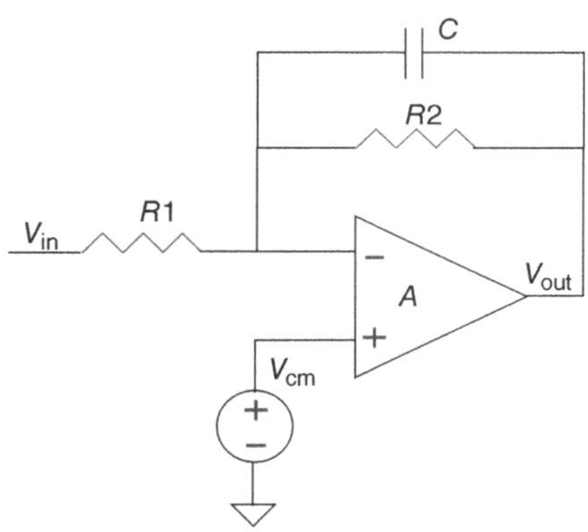

Figure Q16a

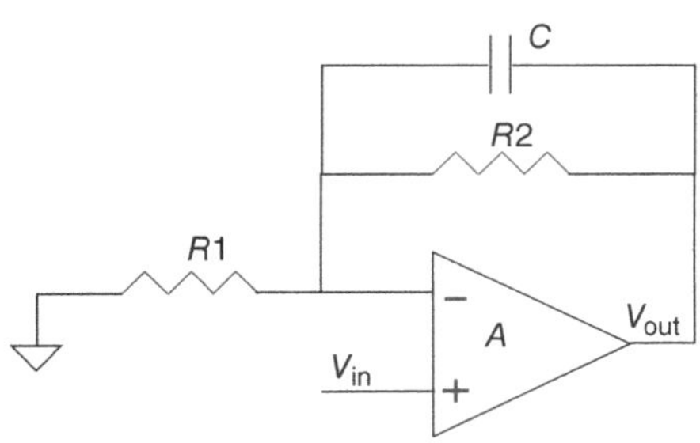

Figure Q16b

Intuitive and Quantitative Analysis: It is clear that the topology in Figure Q16a is an inverting amplifier, while the topology in Figure Q16b is a non-inverting amplifier, both with negative feedback. Without the capacitor C, the gain of the inverting amplifier with ideal op-amp is $-(R_2/R_1)$, and the gain of the non-inverting amplifier with ideal op-amp is $1 + R_2/R_1$ (*see Appendix 4, Section 4, Basic op-amp based closed-loop circuits*).

What is the impact of C on these two amplifiers? As frequency goes up, the impedance of the capacitors reduces, lowering the total impedance at the feedback paths. The gain would drop in both topologies. These two topologies therefore represent a low-pass filtering function, but their frequency responses are not quite the same.

An easy way to obtain the transfer function and frequency response of the amplifiers is to analyze the circuits in the complex frequency domain (s-domain). The s-domain contains information of both transient response and steady-state response. Please refer to *Appendix 3, Circuit Analysis in Complex Frequency Domain (s-domain)*, for detailed explanations on how to do circuit analysis in the s-domain.

The impedance of the basic circuit elements in the s-domain is as follows:

For resistances, s-domain impedance is:

$$Z_R = R$$

For capacitances, s-domain impedance is:

$$Z_C = \frac{1}{C \cdot s}$$

For inductances, s-domain impedance is:

$$Z_L = L \cdot s$$

A primary advantage of s-domain analysis is that the impedances of capacitors and inductors can be treated as algebraic quantities, similar to resistances. For example, if a capacitance C and a resistance R are in parallel, the total s-domain impedance is:

$$Z = \frac{1}{\frac{1}{Z_R} + \frac{1}{Z_C}} = \frac{1}{\frac{1}{R} + C \cdot s} = \frac{R}{1 + R \cdot C \cdot s}$$

Using s-domain impedances, the transfer function, $H(s)$, of the inverting amplifier in Figure Q16a is:

$$H(s) = -\frac{Z_2}{Z_1} = -\frac{\frac{R_2}{1 + R_2 \cdot C \cdot s}}{R_1} = -\frac{R_2}{R_1} \cdot \frac{1}{R_2 \cdot C \cdot s + 1}$$

where Z_2 is the s-domain impedance at the feedback path (R_2 in parallel with C), and Z_1 is the s-domain impedance of R_1. Notice that the input voltage V_{cm} is not in the signal transfer function $H(s)$, as it is a DC voltage used to set the input and output bias voltage.

The s-domain transfer function shows that this is a first-order low-pass system with the DC gain of $-R_2/R_1$ (You can obtain DC gain by substituting s with zero). The time constant of the circuit is $R_2 \cdot C$, and the -3 dB frequency is:

$$f_{-3\text{dB}} = \frac{1}{2 \cdot \pi \cdot R_2 \cdot C}$$

To obtain the frequency response from the transfer function, we need to replace the complex frequency s with $j \cdot \omega$ in the transfer function:

$$s = j \cdot \omega$$

where ω is the angular frequency with the unit of rad/s. This will transform the function from complex s-domain to a steady-state frequency domain. The modulus of the resulting function is the magnitude response, and the angle is the phase response.

The derivation of the frequency response is left to interested readers. To draw the Bode plot, however, we just need the information of the DC gain, the frequency of poles, and the frequency of zeros of the circuit. The Bode plot (magnitude response) is shown in Figure Q16c. Note that the op-amp in this question is an ideal one. In practical circuits, more poles will show up at high frequencies reflecting the frequency characteristics of the op-amp.

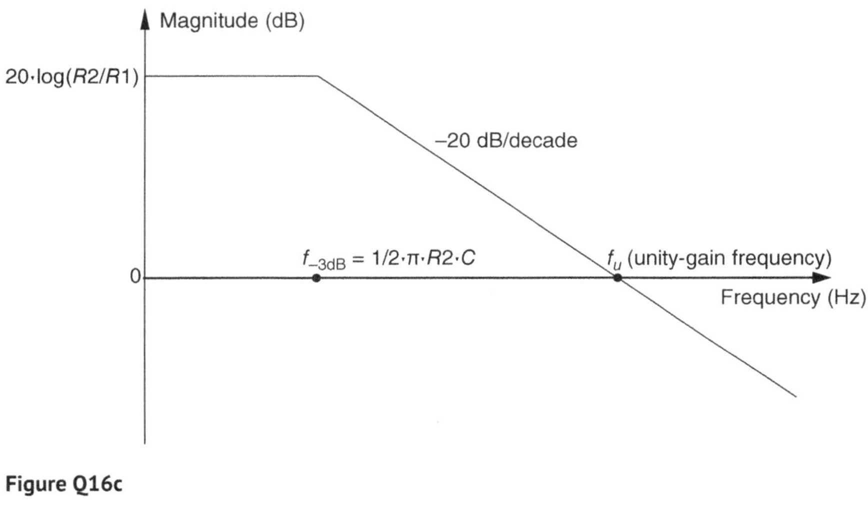

Figure Q16c

Similarly, using s-domain impedances, the transfer function, $H(s)$, of the non-inverting amplifier in Figure Q16b is:

$$H(s) = 1 + \frac{Z_2}{Z_1} = 1 + \frac{\frac{R_2}{1 + R_2 \cdot C \cdot s}}{R_1}$$

$$= 1 + \frac{R_2}{R_1} \cdot \frac{1}{R_2 \cdot C \cdot s + 1} = \frac{R_2 \cdot R_1 \cdot C \cdot s + R_1 + R_2}{R_1 \cdot (R_2 \cdot C \cdot s + 1)}$$

where Z_2 is the s-domain impedance at the feedback path (R_2 in parallel with C), and Z_1 is the s-domain impedance of R_1.

The s-domain transfer function indicates that this is a first-order system with one pole and one zero, and a DC gain of $1 + R_2/R_1$. The frequency of the pole, or the -3 dB frequency of the amplifier, is at:

$$f_{-3\text{dB}} = \frac{1}{2 \cdot \pi \cdot R_2 \cdot C}$$

This is the same as the low-pass inverting amplifier topology in Figure Q16a. However, the system also has a (LHP) zero at the frequency of:

$$f_z = \frac{R_1 + R_2}{2 \cdot \pi \cdot R_1 \cdot R_2 \cdot C} = \frac{1}{2 \cdot \pi \cdot (R_1 \parallel R_2) \cdot C}$$

It is apparent that $f_z > f_{-3\text{dB}}$. The gain of the system starts to drop at $f_{-3\text{dB}}$ with a slope of -20 dB/decade, and at the zero frequency (f_z), the gain drops to 0 dB, and stays at 0 dB from this point. Figure Q16d illustrates the Bode plot (magnitude frequency response) of the non-inverting amplifier.

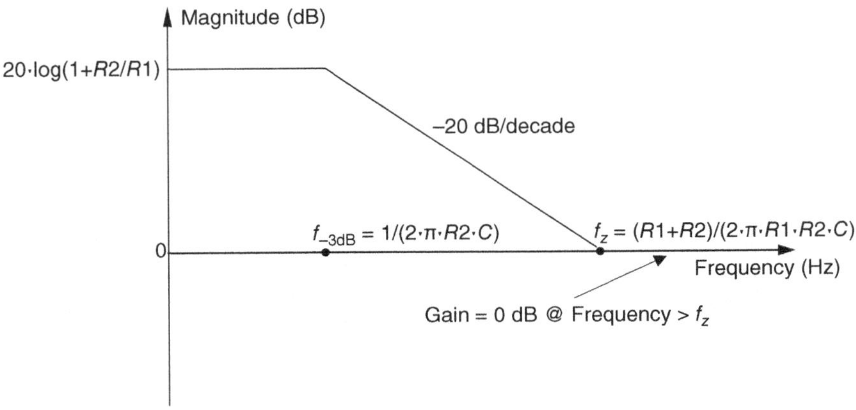

Figure Q16d

The major difference, in terms of frequency response, between the two closed-loop amplifiers is that the non-inverting amplifier has a zero. Intuitively, this is because the capacitance C shorts the feedback resistance R_2 at high frequencies and the amplifier becomes a unity-gain follower in the non-inverting topology.

Again, this assumes the op-amp is ideal and has infinite bandwidth. In practice, the gain will start to drop again from 0 dB when the op-amp runs out of bandwidth.

Short Answer to Question 16

Please describe the transfer function. i.e., the relationship between the input and output of each of the topologies and draw the Bode plot (magnitude frequency response only).

The transfer function of the inverting amplifier in Figure Q16a is:

$$H(s) = -\frac{R_2}{R_1} \cdot \frac{1}{R_2 \cdot C \cdot s + 1}$$

The transfer function of the non-inverting amplifier in Figure Q16b is:

$$H(s) = \frac{R_2 \cdot R_1 \cdot C \cdot s + R_1 + R_2}{R_1 \cdot (R_2 \cdot C \cdot s + 1)}$$

Please refer to Figure Q16c and Figure Q16d for the Bode plots of the magnitude frequency response of the inverting and non-inverting amplifier.

Question 17

An inverting amplifier and a unity-gain buffer are shown in Figure Q17a and Figure Q17b, respectively. Assume the op-amp has a single pole with a DC gain $A(0)$ and a -3 dB bandwidth *fb*. **(a)** What is the feedback factor (β) of the two amplifiers? **(b)** What is the closed-loop -3 dB bandwidth of the two amplifiers? **(c)** Which amplifier has more phase margin?

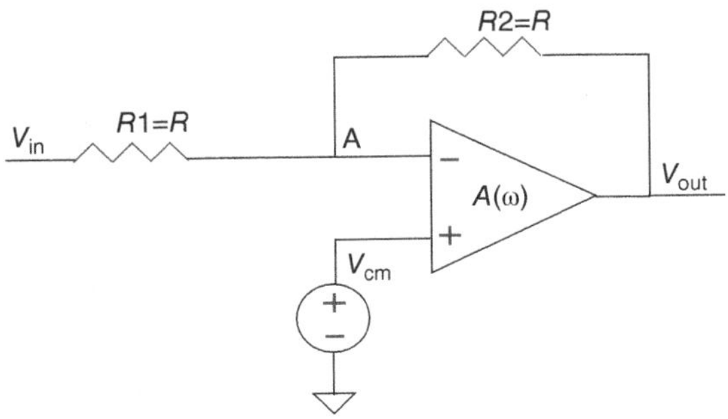

Figure Q17a

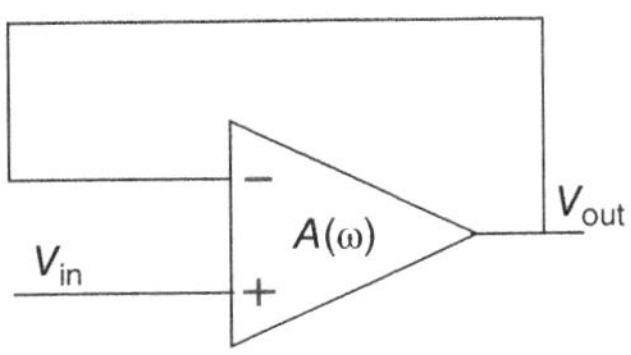

Figure Q17b

Intuitive and Quantitative Analysis: The feedback factor is an important concept in negative feedback circuits, and it refers to the percentage of the circuit output that is fed back to interact with the input. In other words, the feedback factor is the transfer function of the feedback path.

The inverting amplifier in Figure Q17a is often referred to as a feedback type called *voltage-sampling, shunt-mixing*. Voltage-sampling indicates that the output voltage is taken as the feedback signal, while shunt-mixing, also known as current-mixing, means that the output and input are summed by currents. Note that in inverting amplifiers, this is the only way for the two voltages, v_{out} and v_{in}, to be summed at the input of the op-amp (v_A). As the op-amp is a voltage amplifier, we will use voltages as the variables in the derivation of the feedback factor and closed-loop transfer functions.

One way to determine the percentage of the v_{out} that is fed back towards the summing point (node A in Figure Q17a) is to let the input be zero, i.e., $v_{\text{in}}(s) = 0$, disconnect the summing point from the op-amp input (loop opened), and calculate the resulted voltage at the summing point solely due to the v_{out}, as illustrated in Figure Q17c. Note that the op-amp non-inverting input is AC ground.

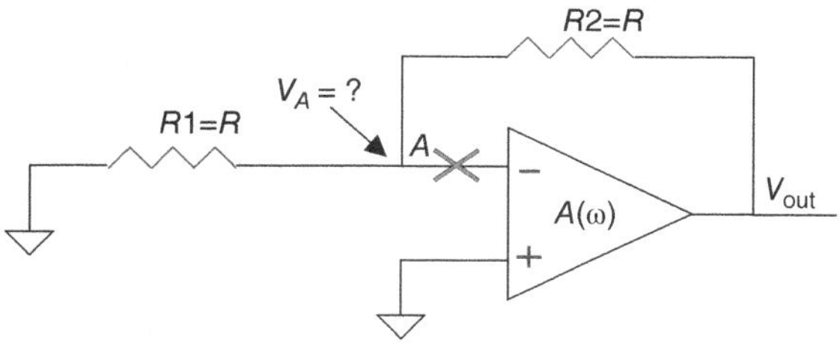

Figure Q17c

Under these conditions, the v_A can be written as:

$$v_A(s) = \frac{R_1}{R_1 + R_2} \cdot v_{out}(s)$$

The feedback factor, $\beta(s)$, therefore is:

$$\beta(s) = \frac{v_A(s)}{v_{out}(s)} = \frac{R_1}{R_1 + R_2}$$

As $R_1 = R_2$, $\beta = 1/2$.

To develop the closed-loop transfer function of the inverting amplifier, we observe the following relationship among v_{in}, v_{out} and the v_A:

$$\begin{cases} v_A(s) = [v_{out}(s) - v_{in}(s)] \cdot \dfrac{R_1}{R_1 + R_2} + v_{in}(s) \\ v_{out}(s) = -A(s) \cdot v_A(s) \\ A(s) = \dfrac{A(0)}{\dfrac{1}{2 \cdot \pi \cdot fb} \cdot s + 1} \end{cases}$$

where *fb* and $A(0)$ are the −3 dB frequency and the DC gain of the op-amp, respectively. Solving the system of equation for $v_{out}(s)/v_{in}(s)$, we obtain the transfer function of the inverting amplifier:

$$H(s) = \frac{v_{out}(s)}{v_{in}(s)} = -\frac{R_2}{R_1} \cdot \frac{1}{\dfrac{1}{2 \cdot \pi \cdot \beta \cdot A(0) \cdot fb} \cdot s + 1 + \dfrac{1}{A(0) \cdot \beta}}$$

$$\approx -\frac{R_2}{R_1} \cdot \frac{1}{\dfrac{1}{2 \cdot \pi \cdot \beta \cdot A(0) \cdot fb} \cdot s + 1}$$

The DC gain and closed-loop −3 dB bandwidth of the (closed-loop) inverting amplifier are therefore:

$$H(0) \approx -\frac{R_2}{R_1}$$

$$f_{-3dB} = \beta \cdot A(0) \cdot fb$$

As $\beta = 1/2$ and $R_1 = R_2 = R$ in this example, the DC gain and −3 dB bandwidth of the closed-loop amplifier become:

$$H(0) \approx -\frac{R_2}{R_1} = -1$$

$$f_{-3dB} = \frac{1}{2} \cdot A(0) \cdot fb = \frac{1}{2} \cdot \text{GBW}_{op}$$

where GBW_{op} is the gain-bandwidth product of the op-amp itself.

The type of negative feedback of the unity-gain buffer in Figure Q17b is voltage-sampling, series-mixing. Similar to the inverting amplifier configuration, the output voltage v_{out} serves as the feedback signal. Because v_{out} is connected to the op-amp's inverting input while the input voltage v_{in} is at the non-inverting input, the two voltages are combined in a series-mixing configuration:

$$v_{net}(s) = v_{in}(s) - \beta(s) \cdot v_{out}(s)$$

where v_{net} is the net input voltage of the op-amp, similar to v_A in the inverting amplifier (Figure Q17a). $\beta(s)$ is the feedback factor.

As 100% of the v_{out} is fed back to the inverting input, the feedback factor is:

$$\beta(s) = 1$$

Combining with the relationship between v_{out} and v_{net}:

$$v_{out}(s) = A(s) \cdot v_{net}(s)$$

The transfer function of the unity-gain buffer is:

$$H(s) = \frac{v_{out}(s)}{v_{in}(s)} = \frac{A(s)}{1 + A(s) \cdot \beta} \approx \frac{1}{\frac{1}{2 \cdot \pi \cdot A(0) \cdot fb} \cdot s + 1}; \text{if } A(0) \gg 1$$

The DC gain and closed-loop −3 dB bandwidth of the unity-gain buffer are therefore:

$$H(0) \approx 1$$
$$f_{-3dB} = A(0) \cdot fb$$

As $\beta = 1$, the −3 dB bandwidth of the closed-loop amplifier becomes:

$$f_{-3dB} = A(0) \cdot fb = \text{GBW}_{op}$$

A common misconception is that since the inverting amplifier and unity-gain buffer both have a closed-loop gain of 1 (magnitude), their feedback factor must be the same. This is simply not true.

The feedback factor has direct impacts to the performances of the closed-loop system: the inverting amplifier's −3 dB bandwidth is half (1/2) of that of the unity-gain buffer (assuming using the same op-amp) due to its smaller feedback factor compared to the latter (1/2 versus 1). What about the phase margin?

The phase margin is judged based on the *open-loop* frequency response. For a unity-gain amplifier, because the feedback factor is 1, the open-loop transfer function is the same as the transfer function of the op-amp itself. But for an inverting amplifier, the open-loop transfer function needs to include the transfer of the feedback loop, or the feedback factor, which is less than 1.

Specifically, the open-loop transfer function of the inverting amplifier is:

$$G(s) = A(s) \cdot \beta = \frac{A(0)}{\frac{1}{2 \cdot \pi \cdot fb} \cdot s + 1} \cdot \frac{R_1}{R_1 + R_2} = \frac{\frac{1}{2} \cdot A(0)}{\frac{1}{2 \cdot \pi \cdot fb} \cdot s + 1}$$

And the open-loop transfer function of the unity-gain buffer is:

$$G(s) = A(s) \cdot \beta = \frac{A(0)}{\frac{1}{2 \cdot \pi \cdot fb} \cdot s + 1}$$

Figure Q17d shows the Bode plot for the two open-loop frequency responses. The two topologies have the same phase response determined by the op-amp −3 dB frequency *fb*, but the unity-gain buffer has an open-loop DC gain and gain-bandwidth product twice that of the inverting amplifier. As a result, the inverting amplifier has a higher phase margin.

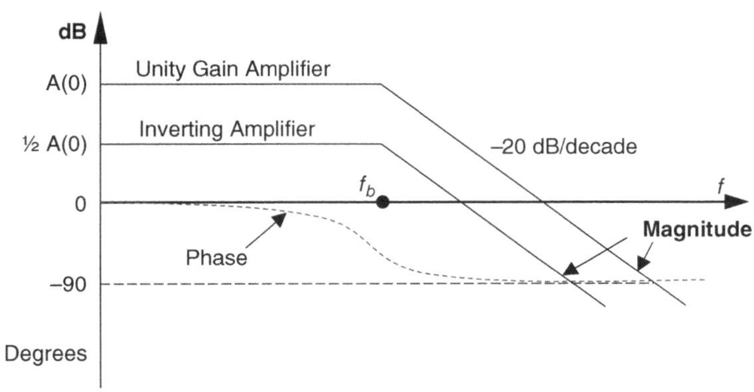

Figure Q17d

Short Answer to Question 17

(a) What is the feedback factor (β) of the two amplifiers? **(b)** What is the closed-loop −3 dB bandwidth of the two amplifiers? **(c)** Which amplifier has more phase margin?

(a) The feedback factor of the inverting amplifier (Figure Q17a) is:

$$\beta(s) = \frac{R_1}{R_1 + R_2} = \frac{1}{2}$$

The feedback factor of the unity-gain buffer (Figure Q17b) is:

$$\beta(s) = 1$$

(b) The closed-loop −3 dB bandwidth of the inverting amplifiers is:

$$f_{-3\text{dB}} = \frac{1}{2} \cdot \text{GBW}_{\text{op}}$$

The closed-loop −3 dB bandwidth of the unity-gain buffer is:

$$f_{-3\text{dB}} = \text{GBW}_{\text{op}}$$

(c) The inverting amplifier has a higher phase margin.

Question 18

A voltage regulator is shown in Figure Q18a. Please draw a diagram showing the relationship between the power supply rejection ratio (PSRR) and the frequency of the VDD noise. $A_1(\omega)$ is an error amplifier with a −3 dB bandwidth $f_{-3\text{dB}}$ and DC gain $A_1(0)$. R_L and C_L are the load resistance and capacitance, respectively. The compensation network is not shown to simplify the PSRR analysis. Assume the reference voltage, V_{ref}, is noise-free. Also assume error amplifier $A_1(\omega)$ has infinite PSRR.

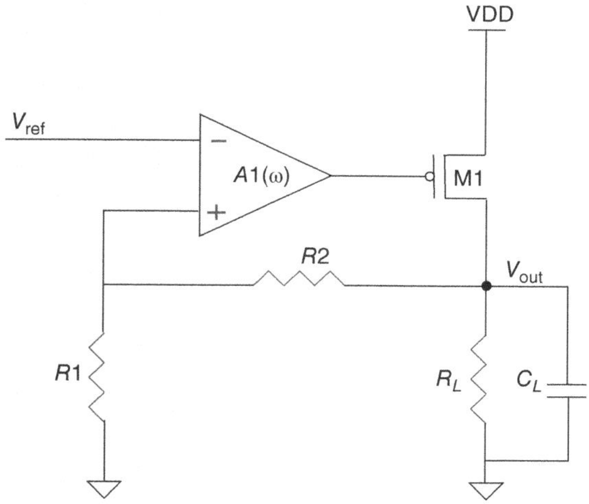

Figure Q18a

Intuitive Analysis: First, let us recall the definition of the PSRR. As there is no single industry-wide standard, several definitions exist. One of the most often used definitions specifies PSRR as:

$$\text{PSRR} = 20 \cdot \log_{10}\left[\frac{v_{\text{supply-noise(RMS)}}}{v_{\text{output-noise(RMS)}}}\right]$$

By this definition, PSRR relates to the reciprocal of the transfer function from the power supply (VDD) to the output of the circuit (V_{out}). Note that power supply noise can exist on both VDD and ground, and in most cases, we treat ground as the reference and assume the noise voltage is on VDD relative to ground.

The circuit diagram in Figure Q18a shows a non-inverting amplifier topology. It has two gain stages – the error amplifier A1 as the first stage, and a common-source amplifier (M1 and R_L) as the output stage. M1 is also known as the *pass transistor* as it directly connects VDD (input supply voltage) and V_{out} (regulated output voltage). Note that although the V_{ref} is connected to the inverting input of the error amplifier, this terminal is in reality the non-inverting input of the overall two-stage amplifier. The load of the amplifier consists of the resistance R_L and the capacitance C_L. The feedback network is of *voltage-sampling, series-mixing* type, and the closed-loop DC gain, A_{CL}, of the non-inverting amplifier is:

$$A_{\text{CL}} = 1 + \frac{R_2}{R_1}$$

When there is noise present on the VDD, there are two paths for it to affect the amplifier output (V_{out}), as shown in Figure Q18b. The major path is from the VDD to the source-to-gate voltage of M1 (V_{sg1}) – the noise of the VDD could affect the V_{sg1} directly. The other path, or minor path, is from the VDD to the output (V_{out}) through the channel of transistor M1 – we will see from an analysis later that the VDD noise can affect the V_{out} through a voltage divider formed by the transistor M1 and the load impedance (R_L in parallel with C_L).

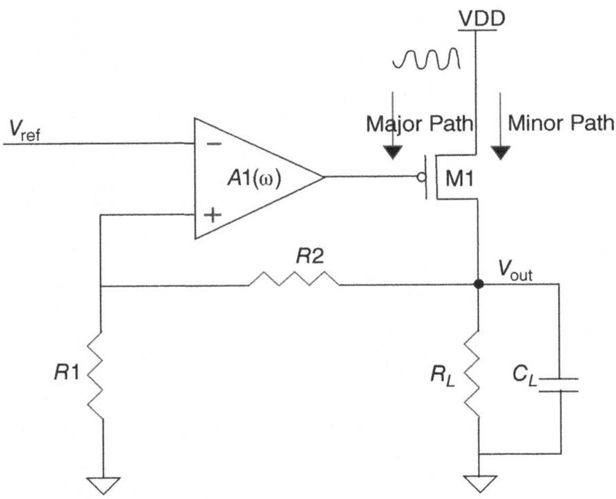

Figure Q18b

Let us start by looking at the major noise path. A common mistake is to assume that since there is no significant capacitance between the source and gate of M1 (at least not specified by the question), and the PSRR of error amplifier $A_1(\omega)$ is infinite, the noise on VDD does not affect the gate voltage of M1 much and therefore it does not affect the PSRR as well. The fact is on the contrary – because the gate voltage of M1 does not track VDD, the source-to-gate voltage of M1 (V_{sg1}) is directly affected by the VDD noise. The source-to-gate voltage matters here, not the gate potential (to ground) itself.

Would this noise at V_{sg1} be amplified by M1 and result in poor PSRR? Another common mistake is overlooking the impact of the feedback loop on the output noise. In reality, for any noise injected at the V_{sg1} with a frequency within the -3 dB bandwidth of the error amplifier A1, the noise is suppressed by the DC gain of the amplifier. This means the VDD noise has little impact on the output voltage and the PSRR is high at noise within the -3 dB bandwidth of the error amplifier.

Once the VDD-noise frequency goes beyond the −3 dB bandwidth of the error amplifier, the gain of the error amplifier starts to decrease, and the noise at the output increases, resulting in a dropping PSRR. The decline in PSRR eventually plateaus because, at sufficiently high frequencies, the feedback loop becomes non-responsive; consequently, the remaining high-frequency noise components are effectively suppressed by the load capacitance. This means that the PSRR typically reaches its minimum (worst-case) value at a specific frequency, and then improves (goes up again) as the frequency increases further.

If we consider the source-to-gate capacitance of M1 (parasitic or purposely designed), when noise frequency goes up, the gate of M1 tracks the VDD better and the noise on V_{sg1} reduces. This means the PSRR is in fact better than the prediction we have made previously, especially at high frequency (depending on the capacitance value). Eventually, at a certain high frequency, the gate of M1 tracks VDD completely (almost) and V_{sg1} is noiseless – the major noise path ceases to exist. But the VDD noise still can feed to the V_{out} through the minor noise path. In fact, the output impedance of M1 and the load impedance form an AC voltage divider between the VDD and ground. This divider ultimately determines the PSRR at high frequencies.

Quantitative Analysis: Let us confirm our intuitive analysis results through some quantitative analysis. First, we check the major noise path. As loop dynamics and transfer function are involved, we model the circuit in the s-domain. Figure Q18c shows the block diagram of the circuit system.

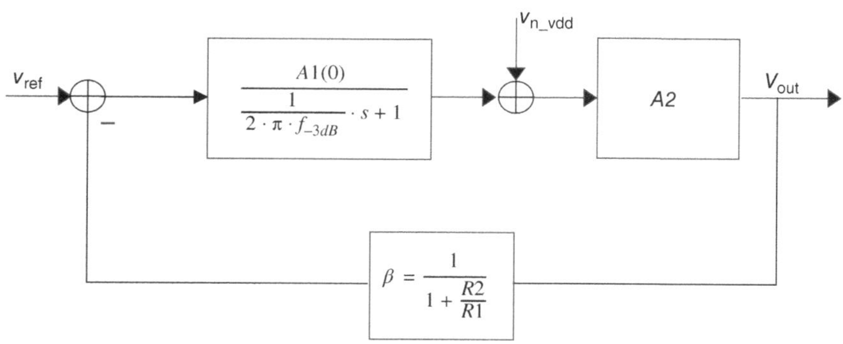

Figure Q18c

The error amplifier $A_1(\omega)$ is modeled as a first-order system with −3 dB cutoff frequency f_{-3dB} and DC gain $A_1(0)$. To make the loop stable, the poles of the second-stage amplifier formed by M1, R_L and C_L are typically set to be significantly higher than the closed-loop bandwidth of the regulator (by the compensation network which is not shown in Figure Q18a). To simplify the derivation (critical in interviews!), we assume the second-stage amplifier has infinite bandwidth and model it as a gain element (A_2). The feedback factor β describes the fraction of the output that is fed back to the input. In this topology, β equals $1/(1 + R_2/R_1)$. V_{n_vdd} is the VDD noise on the source of M1, which is injected to the input of second-stage amplifier A_2.

The signal (v_{ref}) transfer function (STF) of the system is:

$$\mathrm{STF}(s) = \frac{v_{out}(s)}{v_{ref}(s)} = \frac{A_1(s) \cdot A_2}{1 + \beta \cdot A_1(s) \cdot A_2} = \frac{A_1(0) \cdot A_2}{\frac{1}{2 \cdot \pi \cdot f_{-3dB}} \cdot s + A_1(0) \cdot A_2 \cdot \beta + 1}$$

As $A_1(0) \cdot A_2 \cdot \beta \gg 1$, the STF can be simplified:

$$\mathrm{STF}(s) \approx \frac{A_1(0) \cdot A_2}{\frac{1}{2 \cdot \pi \cdot f_{-3dB}} \cdot s + A_1(0) \cdot A_2 \cdot \beta} = \frac{\frac{1}{\beta}}{\frac{1}{A_1(0) \cdot A_2 \cdot \beta \cdot 2 \cdot \pi \cdot f_{-3dB}} \cdot s + 1}$$

The closed-loop STF is a first-order system with DC gain of $1/\beta$, or $1 + R_2/R_1$, and an elevated −3 dB bandwidth boosted by the loop gain, $A_1(0) \cdot A_2 \cdot \beta$.

Importantly, the noise transfer function (NTF) of the system is:

$$\mathrm{NTF}(s) = \frac{v_{out}(s)}{v_{n\_vdd}(s)} = \frac{A_2}{1 + \beta \cdot A_1(s) \cdot A_2}$$

$$\approx \frac{1}{A_1(0) \cdot \beta} \cdot \frac{\frac{1}{2 \cdot \pi \cdot f_{-3dB}} \cdot s + 1}{\frac{1}{A_1(0) \cdot A_2 \cdot \beta \cdot 2 \cdot \pi \cdot f_{-3dB}} \cdot s + 1}$$

Again, we assume $A_1(0) \cdot A_2 \cdot \beta \gg 1$ in the above derivation.

The NTF has one pole and one zero. The zero frequency is at −3 dB bandwidth of the error amplifier, and pole frequency is at the gain-bandwidth product of the loop.

For DC to low frequency noise, s is negligible and the noise gain is:

$$\mathrm{NTF}(0) = \frac{1}{A_1(0) \cdot \beta}$$

The VDD noise is attenuated by a factor of the error amplifier's DC gain multiplied by the feedback factor. The PSRR is:

$$\mathrm{PSRR}(0) = \frac{1}{\mathrm{NTF}(0)} = A_1(0) \cdot \beta$$

Note that the PSRR at DC is called the *line regulation* in many regulator product datasheets.

When the noise frequency approaches the bandwidth of the error amplifier (f_{-3dB}), the zero starts to take effect, and the VDD-noise gain begins to increase at a rate of 20 dB/decade. This simultaneously means the PSRR decreases at a rate of −20 dB/decade. As the VDD-noise frequency further increases beyond the gain-bandwidth product of the loop, the closed-loop pole comes into play and cancels the effect of the zero. Consequently, the VDD-noise gain becomes flat, and the PSRR stops degrading with frequency.

Note that this derivation neglects the second-stage amplifier pole and other non-dominant poles. At higher frequencies, these poles begin to attenuate the noise, causing the output noise gain to decrease and the PSRR to improve (increase) accordingly.

The results of the system transfer function analysis are consistent with those obtained through intuitive analysis. Figure Q18d is the curve of the PSRR versus the VDD-noise frequency. Please note that this does not take the low-pass nature of the VDD to V_{sg1} noise transfer (due to the source-to-gate capacitance of M1) into account, a factor which would further improve the PSRR at high frequency.

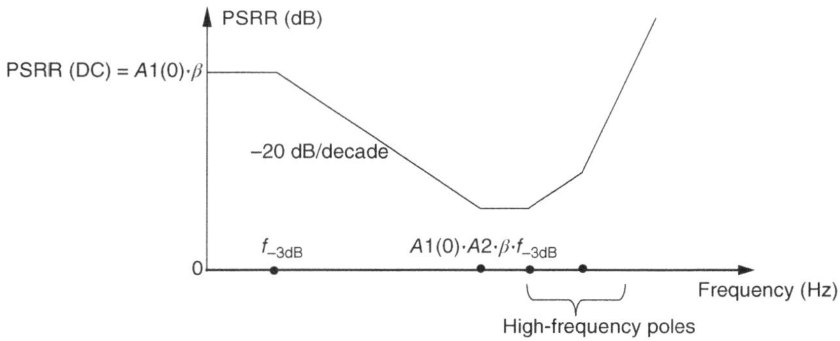

Figure Q18d

Next, let us analyze the minor noise path. As we mentioned previously, this path becomes visible when the noise frequency is beyond the loop bandwidth and the feedback loop ceases to function. M1 behaves like a DC current source (AC current = 0) as the voltage of its gate tracks that of its source (VDD) at high frequency. The noise on VDD, however, could still affect the output through the voltage divider formed by the output impedance of this current source and the load impedance. Figure Q18e is the small-signal model for minor noise path.

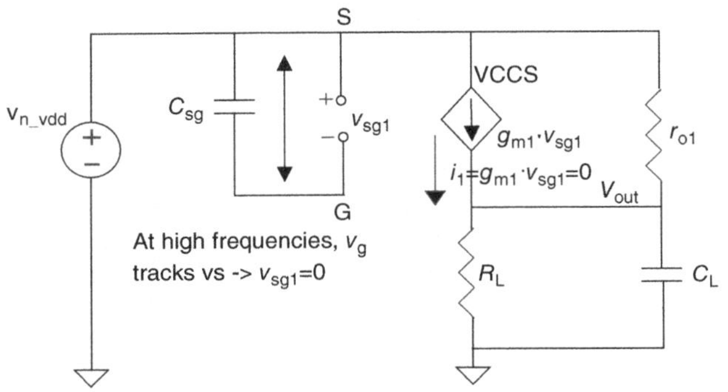

Figure Q18e

Under these conditions, the NTF of the minor path becomes:

$$\mathrm{NTF}(s) = \frac{v_{\mathrm{out}}(s)}{v_{n\_\mathrm{vdd}}(s)} = \frac{\dfrac{1}{\frac{1}{R_L} + C_L \cdot s}}{r_{o1} + \dfrac{1}{\frac{1}{R_L} + C_L \cdot s}} \approx \frac{\dfrac{R_L}{r_{o1}}}{R_L \cdot C_L \cdot s + 1}$$

where r_{o1} is the output impedance of the transistor M1, and we assume $R_L \ll r_{o1}$ in the above derivation.

Because $R_L \ll r_{o1}$ and the NTF is low-pass, this noise contribution is insignificant. Note that there could be some parasitic capacitances between the source and drain of M1, for example, coupling capacitance originating from routing. This will ultimately limit the PSRR to a finite number.

This question assumes that the V_{ref} of the voltage regulator is noise-free. This is important. We know through the previous analysis that the signal (reference) transfer function (STF) is low-pass, so that noise at V_{ref}, if any, within the bandwidth of the STF, will show up at the output of the regulator, and result in poor PSRR if V_{ref} is affected by the power supply noise.

Short Answer to Question 18

Please draw a diagram showing the relationship between the PSRR and the frequency of the VDD noise.

The PSRR is high from DC to the −3 dB bandwidth of the error amplifier, thanks to the regulation of the negative-feedback loop. It starts to degrade at frequencies higher than the bandwidth of the error amplifier. As the noise frequency exceeds the loop bandwidth, PSRR begins to improve again. At these frequencies, the loop becomes inactive, and the pass transistor's gate voltage tracks VDD, thereby mitigating the impact of supply noise on the output. The ultimate PSRR at very high frequency is determined by the impedance divider between the output impedance of the passing transistor and the load.

Please refer to Figure Q18d for the sketch of the PSRR versus VDD-noise frequency.

Question 19

A fully differential amplifier is shown in Figure Q19a. The amplifier has a DC gain of 25, and a −3 dB bandwidth of 100 MHz. The width and length of the input transistors M1 and M2 are 20 μm and 1 μm, respectively. The transconductance parameter of the input transistors (K_{pn}) is 120 $\mu A/V^2$. The gate oxide capacitance density of the input transistors (C'_{ox}) is 1.75 f/μm^2. The amplifier is operated in an open-loop configuration with a load capacitance of =100 fF, as shown in Figure Q19b. Assume small-signal conditions and that the amplifier is not slew-rate-limited. **(a)** Estimate the current consumption of the amplifier. **(b)** Is it possible to reduce power consumption while maintaining the existing gain, bandwidth, and load capacitance specifications?

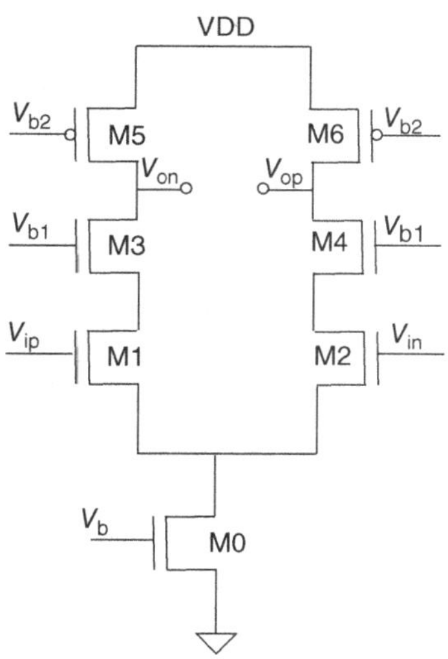

Figure Q19a

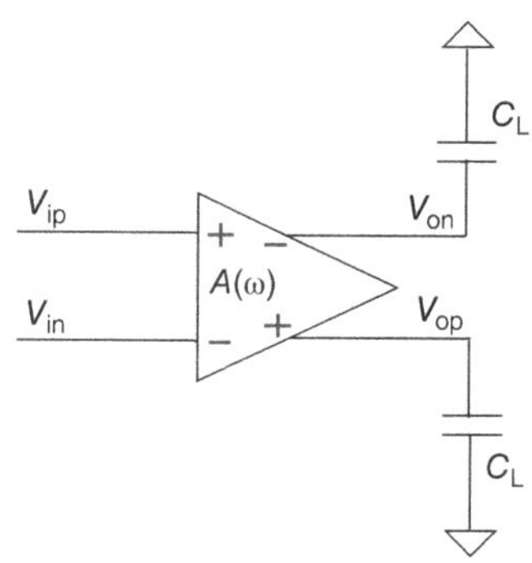

Figure Q19b

Intuitive and Quantitative Analysis: This is an open-loop amplifier application. Driving a capacitive load while maintaining a target bandwidth imposes specific requirements on the input transistors' transconductance, among other design parameters. For a single-pole amplifier as shown in Figure Q19a (ignoring the nondominant poles associated with the cascode nodes), the unity-gain frequency (f_u) equals the gain-bandwidth product (GBW) and is given by the following equation:

$$f_u = \text{GBW} = \frac{g_{m1}}{2 \cdot \pi \cdot C_L}$$

where g_{m1} is the transconductance of the input transistor pair (M1 and M2).

As the DC gain and −3 dB bandwidth of the amplifier are given, the GBW is determined:

$$\text{GBW} = A(0) \cdot f_{-3\text{dB}} = 2.5 \cdot 10^9(\text{Hz})$$

The transconductance requirement for input transistor pair therefore is:

$$g_{m1} = f_T \cdot 2 \cdot \pi \cdot C_L = \text{GBW} \cdot 2 \cdot \pi \cdot C_L = 1.57 \cdot 10^{-3}(\text{A/V})$$

The transconductance of the input transistors relates to the drain current, W/L ratio, and transconductance parameter:

$$g_{m1} = \sqrt{2 \cdot K_{pn} \cdot \frac{W_1}{L_1} \cdot I_1} = 1.57 \cdot 10^{-3}(\text{A/V})$$

Substituting K_{pn}, W_1 and L_1, the required current in M1 is:

$$I_1 \approx 5 \cdot 10^{-4}\text{A} = 0.5 \text{ mA}$$

Therefore, the total current consumption is:

$$I = 2 \cdot I_1 = 1 \text{ mA}$$

This answers part (a). However, this assumes the amplifier is transconductance (g_m) limited, meaning the current consumption primarily depends on the transconductance requirement. Not all amplifiers are transconductance-limited; some can be slew-rate-limited, especially when the output swing is large and the capacitive load is heavy. In practice, it is necessary to verify that the bias current satisfies the slew-rate requirements for a given bandwidth, full-scale output range, and load capacitance (see Q11 for the derivation of the minimum current required to achieve slew-free settling).

How can power consumption be reduced under fixed gain, bandwidth, and load constraints? Since these specifications are fixed, a topological shift may be necessary. In a single-stage amplifier, the current is dictated by the transconductance (g_m) required to meet the target gain-bandwidth product (GBW). For CMOS transistors, the relationship between bias current and transconductance is sub-linear (e.g., quadratic in strong inversion), meaning current increases disproportionately as GBW requirements rise. By distributing the total gain across multiple cascaded stages, power savings can be realized: while the total gain is the product of the individual stages, the total current consumption is merely the sum.

A quick calculation demonstrates this principle. Consider two identical gain stages, each with a gain of 5, cascading to provide a total gain of 25. While the gain multiplies, the bandwidth is compromised. When two identical first-order systems are cascaded, the total magnitude response is −6 dB at the original −3 dB corner frequency. Consequently, the overall −3 dB bandwidth is reduced to approximately 64% of an individual stage's bandwidth. To maintain the original target bandwidth for the complete system, the bandwidth of each individual stage

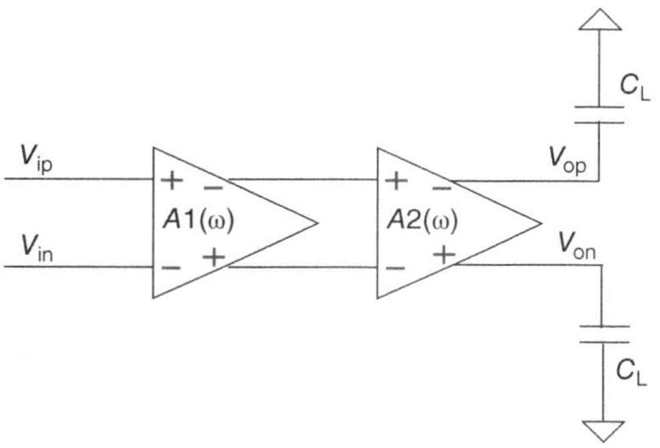

Figure Q19c

must be increased by approximately 1.5 times. The new amplifier is shown in Figure Q19c.

Start with the first amplifier (A_1). The gain-bandwidth product of A_1 now becomes:

$$\text{GBW1} = A'(0) \cdot f'_{-3\text{dB}} = 5 \cdot 1.5 \cdot 100 \cdot 10^6 = 7.5 \cdot 10^8(\text{Hz})$$

where $A'(0)$ and $f'_{-3\text{dB}}$ are the new DC gain and −3 dB bandwidth for each of the two stages, respectively.

Unlike the second-stage A_2, the first stage A_1 drives the input capacitance of A_2 plus its own output parasitics. Since the amplifiers have cascode transistors (M3 and M4) to reduce the Miller effect of the gate-to-drain capacitances, we assume the total load capacitance of the first stage is approximately the gate oxide capacitance of the input transistor of the second stage:

$$C_{L1} = C'_{\text{ox}} \cdot W_2 \cdot L_2 = 35 \cdot 10^{-15} = 35(\text{fF})$$

W_2 and L_2 are the width and length of the second-stage input transistors (assuming the input transistors in both stages are sized identically to those in the original single-stage amplifier.).

The required new transconductance, g'_{m1}, is:

$$g'_{m1} = \text{GBW}_1 \cdot 2 \cdot \pi \cdot C_{L1} = 1.65 \cdot 10^{-4}\left(\frac{\text{A}}{\text{V}}\right)$$

This determines the required bias current for the first amplifier stage (A_1):

$$I'_1 = 2 \cdot 5.6 \cdot 10^{-6}\text{A} = 11\ \mu\text{A}$$

Applying the same procedures to calculate the new current requirement for the second stage, noting that $C_L = 100\,\text{fF}$, we obtain:

$$I'_2 = 2 \cdot 46 \cdot 10^{-6}\text{A} = 92\,\mu\text{A}$$

The total current consumption is then:

$$I' = I'_1 + I'_2 = 103\,\mu\text{A} \approx 0.1\,\text{mA}$$

The current consumption is reduced by a factor of 10.

Note should be taken that to simplify the calculations, we assume that the *W/L* ratios of M1 and M2 remain constant even as their transconductances decrease significantly within the two-stage topology. In practice, however, adjusting these *W/L* ratios is necessary to maintain proper device operating conditions. These adjustments affect the input and output capacitance of the amplifier stages, factors which must be considered during the optimization of a multistage amplifier.

This simplified case demonstrates the potential for power savings when employing a multistage topology in open-loop applications. These savings assume that power consumption is primarily dictated by gain-bandwidth requirements (transconductance-limited). If the design is instead slew-rate-limited, the multistage approach may yield no power benefits. Furthermore, cascading multiple stages introduces significant tradeoffs, most notably the accumulation of noise and increased total offset at the output.

Short Answer to Question 19

(a) Estimate the current consumption of the amplifier. (b) Is it possible to reduce power consumption while maintaining the existing gain, bandwidth, and load capacitance specifications?

(a) Assuming the amplifier is transconductance (g_m) limited, the current consumption is:

$$I \approx 1\,\text{mA}$$

(b) Utilizing multiple gain stages to realize a large gain requirement can potentially lower power consumption. In this example, by using two stages each with a gain of 5, the current consumption can be reduced to about 0.1 m*A*.

Question 20

A CMOS voltage source follower circuit driving a capacitive load is shown in Figure Q20a. What are the DC gain and −3 dB bandwidth of the source follower?

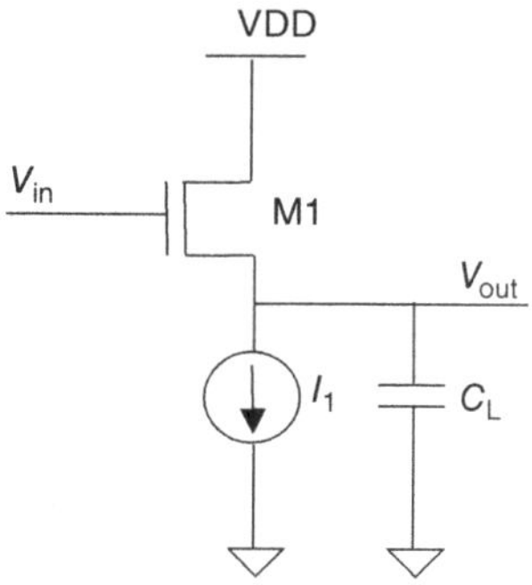

Figure Q20a

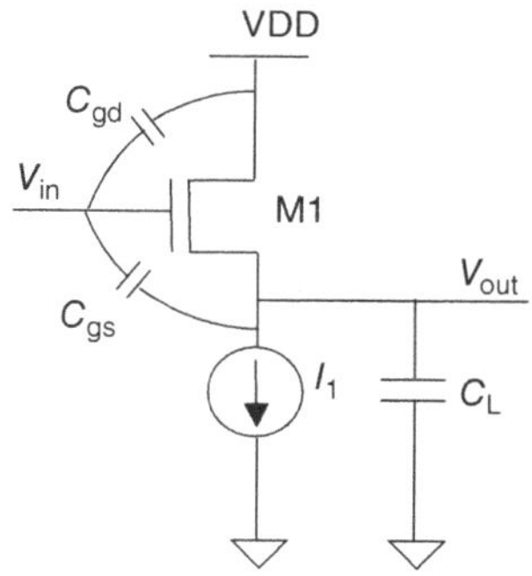

Figure Q20b

Intuitive and Quantitative Analysis: All the questions in this chapter so far are about common-source amplifiers. Yet source followers (common-drain) are another type of amplifiers with a wide range of applications such as voltage buffers and drivers. To facilitate the analysis, the two primary parasitic capacitances of transistor M1 – the gate-to-source capacitance C_{gs} and the gate-to-drain capacitance C_{gd} – are included in Figure Q20b.

In Question 2, we have analyzed the input capacitance of a simple source follower that has a current source as load with *no load capacitance*. We found out that the transfer function of the ideal source follower has a pole and zero almost at identical location:

$$H(s) = \frac{v_{out}(s)}{v_{in}(s)} = \frac{C_{gs} \cdot r_o \cdot s + g_m \cdot r_o}{C_{gs} \cdot r_o \cdot s + g_m \cdot r_o + 1}$$

where g_m and r_o are the transconductance and output impedance of the transistor M1.

The frequency of the pole is:

$$\omega_p = \frac{g_m \cdot r_o + 1}{C_{gs} \cdot r_o}$$

And frequency of the zero (LHP) is:

$$\omega_z = -\frac{g_m \cdot r_o}{C_{gs} \cdot r_o} = \frac{g_m}{C_{gs}}$$

Intuitively, with load capacitance C_L, the pole moves to a lower frequency, because now the amplifier needs to drive this load capacitance. The zero is caused by the high-frequency input feedthrough from the gate to the output via the gate-to-source capacitance (C_{gs}). As long as the impedance at the output is dominated by $1/g_m$, the impact from the load capacitance on the zero is insignificant. Based on this analysis, it is expected that the addition of load capacitance will cause the pole and zero to diverge.

Figure Q20c is the small-signal model for the source follower with the load capacitance C_L.

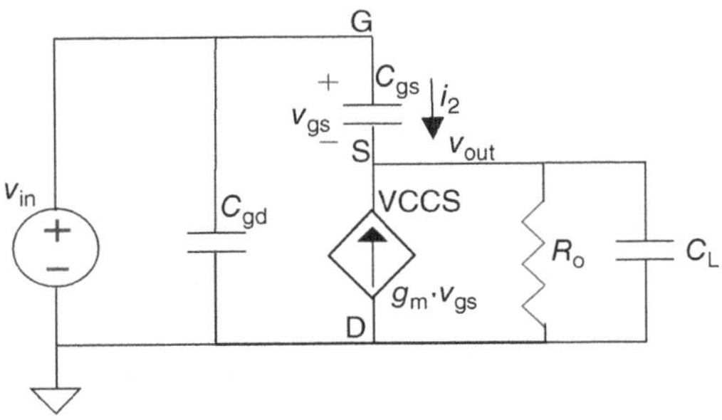

Figure Q20c

The only difference between Figures Q20c and Q2c is that the load capacitance, C_L, is put in parallel with the output impedance of the transistor M1, r_o. The equivalent impedance, r'_o, with these two components in parallel is:

$$r'_o = r_o \parallel \left(\frac{1}{C_L \cdot s}\right) = \frac{r_o}{r_o \cdot C_L \cdot s + 1}$$

Substitute r'_o for r_o in the result of Question 2, we obtain the transfer function of the source follower with the load capacitance:

$$H(s) = \frac{v_{\text{out}}(s)}{v_{\text{in}}(s)} = \frac{C_{gs} \cdot r_o \cdot s + g_m \cdot r_o}{(C_{gs} + C_L) \cdot r_o \cdot s + g_m \cdot r_o + 1} \approx \frac{C_{gs} \cdot s + g_m}{(C_{gs} + C_L) \cdot s + g_m}$$

The DC gain $(s = 0)$ is:

$$H(0) = \frac{v_{\text{out}}(0)}{v_{\text{in}}(0)} = \frac{g_m \cdot r_o}{g_m \cdot r_o + 1} \approx 1$$

The frequency of the pole and zero (LHP zero) are:

$$\omega_p = \frac{g_m}{C_{gs} + C_L}$$

$$\omega_z = \frac{g_m}{C_{gs}}$$

The magnitude frequency response is shown in Figure Q20d:

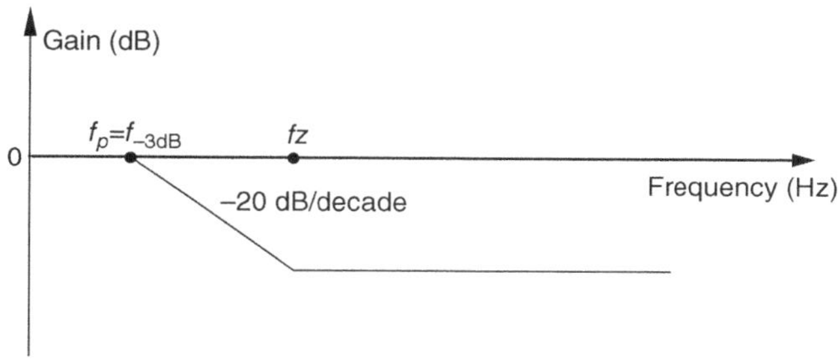

Figure Q20d

The −3 dB bandwidth is at the pole frequency:

$$f_{-3\text{dB}} = f_p = \frac{g_m}{2 \cdot \pi \cdot \left(C_{gs} + C_L\right)}$$

The gain at frequencies exceeding the LHP zero (fz) is given by:

$$H(\infty) \approx \frac{C_{gs}}{C_{gs} + C_L}$$

This behavior is anticipated: at high frequencies, the amplifier's active operation ceases, and the output is determined solely by the passive capacitive voltage divider formed by the gate-to-source capacitance $\left(C_{gs}\right)$ and the load capacitance (C_L).

This derivation neglects several critical impedances inherent to source followers, such as the series gate resistance (R_s) of transistor M1 and the parasitic elements associated with the current source I_1. Incorporating these factors would yield a second-order or higher-order transfer function, potentially introducing frequency response peaking and (steeper) high-frequency roll-off. These effects are left for interested readers to explore. For an interview discussion, it will probably not go to a depth and complexity beyond what we have discussed so far. The point of interviews often is to make sure that candidates possess the right circuit analysis knowledge and approach, so they can perform analysis in depth if needed for real-world projects.

Short Answer to Question 20

What is the DC gain and −3 dB bandwidth of the source follower?

The DC gain of the source follower in Figure Q20a is:

$$H(0) = \frac{g_m \cdot r_o}{g_m \cdot r_o + 1} \approx 1$$

And the −3 dB bandwidth is:

$$f_{-3\text{dB}} = \frac{g_m}{2 \cdot \pi \cdot \left(C_{gs} + C_L\right)}$$

General References

1 Allen PE and Holberg DR. *CMOS Analog Circuit Design*. 3rd ed. New York (NY): Oxford University Press; 2012.
2 Baker RJ. *CMOS: Circuit Design, Layout, and Simulation*. 3rd ed. Hoboken (NJ): Wiley/IEEE Press; 2010.
3 Carusone TC, Johns DA and Martin KW. *Analog Integrated Circuit Design*. 2nd ed. Hoboken (NJ): Wiley; 2012.
4 Franco S. *Design with Operational Amplifiers and Analog Integrated Circuits*. 4th ed. New York (NY): McGraw-Hill Education; 2015.
5 Gray PR, Hurst PJ, Lewis SH and Meyer RG. *Analysis and Design of Analog Integrated Circuits*. 6th ed. Hoboken (NJ): Wiley; 2024.
6 Gregorian R. *Introduction to CMOS OP-AMPs and Comparators*. New York (NY): John Wiley & Sons; 1999.
7 Lee TH. *The Design of CMOS Radio-Frequency Integrated Circuits*. 2nd ed. Cambridge (UK): Cambridge University Press; 2004.
8 Razavi B. *Design of Analog CMOS Integrated Circuits*. 2nd ed. New York (NY): McGraw-Hill Education; 2017.
9 Rincon-Mora GA. *Analog IC Design with Low-Dropout Regulators*. 2nd ed. New York (NY): McGraw-Hill Education; 2014.
10 Sansen WMC. *Analog Design Essentials*. New York (NY): Springer; 2006.

3

Bias and References

Bias circuits are fundamental building blocks in analog circuit systems and are needed wherever DC reference currents are required. Reference voltage generators provide stable reference voltages for bias current generation and serve as references, or "rulers," for circuit systems such as data converters. In high-precision analog circuits, the requirements on references and bias can be quite demanding. Design concerns include matching, noise, temperature coefficient, and drift. Factory calibration and trimming are often required.

For many new college graduates, bias circuits are likely among the first circuit blocks they are asked to work on. A good understanding of bias and references can benefit a designer's work sooner than one might expect.

In this chapter, we collect five frequently asked questions on bias and references, including three questions on current mirrors and two questions on voltage references. They are organized as follows:

We start with a question on how to design a precision current mirror (Question 21). The next challenge is on the effect of degeneration on device matching (Question 22). Question 23 evaluates the impact of various resistances on the matching and noise performance of a current mirror. Questions 24 and 25 are on two popular reference voltage generators – a bandgap reference and a beta-multiplier reference, respectively.

Question 21

A current mirror is shown in Figure Q21a. Transistors M1 and M2 have the same size (W/L). Due to the mismatch between the two transistors, the drain current of M2 deviates by an error term i_{err}. Assume that $\pm 1\sigma$ (standard deviation) of the input-referred offset voltage between the two transistors due to mismatch, v_{os}, can be modeled as follows:

$$|v_{\mathrm{os}}| = \frac{K_m}{\sqrt{W \cdot L}}$$

where K_m is a process technology related constant, W and L are the width and length of the transistor pair. Further define the error (in percentage) of the mirrored current as:

$$\varepsilon\% = \frac{i_{\mathrm{err}}}{I} \cdot 100\%$$

Explain how to design the transistor width (W) and length (L) to keep the error percentage within a specified error band, for example, $\pm 2\%$, assuming transistor mismatch is the sole cause of the error in the M2 drain current.

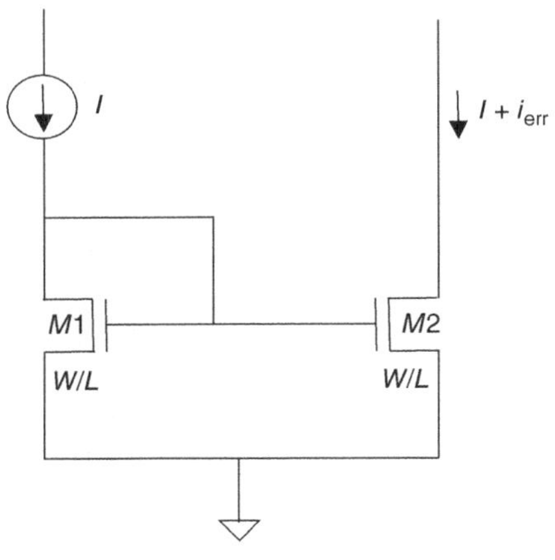

Figure Q21a

Intuitive and Quantitative Analysis: This is a classical current mirror question. First, we need to understand what the input-referred offset voltage (v_{os}) for a matched transistor pair is. As shown in Figure Q21b, the input-referred offset, v_{os}, represents the gate voltage mismatch that would result in the same level of drain current deviation (i_{err}) as that produced by the inherent physical mismatches between the two devices. Note that we assume the two transistors are perfectly matched when we use v_{os}.

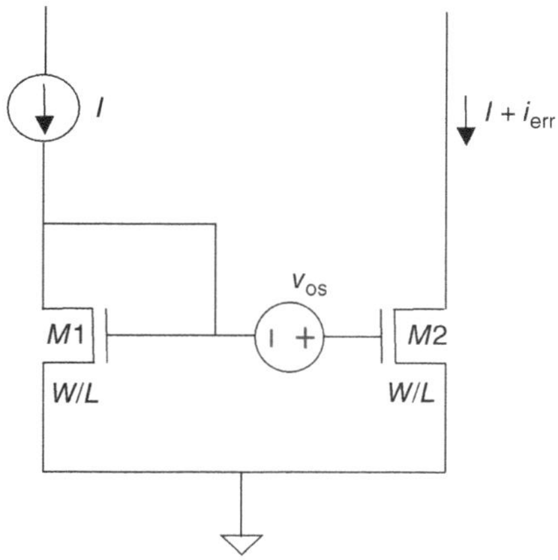

Figure Q21b

To analyze the current mismatch of the two transistors, we need to relate the mismatched current of the transistors to the input-referred offset voltage, v_{OS}. One may apply the square-law equation for the CMOS transistors (channel-length modulation is ignored because the question states that the mismatch is the only concern):

$$I = K_p \cdot \frac{W}{2 \cdot L} \cdot (V_{\mathrm{GS}} - V_{\mathrm{TH}})^2$$

This approach, however, results in a somewhat cumbersome derivation, which may be impractical under the strict time constraints of an interview. A quick way to establish the relationship between the mismatched current of the two transistors and v_{OS} is to consider v_{OS} as a "DC signal," or "DC noise." Then the "error" current of M2 is:

$$i_{\mathrm{err}} = g_m \cdot v_{\mathrm{OS}} = \sqrt{2 \cdot K_p \cdot \frac{W}{L} \cdot I} \cdot v_{\mathrm{OS}}$$

where i_{err} is the error current of M2, g_m is the transconductance of the matched pair, and I denotes the nominal drain current of each transistor in the absence of mismatch. Substituting the relationship for v_{OS} as a function of transistor dimensions and process parameters $(\pm 1\sigma)$:

$$v_{\mathrm{OS}} = \frac{K_m}{\sqrt{W \cdot L}}$$

We obtain:

$$i_{\mathrm{err}} = \sqrt{2 \cdot K_p \cdot \frac{W}{L} \cdot I} \cdot \frac{K_m}{\sqrt{W \cdot L}} = \frac{K_m \cdot \sqrt{2 \cdot K_p \cdot I}}{L}$$

If we adopt a $\pm 3\sigma$ design constraint – corresponding to a 99.7% production yield – the requirement for the error band is expressed as:

$$\varepsilon\% = 3 \cdot \frac{i_{\mathrm{err}}}{I} \cdot 100\% = \frac{3 \cdot \sqrt{2 \cdot K_p} \cdot K_m}{L \cdot \sqrt{I}} \cdot 100\%$$

The mismatch error is inversely proportional to the transistor length and the square root of the current. It does not depend on the width of the transistor – this outcome is due to the fact that increasing W raises g_m (increasing the gain applied to v_{OS}) and simultaneously reduces the magnitude of v_{OS}.

This implies that a specific channel length is required to ensure the mismatch error remains within the target tolerance. Longer channel length helps. A larger current also helps reduce the mismatch error at the cost of increased power consumption.

A common question arises regarding the role of device width: Does this result imply that W has no impact on mismatch performance? Given the assumptions of this problem, it is true that the width of the transistors does not affect the mismatch error. In reality, however, a transistor's width affects its W/L ratio, which in turn impacts the overdrive voltage $(V_{\mathrm{GS}} - V_{\mathrm{TH}})$. Recall the current square law:

$$I = K_p \cdot \frac{W}{2 \cdot L} \cdot (V_{\mathrm{GS}} - V_{\mathrm{TH}})^2$$

The square-law model indicates that as the overdrive voltage decreases, the sensitivity of the drain current to threshold voltage variations increases quadratically. Consequently, a large W/L ratio results in a lower overdrive voltage; when V_{GS} is biased near V_{TH}, any threshold mismatch between the transistors manifests as a significant current error. Conversely, a high overdrive voltage requires a larger drain-to-source voltage V_{DS} to maintain saturation, thereby reducing signal headroom. Therefore, once the channel length is established, the W/L ratio must be carefully balanced: it must be small enough to mitigate mismatch sensitivity, yet large enough to preserve sufficient voltage headroom.

Before concluding this analysis, it is important to note that factors beyond transistor mismatch also contribute to current errors in practical designs. Among them is channel-length modulation – changes in the drain-to-source voltage (V_{DS}) would cause current variations. Cascode current mirrors and regulated drain-to-source voltage are common techniques in practice to deal with the current mirror channel-length modulation problem.

Short Answer to Question 21

Explain how to design the transistor width (W) and length (L) to keep the error percentage within a specified error band, for example, ±2%, assuming transistor mismatch is the sole cause of the error in the M2 drain current.

The error (mismatched) current in a current mirror equals input-referred offset voltage multiplied by the transconductance of the transistor. The following equation (±3σ design constraint) can be used to determine the minimum transistor length for a given precision requirement ($\varepsilon\%$) and process technology (K_m):

$$\varepsilon\% = \frac{3 \cdot \sqrt{2 \cdot K_p} \cdot K_m}{L \cdot \sqrt{I}} \cdot 100\%$$

We also need to ensure a well-balanced W/L ratio – not too large, which would increase the drain current's sensitivity to threshold voltage variations, and not too small, which would reduce the headroom for the current source to stay in saturation.

Question 22

Most of Question 22 is the same as Question 21, except that the transistors M1 and M2 both have a resistor between their source and the ground (R_1 and R_2, respectively), as shown in Figure Q22a. Assume that R_1 and R_2 are perfectly matched to each other with the resistance value of R. Please evaluate the impact of the resistors on the error in the mirrored current, i_{err}.

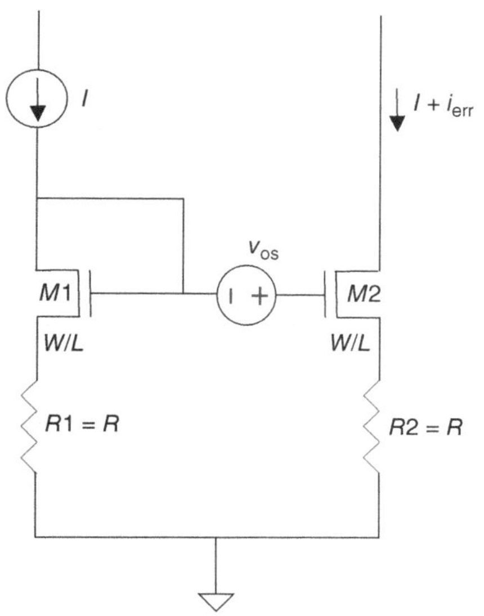

Figure Q22a

Intuitive and Quantitative Analysis: The analysis of Question 21 demonstrates that the mismatch-induced error current in M2 is equal to the product of the input-referred offset voltage (v_{os}) and the transconductance (g_m) of M2. With the degeneration resistance R at the sources of M1 and M2, the effective transconductance of the transistors is reduced. This implies that the error current, i_{err}, would be reduced.

To analyze quantitatively the impact of the degeneration resistance R on the error current, we need to find the equivalent transconductance of the degenerated transistors. As shown in Figure Q22b, the equivalent transconductance, g_{m_eq}, refers to the transconductance of an equivalent transistor, M2_eq, without the degeneration resistance, but with the effect of the degeneration resistance transferred to its transconductance.

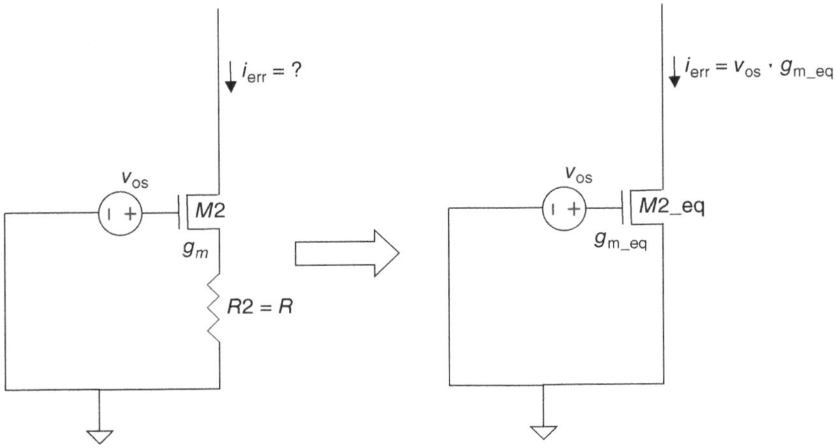

Figure Q22b

Starting with the original degenerated circuit, we obtain:

$$\begin{cases} v_{os} = v_{gs} + i_{err} \cdot R \\ i_{err} = v_{gs} \cdot g_m \end{cases}$$

Note that v_{gs} is an "AC" gate-to-source voltage caused by v_{os}. Solving the system of equations yields:

$$g_{m\_eq} = \frac{i_{err}}{v_{os}} = \frac{g_m}{1 + g_m \cdot R} \approx \frac{1}{R}; \text{ if } g_m \cdot R \gg 1$$

With a large loop gain, the equivalent transconductance depends primarily on the degeneration resistance (R). The error current ($\pm 1\sigma$), i_{err}, can be calculated from the following equation:

$$i_{err} = g_{m\_eq} \cdot v_{os} \approx \frac{1}{R} \cdot \frac{K_m}{\sqrt{W \cdot L}}$$

The $\pm 3\sigma$ error (in percentage) is:

$$\varepsilon\% = \frac{3 \cdot i_{err}}{I} \cdot 100\% = \frac{3}{R \cdot I} \cdot \frac{K_m}{\sqrt{W \cdot L}} \cdot 100\%$$

Evidently, the source degeneration resistors provide an additional design parameter; the current mismatch can be mitigated by increasing the resistance value R. When the degeneration is sufficiently large, the influence of transistor nonidealities on the total current error diminishes.

What about the mismatch of the degeneration resistors themselves? As a matter of fact, the use of source degeneration shifts the dominant source of current mirror mismatch from the transistors to the resistors. Matching of resistors is typically significantly better than that of transistors for most CMOS process technologies. This renders resistive degeneration a viable solution for the reduction of mismatches in CMOS current mirrors. For example, degeneration resistors can be used in current sources in a digital-to-analog converter to improve matching and precision. Keep in mind that large-value resistors might reduce the voltage headroom due to IR drops, and well-matched resistors could take up a large layout area as well.

Short Answer to Question 22

Please evaluate the impact of the resistors on the error in the mirrored current, i_{err}.

The current mirror's sensitivity to transistor mismatch is reduced by the source degeneration resistance:

$$\varepsilon\% = \frac{3}{R \cdot I} \cdot \frac{K_m}{\sqrt{W \cdot L}} \cdot 100\%$$

We can reduce current mismatch by increasing the degeneration-resistance value. When the degeneration resistance is sufficiently large, the impact of the transistor mismatch becomes insignificant. Matching of resistors is typically significantly better than that of transistors for most CMOS process technologies.

Question 23

A current mirror consisting of transistors M1 and M2 is shown in Figure Q23a. M3 is the current source providing the reference current, I_1, to the current mirror. R_1, R_2, and R_3 are resistors at three locations in the current mirror, as shown in Figure Q23a. Please comment on **(a)** the impact of these resistances on the current mirror in terms of the accuracy of the DC current I_2. **(b)** the R_2's noise contribution to I_2. For questions in part **(b)**, we assume M3 is an ideal current source (i.e., M3 has an infinite output resistance) and $R_3 = 0$. Also assume $R_2 \gg 1/g_m$, where g_m is the transconductance of M1 and M2.

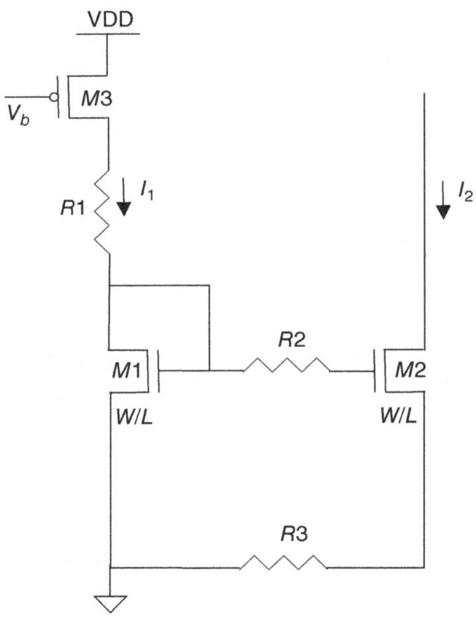

Figure Q23a

Intuitive and Quantitative Analysis: Let us first consider the DC accuracy of I_2. The question we ask ourselves is: "What impacts can resistances have on the DC behavior of a current mirror?" For one, a resistor can have a voltage drop (IR drop) across it if there is current flowing through it. Such a voltage drop can potentially cause a mismatch problem in a current mirror depending on where it occurs.

R_1 has a voltage drop across it due to the reference current, I_1, from the current source M3. The V_{GS} of the transistor M1 is determined by the current I_1 and the parameters of M1, such as its transconductance parameter, threshold voltage, width and length, and does not depend on the voltage drop across R_1. However, if the voltage across R_1 is significantly large, it can cause voltage headroom issues for the current source providing the reference current. The reference current, I_1, would decrease due to insufficient source-to-drain voltage (V_{SD}) of M3. Voltage headroom issues are more often found in cascode current sources. Also, in some circuit systems, the reference currents from a master bias unit must travel long distances in the routing and experience large parasitic resistances before reaching the local bias units, so the IR drops of the routing need to be checked.

Unlike R_1, R_2 has no DC current flowing through it and therefore has no DC voltage drop. The resistance at the gate therefore affects neither the gate voltage of the transistor pair nor the current I_2. Note that we assume the gate leakage current is negligible.

What about R_3? A common mistake is to think that the impact of R_3 is the same as that of R_2 by assuming there is no current flowing through R_3. In fact, according to the circuit diagram in Figure Q23a, current I_2 must flow through R_3 to reach the ground potential (ground tap), as illustrated in Figure Q23b.

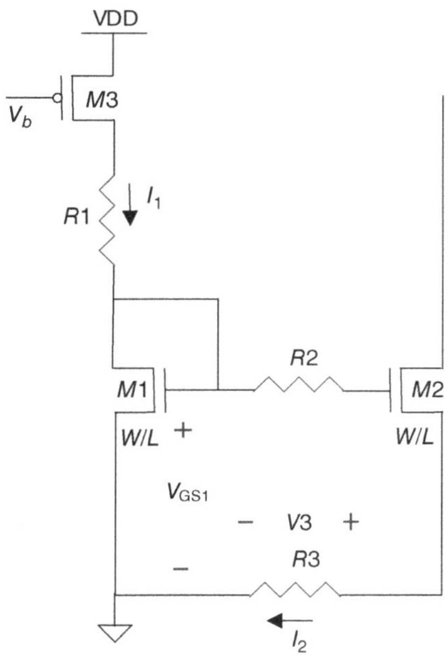

Figure Q23b

The IR drop V_3 (Figure Q23b), caused by the ground current I_2 flowing through the ground resistance R_3, offsets the gate-to-source voltage V_{GS} between M1 and M2. This would in turn cause a mismatch between currents I_1 and I_2.

In circuit systems, the distribution of currents in ground lines is rather complex, depending on the topology of the circuits and their ground network. Most, if not all, ground branches carry some current, large or small, resulting in voltage drops along ground lines due to routing parasitic resistances. For this reason, we do not distribute a bias voltage (V_{GS1} in Figure Q23b) over a long routing distance to the mirrored transistors (M2 in Figure Q23b), as ground voltage drops would cause bias current mismatches. Instead, we distribute the bias current (I_1 in Figure Q23a) over long distances and generate the bias voltage at the destination.

The question in part (b) asks about the noise contribution of R_2 to the mirrored current I_2, assuming I_1 comes from a current source, $R_3 = 0$, and $R_2 \gg 1/g_m$. R_2 generates thermal noise, and we add this noise source to the simplified current mirror as shown in Figure Q23c.

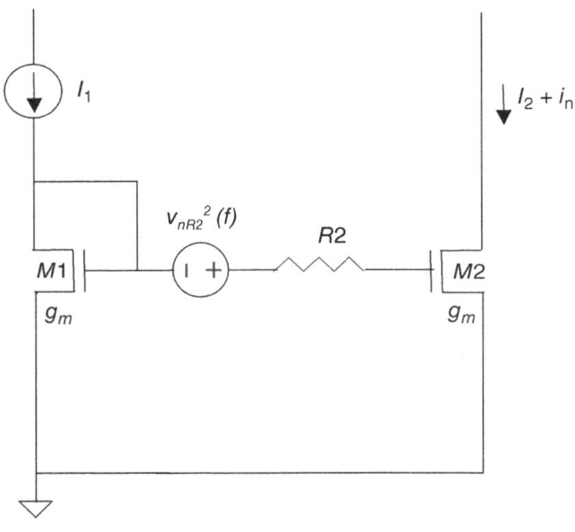

Figure Q23c

The thermal noise of a resistor can be modeled as either a current noise source in parallel with the resistor or a voltage noise source in series with the resistor. The power spectral densities (PSDs) of the current noise and voltage noise are:

$$i_{\text{thermal}}^2(f) = \frac{4 \cdot K \cdot T}{R};\ v_{\text{thermal}}^2(f) = 4 \cdot K \cdot T \cdot R$$

It is important to realize that these two noise models describe the same thermal noise and are in fact equivalent. If we apply Thévenin's theorem to the current noise model, we can obtain the voltage noise model. Usually, we use a voltage noise model if the noise source drives a high-impedance load, and use a current noise model if the noise source drives a low-impedance load. In

Figure Q23c, we model the noise source of R_2 as a voltage noise source because it drives a high-impedance load (gate of M2).

Because I_1 is from a DC current source, the gate voltage of M1 is at AC ground and the resistance looking into the gate of M1 is $1/g_m$. We will ignore this resistance in the analysis since $R_2 \gg 1/g_m$. The circuit can be further simplified as shown in Figure Q23d. The PSD of the noise voltage source of R_2 is:

$$v_{nR2}^2(f) = 4 \cdot K \cdot T \cdot R_2$$

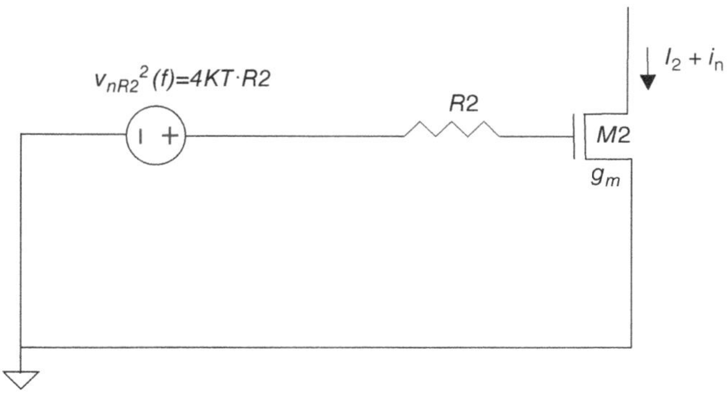

Figure Q23d

The resulting noise current (RMS) at the drain of M2, i_n, is:

$$i_n(\text{RMS}) = \sqrt{\int_{f_{\min}}^{f_{\max}} v_{nR2}^2(f) \cdot g_m^2 \cdot df}$$

This brings up the question of what the current mirror's equivalent noise bandwidth is for the thermal noise presented at the gate of M1. The noise bandwidth is related to the *RC* time constant at the gate of M2 and the bandwidth limitation of the transconductance of M2. Usually, the bandwidth of the transconductance is significantly higher so that the time constant at the gate of M2 (i.e., the time constant set by R_2 and the gate parasitic capacitance of M2) dominates the noise bandwidth. We can clearly identify this as KT/C noise, the thermal noise power sampled onto a capacitor (see *Appendix 4, Section 6, Circuit noise fundamentals*). The noise current (RMS) at the drain of M2 is:

$$i_n(\text{RMS}) = g_m \cdot \sqrt{\frac{K \cdot T}{C_g}}$$

where C_g is the capacitance at the gate of M2.

Because this is KT/C noise, the value of R_2 does not affect the RMS value of thermal noise in the mirrored current. A small series resistance at the gate of M2 is usually preferred over a large resistance for thermal noise reduction, however, since quite often the subsequent circuit may have a larger time constant than that at the gate of M2. Also note that we have not taken the noise of M1 and M2 into consideration so far. In fact, the noise sources associated with M2 often dominate the noise contribution to I_2.

In practice, a decoupling capacitor is often added to the gate of the mirrored transistors, as shown in Figure Q23e. This effectively lowers the thermal noise at the gate. The decoupling capacitor also improves the Power Supply Rejection of the current mirror by filtering away the high-frequency noise at the gate of the mirrored transistors caused by noise on the power rails.

In many circuit applications, the bias current noise due to the device intrinsic noise sources, such as thermal noise, is negligible. In low-noise applications, such as low-noise amplifiers and high-precision data converters, however, these noise contributions can pose a threat to the noise performance of the system.

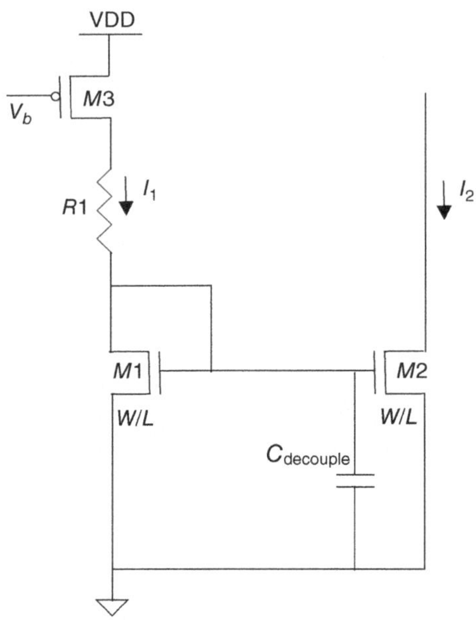

Figure Q23e

Short Answer to Question 23

(a) the impact of these resistances on the current mirror in terms of the accuracy of the DC current I_2. **(b)** the R_2's noise contribution to I_2. For questions in part (b), we assume M3 is an ideal current source (i.e., M3 has an infinite output resistance) and $R_3 = 0$. Also assume $R_2 \gg 1/g_m$, where g_m is the transconductance of M1 and M2.

(a) R_3 has the largest and most direct impact on the accuracy of DC current I_2 due to V_{GS} mismatch caused by ground/return voltage drop. R_2 has no impact (assuming the gate leakage current is negligible). R_1 does not have a direct impact unless the voltage drop across it is large enough to cause voltage headroom issues for M3.

(b) The noise current (RMS) at the drain of M2 is:

$$i_n(\text{RMS}) = g_m \cdot \sqrt{\frac{K \cdot T}{C_g}}$$

where C_g is the capacitance at the gate of M2.

Question 24

A CMOS bandgap voltage reference is shown in Figure Q24a (the startup circuit is not shown). An amplifier A is used to regulate node $A(V_{\text{REF}})$ and node B so that their voltages are equal. The polarity of the two inputs of the amplifier is intentionally left unlabeled. Resistors R_1 and R_2 are scaled by a factor of L relative to R_3. Diode D2 (a parasitic PNP) consists of K devices equivalent to D1 in parallel. **(a)** Please determine the sign (inverting or non-inverting) of the two inputs for the amplifier. **(b)** How should the parameters be set so that the temperature coefficient of V_{REF} is close to zero? **(c)** Explain why it is not practically feasible to maintain a zero temperature coefficient over a wide temperature range?

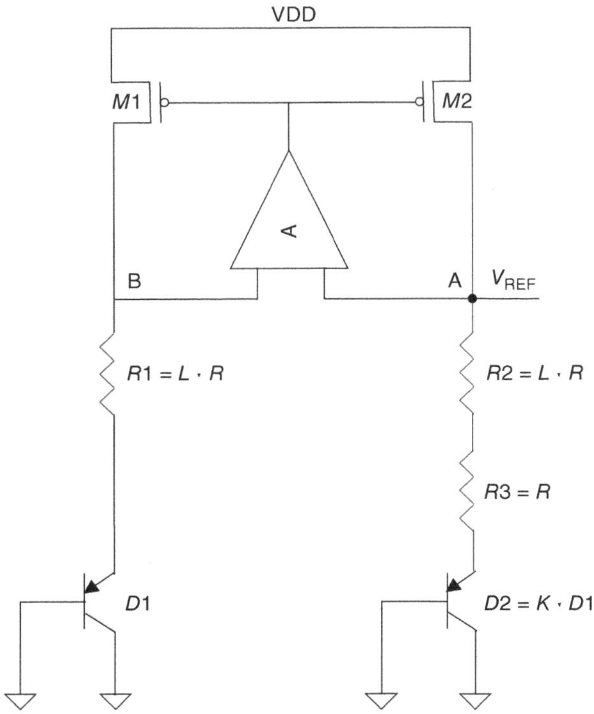

Figure Q24a

Intuitive and Quantitative Analysis: Let us start with the question in (a). To some readers, this can be a somewhat unexpected question for which they are unprepared. A common confusion is that the amplifier has two feedback paths (the left-side path and the right-side path) – no matter which input is inverting there is always one with positive feedback and the other one with negative feedback. The key fact to notice about this type of bandgap circuit is that the gains of the two feedback loops are not equal. On both sides, the feedback loop consists of amplifier A and a common-source amplifier formed by M1 or M2. The common-source amplifier formed by M2 has a larger load resistance than that formed by M1 and therefore larger loop gain. This is the path that needs to be configured as negative feedback so that it overrides the effect of the other positive-feedback loop. By this reasoning, the input on the right side (i.e., the input at node A) should be the non-inverting input (Figure Q24b). We can perform a sanity check by investigating the feedback response in the event that the voltages at the two nodes become unequal.

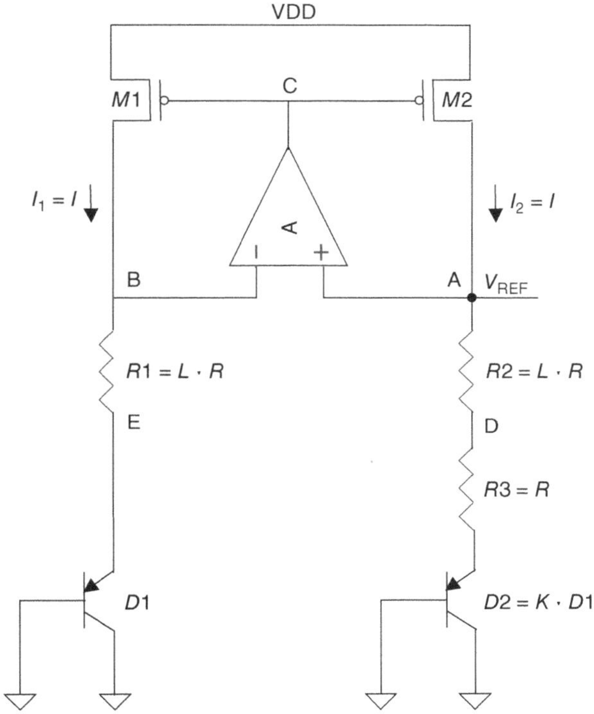

Figure Q24b

At the steady state, V_A is equal to V_B. If V_A increases relative to V_B, the output voltage V_C of amplifier A increases, meaning that the source-to-gate voltage of transistors M1 and M2 decreases. This subsequently causes I_1 and I_2 to decrease (note that $I_1 = I_2$). As the impedance at node A is larger than that at node B, V_A decreases more than V_B, which in turn reduces the difference between the two voltages. Note that in this process, V_B also decreases, which is a positive-feedback effect, but it is offset by the larger changes on V_A which is a stronger negative-feedback action.

To answer the question in (b), we need to derive the relationship between V_{REF} and the parameters of the bandgap circuit. As discussed earlier, the amplifier A forces $I_1 = I_2$ and $V_B = V_{REF}(V_A)$. Note that $V_E = V_D$ because $R_1 = R_2$, and $I_1 = I_2$. This yields,

$$V_E = V_{d1} = V_{d2} + I \cdot R_3 = V_{d2} + I \cdot R$$

The *I-V* relationship of a diode (parasitic PNP on CMOS Process) is:

$$I_d = I_s \cdot e^{\frac{V_d}{n \cdot V_T}}; \text{ or } V_d = n \cdot V_T \cdot \ln\left(\frac{I_d}{I_s}\right)$$

where V_d is the diode forward bias voltage, I_d is the diode forward bias current, I_s is the diode's scale current, n is the emission coefficient, T is absolute temperature, and V_T is the thermal voltage:

$$V_T = \frac{k \cdot T}{q}$$

$$k = 1.38 \cdot 10^{-23}\ \text{J/K}(\text{Boltzmann Constant})$$

$$q = 1.6 \cdot 10^{-19}\ \text{C}(\text{elementary charge})$$

By applying the diode *I-V* equation to diode D1 and D2, we obtain:

$$V_{d1} = n \cdot V_T \cdot \ln\left(\frac{I}{I_s}\right)$$

$$V_{d2} = n \cdot V_T \cdot \ln\left(\frac{I}{K \cdot I_s}\right)$$

Combining these equations together, we have:

$$I_2 = I_1 = I = \frac{n \cdot k \cdot \ln(K)}{q \cdot R} \cdot T$$

This is a PTAT (proportional to absolute temperature) current.

The reference voltage (V_{REF}) is:

$$V_{\text{REF}} = V_B = \frac{L \cdot n \cdot k \cdot \ln(K)}{q} \cdot T + V_{d1} = L \cdot n \cdot \ln(K) \cdot V_T + V_{d1}$$

Taking the derivative with respect to temperature at both sides of the equation, we obtain:

$$\frac{\partial V_{\text{REF}}}{\partial T} = L \cdot n \cdot \ln(K) \cdot \frac{\partial V_T}{\partial T} + \frac{\partial V_{d1}}{\partial T}$$

where

$$\frac{\partial V_T}{\partial T} = \frac{k}{q} \approx 0.086\ \text{mV/°C}$$

$$\frac{\partial V_{d1}}{\partial T} \approx -2\frac{\text{mV}}{\text{°C}}; \text{depends on process, bias, and temperature}$$

The first part of the V_{ref} is a PTAT voltage and has a positive temperature coefficient, and the second part of the V_{ref} is a diode voltage, which is a CTAT (complementary to absolute temperature) voltage with a negative temperature coefficient.

The negative temperature coefficient of diodes is because the increasing temperature provides more energy to the charge carriers and it becomes easier for them to overcome the energy barrier (bandgap). By properly designing the two ratio parameters, L and K, a temperature coefficient close to zero can be achieved.

Maintaining a zero temperature coefficient over a wide temperature range (e.g., −40 °C to +125 °C) is not practically feasible. This is because the temperature coefficient of the diode is influenced not only by the process and manufacturing variations, but also by the bias current and the operating temperature itself. In other words, the diode voltage has a nonlinear temperature dependency. Even if perfect cancellation of the PTAT and CTAT components is achieved at a specific temperature point, the diode's effective temperature coefficient shifts as the temperature deviates from that point. Furthermore, other components within the bandgap reference circuit introduce additional nonidealities, such as: DC offset, low-frequency noise, finite gain of the operational amplifier, and resistor mismatch. These factors introduce errors into the bandgap output voltage. Mechanical stress during IC packaging and reflow processes further increases complexity, as it can induce shifts in both the bandgap output voltage and its temperature characteristics. Modern state-of-the-art bandgap references employ additional circuits to alleviate some of the circuit nonidealities and sophisticated trimming methods to compensate for the variations during the manufacture process.

Short Answer to Question 24

(a) Please determine the sign (inverting or non-inverting) of the two inputs for the amplifier. **(b)** How to set the parameters so that the temperature coefficient of V_{REF} is close to zero? **(c)** Explain why it is not practically feasible to keep a zero temperature coefficient over a wide temperature range?

(a) Both positive and negative feedback exist around the op-amp, and it is necessary that negative feedback dominate. The inverting input of the op-amp should be connected to node B in Figure Q24b.

(b) The temperature coefficient of V_{REF} is:

$$\frac{\partial V_{\text{REF}}}{\partial T} = L \cdot n \cdot \ln(K) \cdot \frac{\partial V_T}{\partial T} + \frac{\partial V_{d1}}{\partial T}$$

The first term involves a PTAT voltage with a positive temperature coefficient, and the second term is a CTAT voltage with a negative temperature coefficient. By properly designing the two ratio parameters, L and K, a temperature coefficient close to zero can be achieved.

(c) This is because the temperature coefficient of the diode is influenced not only by the process and manufacturing variations, but also by the bias current and the operating temperature itself.

Question 25

A beta-multiplier voltage reference is shown in Figure Q25a (the startup circuit is not shown). M3 and M4 are a matched pair with the same width and length, and M2 is larger than M1 by a multiplier of K (i.e., Transistor M2 is composed of K parallel instances of M1). V_{REF} is the output reference voltage. Assume that the resistor R has a positive temperature coefficient. **(a)** Explain intuitively why the beta multiplier can achieve a stable reference voltage over temperature. **(b)** The beta multiplier is a circuit that uses feedback. Please identify the nature of this feedback, i.e., negative or positive feedback, and analyze the stability of the circuit. **(c)** Explain the merits and drawbacks of beta-multiplier voltage references versus bandgap voltage references.

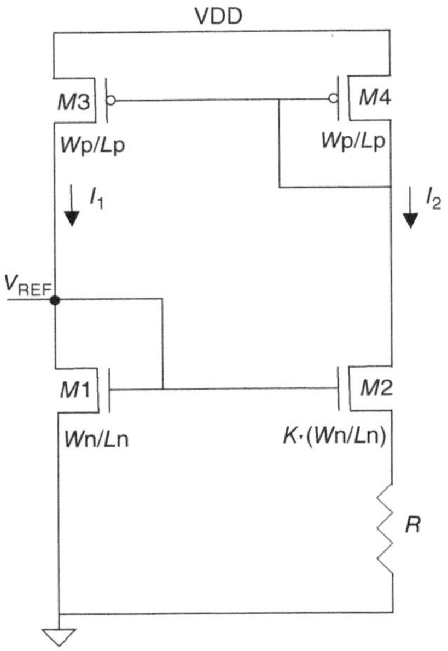

Figure Q25a

Intuitive and Quantitative Analysis: The output reference voltage, V_{REF}, equals the gate-to-source voltage of M2 (V_{GS2}) plus the voltage drop across the resistor R.

$$V_{REF} = V_{GS2} + I \cdot R$$

When the temperature increases, the resistance of R also increases due to its positive temperature coefficient. The gate-to-source voltage, V_{GS2}, decreases because of the negative temperature coefficient of the threshold voltage of

M2 (V_{TH2}). As such, it is possible to achieve a stable voltage over temperature with the beta multiplier.

The quantitative derivation of the temperature coefficient of the beta multiplier is a little involved, but we illustrate it below for interested readers – and some results turn out to be handy for answering the questions in (b) and (c). We notice that currents I_1 and I_2 are forced to be equal by the PMOS current mirror on the top of the circuit. Neglecting body effect of M2, and assuming:

$$I_1 = I_2 = I$$

$$V_{TH1} = V_{TH2} = V_{THN}$$

$$\beta = K_{pn} \cdot \frac{W_N}{L_N}; K_{pn}\text{: Transconductance parameter of NMOS}$$

We have the following system of equations:

$$\begin{cases} V_{GS1} = V_{GS2} + I \cdot R \\ V_{GS1} = \sqrt{\dfrac{2 \cdot I}{\beta}} + V_{THN} \\ V_{GS2} = \sqrt{\dfrac{2 \cdot I}{K \cdot \beta}} + V_{THN} \end{cases}$$

Solving the system of equations yields:

$$I = \frac{2}{R^2 \cdot \beta} \cdot \left(1 - \frac{1}{\sqrt{K}}\right)$$

$$V_{REF} = V_{GS1} = \frac{2}{R \cdot \beta} \cdot \left(1 - \frac{1}{\sqrt{K}}\right) + V_{THN}$$

And the temperature coefficient of V_{REF} is:

$$\frac{\partial V_{REF}}{\partial T} = \frac{\partial V_{THN}}{\partial T} - \frac{2}{R^2 \cdot \beta} \cdot \left(1 - \frac{1}{\sqrt{K}}\right) \cdot \left(\frac{1}{R} \cdot \frac{\partial R}{\partial T} + \frac{1}{K_{pn}} \cdot \frac{\partial K_{pn}}{\partial T}\right)$$

Notice that there are three temperature sensitive circuit parameters in the equation: besides the resistance and threshold voltage of NMOS transistors, the NMOS transconductance parameter K_{pn} also plays a role. The temperature coefficient of the transconductance parameter is negative due to the reduced carrier mobility at higher temperatures. By properly choosing the resistance value R, the transistor size (therefore the β), and the ratio K, a low temperature coefficient at certain temperatures can be achieved.

Now let us turn to part (b) and identify the feedback path in the beta-multiplier circuit. An obvious one is the degeneration resistor R, which is negative feedback in nature. However, this is not the only feedback path in the circuit. As illustrated in Figure Q25b, M2 and M4 form a common-source amplifier with its voltage output (V_B) fed to the input of a second common-source amplifier formed by M3 and M1. The output voltage of this second amplifier, V_{REF}, is fed back again to the input of the first amplifier (M3/M4). By tracing this feedback path and noting that both amplifiers have an inverting voltage gain, we conclude that this is positive feedback similar to those found in circuits such as latches.

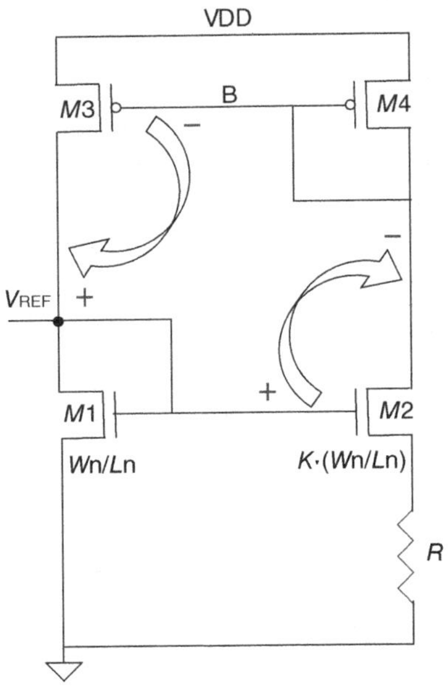

Figure Q25b

Would the circuit be stable with this positive feedback? To answer this question, we need to analyze the loop gain (A_{loop}) of this positive feedback, which must be less than 1 to keep the circuit from latching up. Intuitively, the degeneration of M2 significantly reduces the gain from the gate to the drain of M2 (and therefore the feedback loop gain), and helps the stability of the beta-multiplier reference.

Quantitatively, the feedback loop gain is the gain from gate of M2 to drain of M2 (notice the degeneration resistor R) multiplied by the gain from gate of M3 to drain of M3:

$$A_{\text{loop}} = \left[\frac{\frac{1}{g_{m4}}}{\left(\frac{1}{g_{m2}} + R\right)}\right] \cdot \left(\frac{g_{m3}}{g_{m1}}\right)$$

Assuming $g_{m3} = g_{m4}$, and utilizing the following relationships derived from, or implied by, the analysis in part (a):

$$g_{m2} = \sqrt{K} \cdot g_{m1}$$

$$I = \frac{2}{R^2 \cdot \beta} \cdot \left(1 - \frac{1}{\sqrt{K}}\right)$$

The loop gain is determined as:

$$A_{\text{loop}} = \frac{1}{2 - \frac{1}{\sqrt{K}}}$$

For stable operation of the circuit, the loop gain needs be less than 1. The condition for ensuring the stability of the beta multiplier, therefore is:

$$K > 1$$

This is guaranteed as M2 must have a larger W/L ratio than M1 to accommodate its degeneration resistor R.

The primary advantages and disadvantages of beta-multiplier references relative to bandgap references are as follows. CMOS bandgap voltage references need to utilize parasitic PNP which causes current to be injected into the substrate. Beta-multiplier references do not have this problem. On the other hand, the reference voltage of beta-multiplier references depends on the value of the degeneration resistor, MOSFET threshold voltage, and transconductance parameter, while the reference voltage of bandgap references primarily relies on the diode forward bias voltage. The former typically has larger variations than the latter and this makes the beta-multiplier reference less accurate than the bandgap reference.

Short Answer to Question 25

(a) Explain intuitively why the beta multiplier can achieve a stable reference voltage over temperature (b) The beta multiplier is a circuit that uses feedback. Please identify the nature of this feedback, i.e., negative or positive feedback, and analyze the stability of the circuit. (c) Explain the merits and drawbacks of beta-multiplier voltage references versus bandgap voltage references.

(a) The output reference voltage (V_{REF}) equals the gate-to-source voltage of M2 (V_{GS2}) plus the voltage drop across resistor R. Resistor R has a positive temperature coefficient, while V_{GS2} has a negative temperature coefficient. As such, it is possible to achieve a stable voltage over temperature with the beta multiplier.

(b) Transistors M1, M2, M3, and M4 form a positive-feedback loop. The loop is stable because the loop gain is less than one due to the source degeneration of M2.

(c) Merits of beta-multiplier voltage references: no current injection to the substrate.

Drawbacks: the reference voltage depends on the value of the degeneration resistor and is less accurate than that of bandgap references.

General References

1 Allen PE and Holberg DR. *CMOS Analog Circuit Design*. 3rd ed. New York (NY): Oxford University Press; 2012.
2 Baker RJ. *CMOS: Circuit Design, Layout, and Simulation*. 3rd ed. Hoboken (NJ): Wiley/IEEE Press; 2010.
3 Carusone TC, Johns DA and Martin KW. *Analog Integrated Circuit Design*. 2nd ed. Hoboken (NJ): Wiley; 2012.
4 Gray PR, Hurst PJ, Lewis SH and Meyer RG. *Analysis and Design of Analog Integrated Circuits*. 6th ed. Hoboken (NJ): Wiley; 2024.
5 Razavi B. *Design of Analog CMOS Integrated Circuits*. 2nd ed. New York (NY): McGraw-Hill Education; 2017.
6 Rincon-Mora GA. *Voltage References: From Diodes to Precision High-Order Bandgap Circuits*. Piscataway (NJ): IEEE Press; 2002.
7 Sansen WMC. *Analog Design Essentials*. New York (NY): Springer; 2006.

4

Switched-capacitor Circuits and Comparators

Mathematically, analog signals are considered continuous and digital logic signals discrete, both in time and in magnitude. Switched-capacitor circuits are a mix of the two. The sampling nature of these circuits makes them *discrete-time* analog circuits: it is discrete with respect to time, but magnitude can still reach any real numbers within the range.

Switched-capacitor circuits are widely used in data converters, chopper amplifiers, and switched-capacitor filters. Here, the switch does the sampling and the capacitor stores the information (charge). In switched-capacitor circuits, clock timing defines the key events and the final settled amount of charge on capacitors determines the circuit performance.

Many nonidealities affect the performance of switched-capacitor circuits. Among them are switch nonlinearity, charge injection, clock feedthrough, amplifier nonidealities (gain, bandwidth, distortion, slew, etc.), clock integrity, device noise, and leakage. Understanding these nonidealities is key to the successful design of switched-capacitor circuits.

Another important type of discrete-time analog circuits is the (clocked) comparator. Comparators are the core building block of the analog-to-digital converters (ADCs), quantizing the continuous analog input into discrete and finite output levels. Unlike precision closed-loop amplifiers, comparators often use open-loop preamplifiers and regenerative latches with positive feedback for fast transitions.

This chapter is organized as follows. The first question (Question 26) investigates the input impedance of switched-capacitor circuits. Question 27 explores the optimization of sampling switch size to reduce nonlinearity and charge injection. Question 28 is about bottom plate sampling. The next three questions (Questions 29, 30, and 31) focus on charge-transfer track-and-hold circuits, covering topics from estimating the amplifier gain and bandwidth requirements to input-referred thermal noise. Switched-capacitor common-mode feedback control is studied in Question 32. The next three questions (Questions 33, 34, and 35) are about comparators, from regenerative latches to preamplifiers and offset cancellation. The last question (Question 36) is about switched-inductor circuits – it is not only interesting to swap capacitors with inductors; the resulting circuits are also widely used in practice.

Question 26

A switched-capacitor circuit is shown in Figure Q26a. The input is an ideal DC voltage source, $V_{\text{in}} = V$, with a source resistance of R_S. Switch S1 and S2 have an on-resistance of R, and zero, respectively, and both have an off-resistance of infinity. V_{S1} and V_{S2} are the control signals for switches S1 and S2, respectively, as shown in Figure Q26b. V_{S1} and V_{S2} are nonoverlapping clocks with a duty cycle close to 50% and period of T_S. Assume initially ($t = 0$) the voltage across capacitor C is zero. **(a)** Sketch the voltage waveform at the input of the switched-capacitor circuit (V_s) and voltage waveform at capacitor (V_{out}). **(b)** For V_{out} to achieve a settling accuracy of 98.2% of V_{in} (equivalent to a settling time of $4 \cdot \tau$, where τ is the time constant of the capacitor charging circuit), what is the maximum R_s allowed, given the value of the switch on-resistance and sampling capacitance C? **(c)** What is the equivalent input resistance of the switched-capacitor circuit (excluding R_s)?

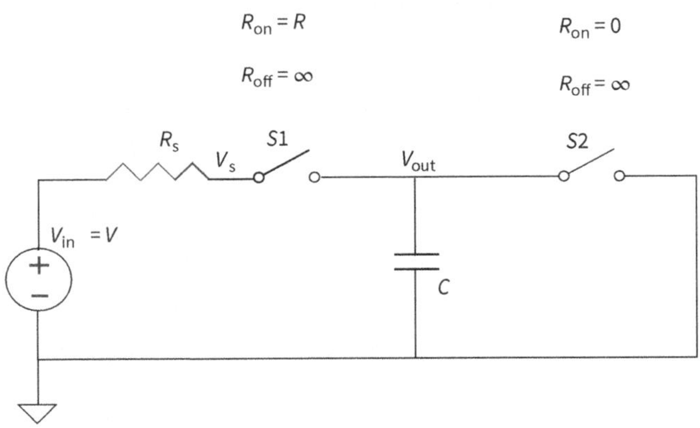

Figure Q26a

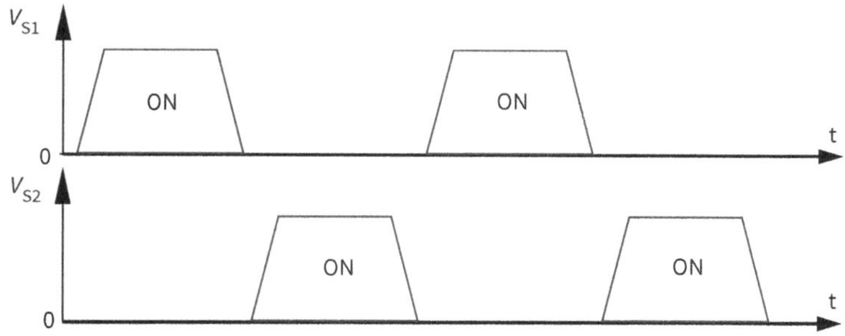

Figure Q26b

Intuitive and Quantitative Analysis: For the question in (a), we need to analyze the transient response of the switched-capacitor circuit. Because the switch S1 has an on-resistance of R, we add this resistance into the circuit diagram as shown in Figure Q26c.

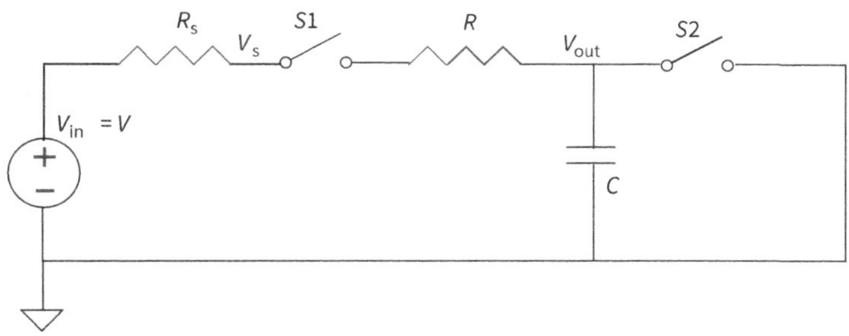

Figure Q26c

When switch S1 closes (i.e., the switch is on) at $t = 0$, the voltage source V_{in} starts to charge the capacitor C. The key for intuitively analyzing a capacitor in a transient response is that the voltage across the capacitor cannot change instantly. At the moment S1 closes, the voltage on the capacitor (V_{out}) is at its initial value which is zero. What about the voltage at node V_s? As the voltage across capacitor C is zero, source voltage V drops entirely on the source resistance R_s and the switch resistance R (per KCL). R_s and R form a voltage divider at that moment, and V_s can be calculated as:

$$V_s = \frac{R}{R_s + R} \cdot V$$

This means the terminal voltage V_s drops at the moment when the switch S1 closes. This is sometimes referred to as "kickback" or "kickback noise."

During the period at which S1 is on, the voltage source V_{in} charges the capacitor C continuously. The time constant is $(R + R_s) \cdot C$. Both V_s and V_{out} go up towards V (the source voltage) following a first-order exponential curve but with different initial voltage – the R_s and R continue to form a voltage divider for the remaining voltage not "occupied" by the capacitor yet. The transient response of V_{out} therefore is:

$$V_{out} = V \cdot \left[1 - e^{-\frac{t}{(R_s+R)\cdot C}}\right]$$

After about half of the clock cycle ($T_s/2$), the switch S1 opens (i.e., the switch is off), and the switch S2 closes. The voltage on the capacitor, V_{out}, starts to discharge (or reset) from its final voltage of the previous charging process. Because the on-resistance of S2 is zero, this instantly brings down the voltage of the capacitor to zero. Note that the control signals of the switch S1 and S2 are "nonoverlapping," meaning that the two switches do not close (turn on) at the same time. This is to prevent the *charge sharing* in which the capacitor is connected to more than one charging or discharging path. Notice that during the short time when both switches are off, the V_s goes back to V, while the V_{out}, or the voltage on the capacitor C, remains its final value at the end of the charging phase.

After another half clock cycle S2 opens and S1 closes again. From this point, the process repeats. The voltage waveform of V_s and V_{out}, is shown in Figure Q26d below. Note that the switching thresholds for both switches are assumed to be at the midpoint of the control signal swing.

For question in (b), we can use the transient response equation of the capacitor voltage that was derived in the analysis of the question in (a). Or even simpler, since the required settling accuracy corresponds to a settling time of $4 \cdot \tau$, or four time constants of the capacitor charging circuit, we have:

$$4 \cdot \tau = 4 \cdot (R + R_s) \cdot C \leq \frac{T_s}{2}$$

The maximum R_s allowed is:

$$R_s(\max) = \frac{T_s}{8 \cdot C} - R$$

Note that due to the nonoverlapping nature of the V_{S1} and V_{S2}, the charging time for capacitor C is in fact less than $T_s/2$.

From the above analysis, we conclude that to reduce the kickback and improve the settling accuracy of the switched-capacitor circuit in Figure Q26a, the source resistance, R_s, should be sufficiently small.

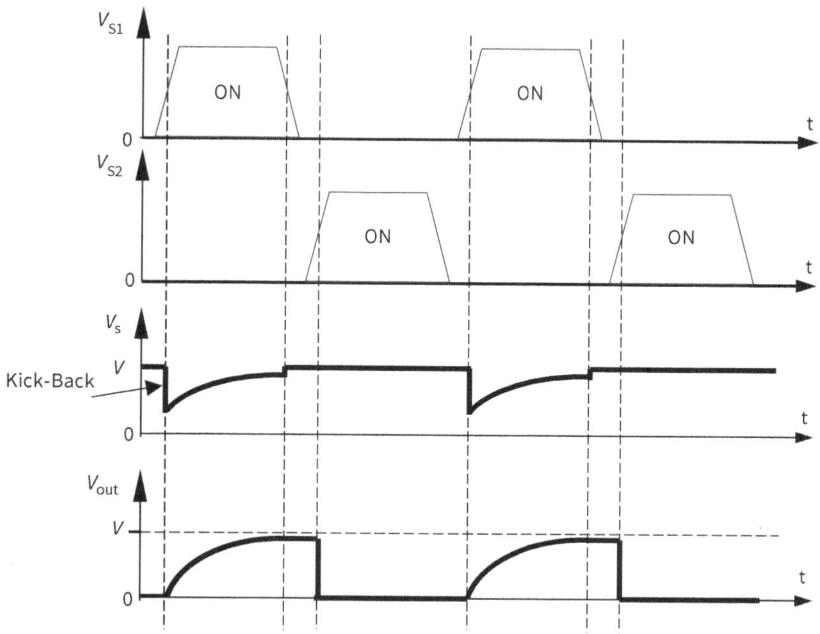

Figure Q26d

Finally, what is the equivalent input resistance of the switched-capacitor circuit? The equivalent input resistance of a switched-capacitor circuit is defined as the resistance that would result in the same average current – or net charge transfer – per clock period as the switching circuit.

The charge, Q, transferred to capacitor C during one clock period is:

$$Q = C \cdot V$$

Note that we assume C is fully charged in each cycle.

The average current flowing in the switched-capacitor circuit during one period is:

$$I_{\text{ave}} = \frac{Q}{T_s} = C \cdot V \cdot f_s$$

where f_s is the clock frequency of the switched-capacitor circuit.

The equivalent input resistance therefore is:

$$R_{\text{eq}} = \frac{V}{I_{\text{ave}}} = \frac{1}{C \cdot f_s} = \frac{T_s}{C}$$

The concept of the equivalent input resistance of a switched-capacitor circuit can offer circuit designers an idea about the average current that would flow into the circuit from a voltage source. It does not, however, give information about the dynamics of the switched-capacitor circuit. The impedance of a capacitor in the s-domain (complex frequency domain) is:

$$Z(s) = \frac{1}{C \cdot s}$$

We can use this impedance for calculations regarding the transient response of the circuit – we have essentially used it already by writing the voltage settling equation and time constant in the analysis of questions in (a) and (b).

Short Answer to Question 26

(a) Sketch the voltage waveform at the input of the switched-capacitor circuit (V_S) and voltage waveform at capacitor (V_{out}). **(b)** For V_{out} to achieve a settling accuracy of 98.2% of V_{in} (equivalent to a settling time of $4 \cdot \tau$, where τ is the time constant of the capacitor charging circuit), what is the maximum Rs allowed, given the value of the switch on-resistance and sampling capacitance C? **(c)** What is the equivalent input resistance of the switched-capacitor circuit (excluding R_S)?

(a) Please refer to Figure Q26d.

(b) Because the setting accuracy requires a settling time of $4 \cdot \tau$, we obtain the following equation:

$$4 \cdot \tau = 4 \cdot (R + R_S) \cdot C \leq \frac{T_S}{2}$$

The maximum R_S allowed is:

$$R_S(\max) = \frac{T_S}{8 \cdot C} - R$$

(c) The equivalent input resistance is:

$$R_{\text{eq}} = \frac{1}{C \cdot f_S} = \frac{T_S}{C}$$

Question 27

A switched-capacitor circuit is shown in Figure Q27a. M1 and M2 form a transmission gate (TG). The control clocks, S1 and $\overline{\text{S1}}$, are complementary to each other to ensure the NMOS and the PMOS transistors turn on and off at the same time.

Explain how to size the two transistors of the TG for **(a)** best switch linearity. **(b)** least charge injections. And **(c)** Would a fully differential topology help to reduce circuit nonlinearity caused by switch nonlinearity and signal-dependent charge injection?

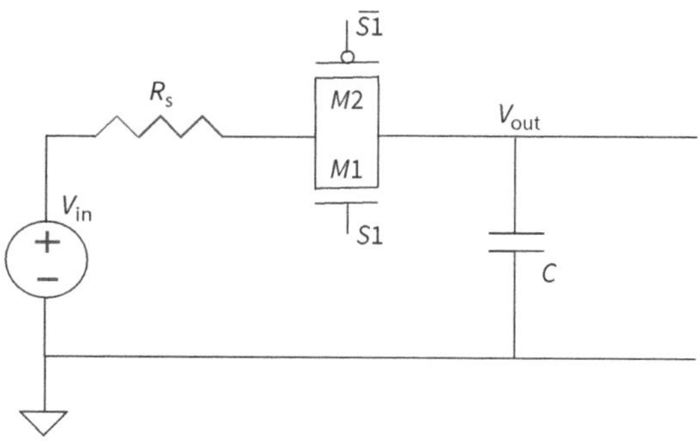

Figure Q27a

Intuitive and Quantitative Analysis: Let us look at the switch linearity first. In switched-capacitor circuits, transistors serving as switches work in the triode region when they are conducting, that is, they behave like resistors. The resistances, however, depend not only on transistor sizes and parameters such as threshold voltage and transconductance parameter, but also on their gate-to-source voltages and therefore the input voltages. This makes the resistances of these switches *signal-dependent*, which results in nonlinearity and harmonic distortions.

In a TG, the on-resistances of the NMOS and PMOS transistors exhibit complementary dependence on the input voltage. When properly designed, this allows the circuit to achieve a relatively constant (flat) equivalent resistance across the input signal range. The on-resistance of a CMOS transistor in triode region (also known as linear region) can be calculated by:

$$R_{\mathrm{CH}} \approx \frac{1}{K_p \cdot \frac{W}{L} \cdot (V_{\mathrm{GS}} - V_{\mathrm{TH}})}$$

where K_p and V_{TH} are the transconductance parameter and threshold voltage of the transistor, respectively.

The resistance of M1 (NMOS) and M2 (PMOS) in Figure Q27a is:

$$R_{M1} \approx \frac{1}{K_{pn} \cdot \frac{W_n}{L_n} \cdot (\mathrm{VDD} - V_{\mathrm{in}} - V_{\mathrm{THN}})}$$

$$R_{M2} \approx \frac{1}{K_{pp} \cdot \frac{W_p}{L_p} \cdot (V_{\text{in}} - V_{\text{THP}})}$$

where K_{pn} and K_{pp} are the transconductance parameters of the NMOS and PMOS, respectively, and V_{THN} and V_{THP} are the threshold voltage of the NMOS and PMOS, respectively.

These equations suggest that R_{M1} and R_{M2} would change in opposite directions in response to a change of V_{in}. Typically, the transconductance parameter of an NMOS transistor is a factor of two to three times that of a PMOS transistor. Assuming $K_{pn} = 3 \cdot K_{pp}$, we can achieve a flat resistance over V_{in} if we make $W_p/L_p = 3 \cdot W_n/L_n$:

$$R_M = R_{M1} \parallel R_{M2}$$
$$\approx \frac{1}{K_{pn} \cdot \frac{W_n}{L_n} \cdot (\text{VDD} - V_{\text{THN}} - V_{\text{THP}})}; V_{\text{THP}} < V_{\text{in}} < \text{VDD} - V_{\text{THN}}$$

The resistance is independent of the input signal V_{in}. Note that this is only true when both transistors are turned on, i.e., $V_{\text{THP}} < V_{\text{in}} < \text{VDD} - V_{\text{THN}}$. These are sometimes referred to as "β-matched" NMOS and PMOS transistors because their device transconductance parameters β, defined as $K_p \cdot W/L$, match each other. Figure Q27b plots the total switch on-resistance as a function of the input voltage:

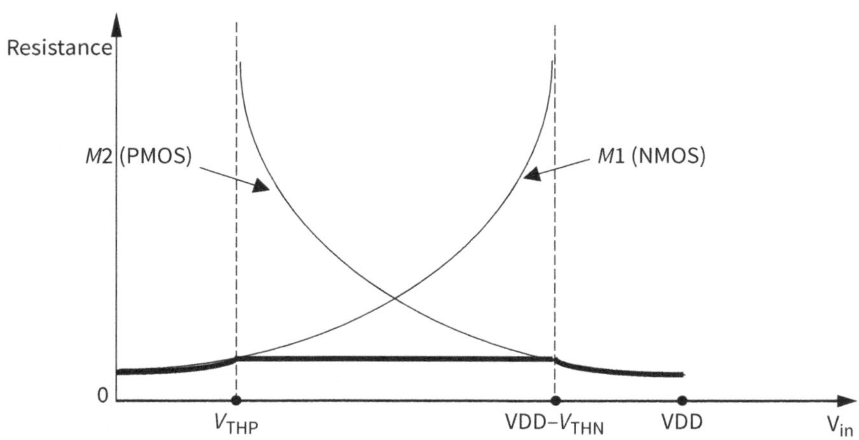

Figure Q27b

Please note that the "β-matched" TG has its limitations on linearization of the switch resistance. First, the resistance compensation only happens when both transistors are on. At input voltages close to power rails (i.e., VDD and ground), only one of the transistors is on and the resistance is highly input dependent. Second, in the previous analysis we ignored the bulk (body) effect of the NMOS. In reality, because of the bulk effect, the threshold voltage of the NMOS changes as V_{in} changes, causing nonlinearity to the switch. Finally, it is not possible to perfectly match the device transconductance parameter β over variations of process and temperature. Because of the last two reasons, the switch resistance is not perfectly constant even when the signal is at the middle section. Signal-dependent resistance results in variations of the time constant and causes signal-dependent settling error. Both odd-order (e.g., 3rd) and even-order (e.g., 2nd) harmonic distortions exist with systems using TG switches.

Consider now the charge injection reduction problem described in section (b). Recall that charge injection occurs when the transistor transitions from the on to off state. At that moment, the charge trapped under the gate oxide is injected back into the circuit, splitting between the source and drain terminals. The charge under the gate oxide resulting from the inverted channel is:

$$Q_{\text{CH}} = C'_{\text{ox}} \cdot W \cdot L \cdot (V_{\text{GS}} - V_{\text{TH}})$$

where C'_{ox} is the unit oxide capacitance.

In the TG, charges under the channel of NMOS and PMOS transistors are:

$$Q_{M1} = -C'_{\text{ox}} \cdot W_n \cdot L_n \cdot (\text{VDD} - V_{\text{in}} - V_{\text{THN}})$$

$$Q_{M2} = C'_{\text{ox}} \cdot W_p \cdot L_p \cdot (V_{\text{in}} - V_{\text{THP}})$$

As illustrated in Figure Q27c, charge injection occurs bidirectionally when the switch turns off. The charge injected toward the input (voltage source) typically has a negligible impact, as this node is driven by a signal source with low output impedance. Conversely, the charge injected onto the capacitor side introduces a significant sampling error in both stored charge and output voltage V_{out}. The distribution of this injected charge depends on the relative impedances seen looking into the source and capacitor directions, as well as the transition times (rise/fall) and the precise timing of the gate control signals.

Assuming half of the charge under the gates is injected onto the capacitor, the total injected charge from both M1 and M2 is:

$$Q_{\text{total}} = \frac{1}{2} \cdot \left[-C'_{\text{ox}} \cdot W_n \cdot L_n \cdot (\text{VDD} - V_{\text{in}} - V_{\text{THN}}) + C'_{\text{ox}} \cdot W_p \cdot L_p \cdot (V_{\text{in}} - V_{\text{THP}})\right]$$

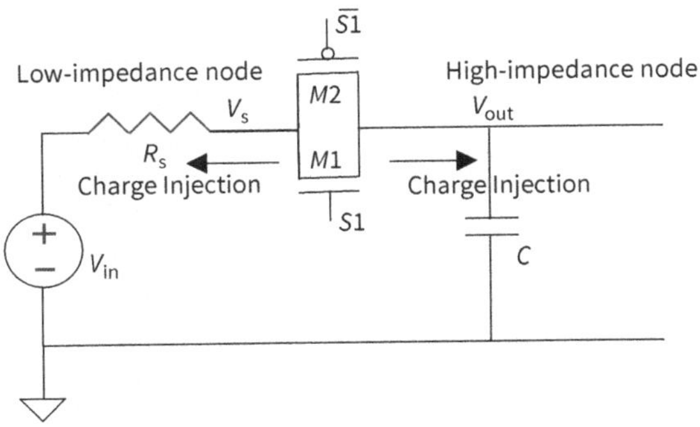

Figure Q27c

The total injected charge from M1 and M2 depends on the size and threshold voltage of the two transistors, as well as the input voltage. To minimize the charge injection, the size of M1 and M2 should be matched. Assume the size (W and L) of M1 and M2 are equal, and let $W_n = W_p = W$, and $L_n = L_p = L$. Also assume $V_{THN} = V_{THP}$. The total charge injected into the capacitor is:

$$Q_{total} = \frac{1}{2} \cdot C'_{ox} \cdot W \cdot L \cdot (2 \cdot V_{in} - \text{VDD})$$

Although matching the NMOS and PMOS device sizes mitigate charge injections in a TG, the net injection remains signal-dependent. Charge injection is minimized when the input voltage V_{in} is approximately VDD/2. Significant deviations from this mid-supply level exacerbate signal-dependent charge injection. Similar to switch nonlinearity, this signal-dependent behavior introduces both odd- and even-order harmonic distortion into the output.

It is important to distinguish between charge injection and *clock feedthrough*, both of which occur during the switch's turn-off transition. As illustrated in Figure Q27d, clock feedthrough is the capacitive coupling of the gate control signal to the output node. When the gate transitions to disable the switch (high-to-low for NMOS or low-to-high for PMOS), the voltage swing is coupled to V_{out} through the parasitic gate-to-source and gate-to-drain overlap capacitances.

The gate parasitic overlap capacitance, C_{ovl}, and the capacitor C form a voltage divider, and the error voltage on the capacitor C caused by clock feedthrough is:

$$V_{err} = \frac{C_{ovl}}{C_{ovl} + C} \cdot \text{VDD}$$

where C_{ovl} is the gate-to-source (or gate-to-drain) overlap capacitance.

"Size-matched" TG can cancel the clock feedthrough to first order as the control signals for NMOS and PMOS are complementary. The cancellation is not perfect,

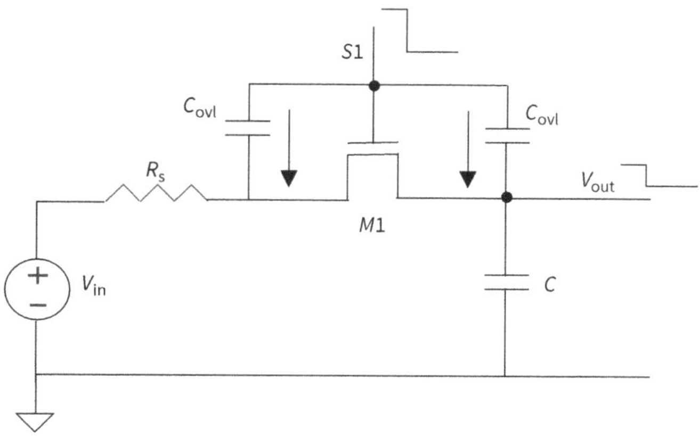

Figure Q27d

however, subject to factors such as timing mismatch between the two complementary control signals.

Question in (c) asks whether differential topology can help to reduce circuit nonlinearity caused by switch nonlinearity and signal-dependent charge injection. As both switch nonlinearity and signal-dependent charge injection cause odd and even-order harmonic distortions, a fully differential topology can eliminate the even-order (e.g., second order) harmonic distortions, but cannot reduce odd-order (e.g., third order) harmonic distortions.

Short Answer to Question 27

Explain how to size the two transistors of the TG for **(a)** best switch linearity. **(b)** least charge injections. And **(c)** Would a fully differential topology help to reduce circuit nonlinearity caused by switch nonlinearity and signal-dependent charge injection?

(a) A relatively constant resistance over V_{in} can be achieved by making the device transconductance parameter (β) of M1 (NMOS) and M2 (PMOS) equal, for best switch linearity. For example, if $K_{pn} = 3 \cdot K_{pp}$, then make $W_p/L_p = 3 \cdot W_n/L_n$.

(b) Charge injection can be mitigated, but not eliminated, by making the size of M1 (NMOS) and M2 (PMOS) the same: $W_n = W_p$, and $L_n = L_p$. Note that the net injection remains signal-dependent.

(c) Switch nonlinearity and signal-dependent charge injection both cause odd and even-order harmonic distortions. A fully differential topology can eliminate the even-order (e.g., second order) harmonic distortions, but cannot reduce odd-order (e.g., third order) harmonic distortions.

Question 28

A switched-capacitor sampling circuit is shown in Figure Q28a. M1 and M1a are sampling switches allowing the input (V_{in}) to charge the sampling capacitor C during the sampling phase when both switches are turned on (controlled by S1/S1a). V_{cm} (equal to VDD/2) is generated by a voltage source. When the sampling is complete, both switches are turned off and the sampled charge is held on the sampling capacitor. Should M1 and M1a be turned off at the same time or at different times at the end of the sampling phase? If they should not be turned off simultaneously, which one should be turned off first?

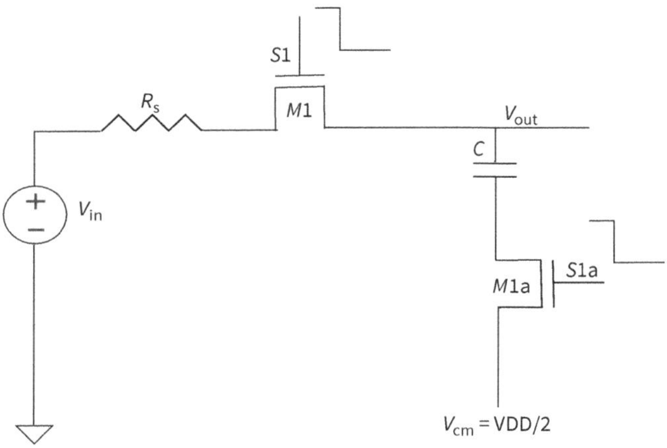

Figure Q28a

Intuitive Analysis: The question focuses on the end of the sampling (charging/discharging) phase. When switches open, they inject charges onto the sampling capacitor C. A common misconception is that M1 and M1a should be disabled simultaneously to allow their injected charges to cancel. However, this argument ignores the fact that perfect cancellation is rarely achieved due to differences in timing, signal levels, and device mismatch. Importantly, in a sampling system, harmonic distortion is a result of the signal-dependent errors, such as errors caused by the signal-dependent charge injections. If the amount of charge injection is signal-independent, it would only cause a DC offset, which can be cancelled out by differential circuits or system calibration, and therefore is not considered a real concern.

It is therefore important to identify charge injection that is signal-dependent. Charge under the gate oxide of M1 from the inverted channel depends on the input signal (V_{in}). On the other hand, charge under the gate of M1a is determined by a fixed DC voltage (V_{cm}). It is clear that the charge injection of M1 is signal-dependent, while charge injection of M1a is not. When M1 and M1a are turned off at the same time, their charge injection cannot cancel each other and is still signal-dependent.

If M1 is turned off earlier than M1a, the signal-dependent charge would store on the sampling capacitor *C* due to the charge injection of M1. After M1 is turned off, the top plate of the sampling capacitor in Figure Q28a (i.e., the terminal of the sampling capacitor that is connected to V_{out}) becomes a floating node. This means the charge injection of M1a when it is turned off subsequently does not affect the sampling capacitor *C*. The charge injected is all from M1 and is signal-dependent.

If M1a is turned off ahead of M1, the signal-independent charge (i.e., a DC offset charge) is injected into the sampling capacitor. Once M1a is open, the terminal of the sampling capacitor that is connected to the ground, or the bottom plate in Figure Q28a, becomes floating. This renders the signal-dependent charge injection from M1 a moment later inconsequential to the sampling capacitor. It is for this reason in the sampling circuit as shown in Figure Q28a, the falling edge of S1a is designed to happen earlier than that of S1, to eliminate the signal-dependent charge injection. The timing diagram of the two switch control signals is shown in Figure Q28b.

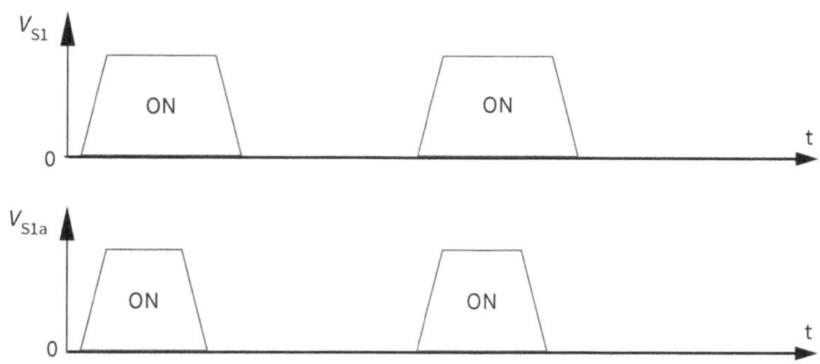

Figure Q28b

This method is also referred to as *bottom plate sampling*, as the switch at the bottom plate opens first and marks the time instant of sampling. Note that the bottom plate refers to the capacitor terminal position in the circuit diagram (as shown in Figure Q28a). It might not represent the physical bottom plate of the capacitor (for example, the bottom plate of a metal-insulator-metal, or MIM) in the layout implementation. In physical implementation, factors such as parasitic capacitances and noise coupling are part of the design considerations.

Note that the effect of the bottom plate sampling is restricted by factors such as the parasitic capacitances at the floating node (e.g., depletion capacitance of switch M1a) and sometimes imperfect timing. Differential sampling circuits are often used to cancel the signal-independent charge injection, clock feedthrough, and even-order harmonic distortions.

There are other methods to reduce the impact of charge injections, such as dummy switches and bootstrap switches. For the sake of brevity, these topics are not discussed further. However, interested readers are encouraged to study these methods due to their importance in practical applications.

Short Answer to Question 28

Should M1 and M1a be turned off at the same time or at different times at the end of the sampling phase? If they should not be turned off simultaneously, which one should be turned off first?

M1 and M1a should be turned off at different times to avoid signal-dependent charge injection. Specifically, M1a should be turned off before M1 is turned off. This is referred to as "bottom plate sampling."

Question 29

A fully differential track-and-hold (T/H) circuit is shown in Figure Q29a. S1/S1a and S2 are nonoverlapping clocks and their timing sequences are shown in Figure Q29b. V_{cm} is a voltage source used to set the common-mode voltage for the differential input signals. Assume that $A(\omega)$ is a single-pole (first-order) op-amp. **(a)** What is the relationship between the output voltage and the input voltage? **(b)** What is the DC gain requirement for the op-amp if the gain error of the T/H needs to be less than 0.1%? **(c)** What is the requirement for the op-amp gain-bandwidth product if the output settling error needs to be less than 0.1% of its steady-state value (The required settling time is approximately seven times the track-and-hold amplifier's closed-loop time constant). Assume that the sampling rate (clock frequency of S1 and S2) is 100 MS/s (100 MHz) and the resistance of the switches is negligible.

Intuitive and Quantitative Analysis: The amplifier in the T/H circuit is closed-loop with negative feedback, similar to that of the single-ended inverting amplifier, but in a fully differential configuration. Recall that in the analysis of the continuous-time closed-loop amplifiers, we use KCL/KVL to determine the relationship between the output and input. In the analysis of the

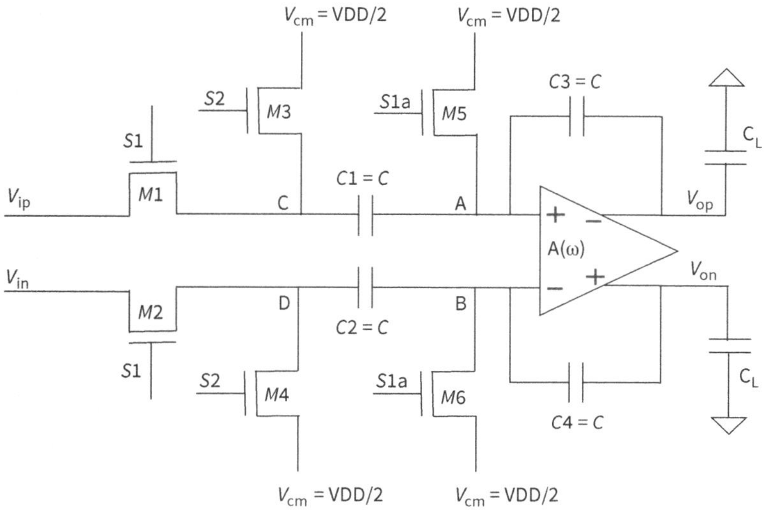

Figure Q29a

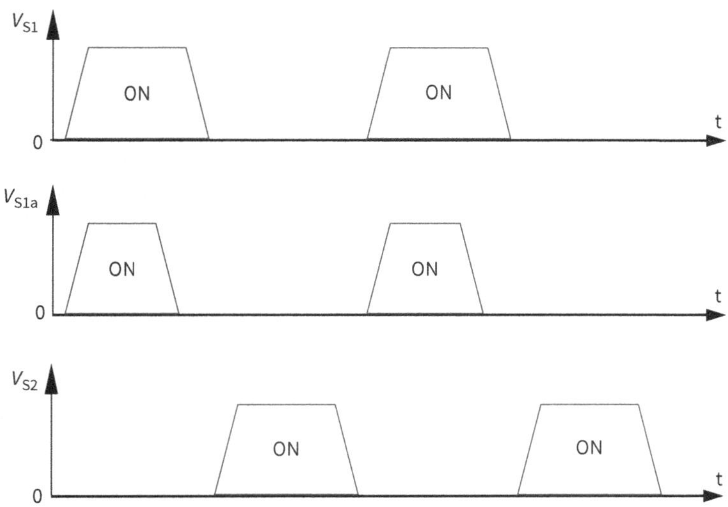

Figure Q29b

switched-capacitor circuit, principle of charge conservation is most often used. *Charge conservation law* says that the total charge of an isolated system remains constant. In a switched-capacitor circuit, this means that the total charge at a circuit node remains constant unless there is a DC path for charge to transfer.

The key nodes in Figure Q29a for the charge conservation analysis are node A and B. Due to the symmetry of the fully differential amplifier, the analysis may be restricted to a single side. The following derivation focuses on the upper (v_{ip}) path. In the tracking phase M1 and M5 are turned on, and M3 is turned off, the sampling capacitor C_1 is connected between two voltage sources, V_{ip} and V_{cm}. Charge at node A is not conserved and the capacitor C_1 is charged by the input signal V_{ip}. Assuming charging time is sufficient, at the end of the tracking phase (the falling edge of S1a), the charge on node A (or on capacitor C_1 and C_3) is:

$$q_A = C_1 \cdot \left(0 - v_{ip}\right) + 0 = -C \cdot v_{ip}$$

Note that, like our small-signal analysis, we only focus on AC components. v_{ip} here stands for the input AC signal at the positive input, and V_{cm} is considered AC ground. Be aware that the inputs of the op-amp are forced to V_{cm} at the tracking phase and the negative feedback is therefore not in effect. This means that the (AC) charge stored on capacitor C_3 (and C_4) during the tracking phase is zero (neglecting the offset voltage of the op-amp).

At the gap between the falling edge of S1a and the rising edge of S2, node A is isolated (no DC path), and the charge on the node is conserved. The op-amp transitions to a closed-loop state as its inputs are not tied to V_{cm} anymore.

In the hold phase, M1 and M5 remain off, and M3 is turned on. There is still no DC path at node A, and charge continues to be conserved. The negative-feedback loop of the amplifier, however, forces the voltage of its two inputs equal and node A (and node B as well) becomes virtual (AC) ground. In the meantime, the voltage of the other side of the capacitor C_1 (node C in Figure Q29a) is forced to become V_{cm}, or AC ground, through the switch M3. The charge on capacitor C_1 is transferred to capacitor C_3, keeping the total charge at node A unchanged. Applying charge conservation at the end of hold phase, we obtain:

$$q_A = -C_1 \cdot v_{ip} + 0 = 0 + C_3 \cdot \left(0 - v_{op}\right) = -C \cdot v_{op}$$

or

$$v_{op} = v_{ip}$$

In differential form, this yields:

$$v_o = v_{op} - v_{on} = v_i = v_{ip} - v_{in}$$

The question in (b) and (c) involves the dynamic characteristics of the closed-loop amplifier during the hold phase, and we need to obtain the transfer function of the circuit. Figure Q29c illustrates the simplified circuit during the hold phase.

At the beginning of the hold phase, voltage at node C and D becomes V_{cm} (AC ground). The voltage across the sampling capacitors (C_1 and C_2) cannot change instantaneously, resulting in an instant voltage change (glitch) at the inputs of the amplifier at the moment M3 and M4 are turned on, as shown in Figure Q29c. The negative feedback would then play its role, voltage of node A goes towards AC ground, charge of the sampling capacitor C_1 transfers to the feedback capacitor C_3, and output settles towards the input voltage last sampled.

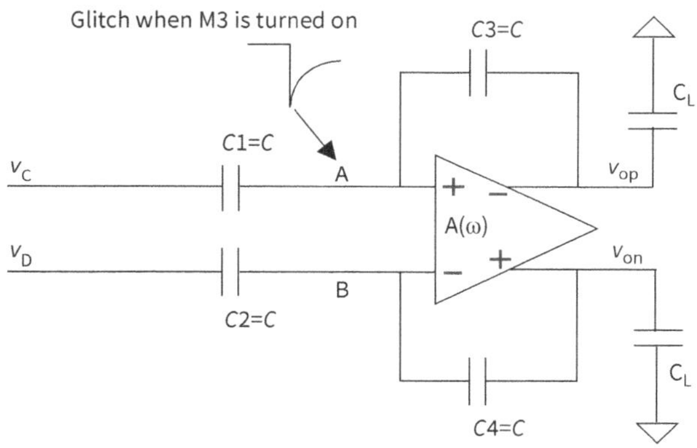

Figure Q29c

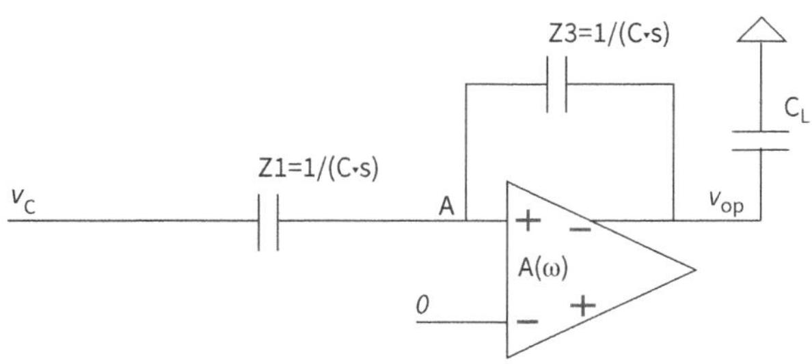

Figure Q29d

To derive the transfer function of the circuit in Figure Q29c, we use the s-domain (complex frequency domain) impedance of the capacitors, and focus on the half-circuit due to its symmetry (Figure Q29d):

The circuit in Figure Q29d is similar to that of a continuous-time resistive inverting amplifier (*voltage-sampling, shunt-mixing*). Applying KCL, we have the following system of equations:

$$\begin{cases} \dfrac{v_C - v_A}{Z_1} = \dfrac{v_A - v_{\text{op}}}{Z_3} \\ v_A \cdot \dfrac{A(0)}{\dfrac{1}{2 \cdot \pi \cdot f_{-3\text{dB}}} \cdot s + 1} = -v_{\text{op}} \\ Z_1 = Z_3 = \dfrac{1}{C \cdot s} \\ A(0) \cdot f_{-3\text{dB}} = \text{GBW} \end{cases}$$

where $f_{-3\text{dB}}$ is the −3 dB frequency of the first-order op-amp. $A(0)$ and GBW are the DC gain and gain-bandwidth product of the op-amp, respectively. Solving the system of equations and expanding to a fully differential configuration lead to:

$$H(s) = \frac{v_o(s)}{v_i(s)} = \frac{v_{\text{op}} - v_{\text{on}}}{v_C - v_D} = -\frac{A(0)}{\dfrac{2}{2 \cdot \pi \cdot f_{-3\text{dB}}} \cdot s + 2 + A(0)}$$

Substituting $s = 0$ yields the DC gain of the T/H circuit:

$$H(0) = \frac{v_o(0)}{v_i(0)} = -\frac{A(0)}{2 + A(0)}$$

The closed-loop gain is less than 1. To keep the gain error less than 0.1%, we have:

$$1 - \frac{A(0)}{2 + A(0)} < 0.001$$

which yields,

$$A(0) > 1998$$

The DC gain of the op-amp needs to be higher than about 66dB.

For question (c) we need to relate the closed-loop time constant to the GBW of the op-amp:

$$H(s) = \frac{v_o(s)}{v_i(s)} = \frac{v_{\text{op}} - v_{\text{on}}}{v_C - v_D} = -\frac{A(0)}{\dfrac{2}{2 \cdot \pi \cdot f_{-3\text{dB}}} \cdot s + 2 + A(0)} \approx \frac{-1}{\dfrac{1}{2 \cdot \pi \cdot \dfrac{\text{GBW}}{2}} \cdot s + 1}$$

Assuming A(0) >> 2.

This is a standard first-order system, and the closed-loop time constant, τ_{cl}, is:

$$\tau_{\text{cl}} = \frac{1}{2 \cdot \pi \cdot \dfrac{\text{GBW}}{2}}$$

To review how the time constant is derived, recall the standard transfer function of a first-order system, presented below:

$$H(s) = \frac{v_o(s)}{v_i(s)} = \frac{K}{\tau \cdot s + 1}; K \text{ is DC gain}, \tau \text{ is time constant}$$

Because the sampling frequency is 100 MHz, the sampling period is 10 ns. According to the timing diagram in Figure Q29b, the hold phase lasts about half of the clock period, making the settling time around 5 ns.

For a settling error less than 0.1% of the final steady-state value, the amplifier needs $7 \cdot \tau_{\text{cl}}$ (closed-loop time constant). That is:

$$\tau_{\text{cl}} = \frac{1}{2 \cdot \pi \cdot \frac{\text{GBW}}{2}} = \frac{5 \text{ ns}}{7}$$

The GBW of the amplifier needs to be:

$$\text{GBW} \geq \frac{7}{5 \cdot 10^{-9} \cdot \pi} \approx 446 \cdot 10^6$$

A DC gain of over 66dB and GBW of about 446 MHz is necessary for the T/H to achieve a DC gain error of less than 0.1% and settling error less than 0.1% at 100 MSPS sampling rate. The total absolute accuracy considering both DC gain error and settling error is about 0.2%. For higher accuracy and higher sampling rate, the GBW requirement increases, which has a direct impact on the power consumption of the T/H circuit.

The topology in Figure Q29a is known as the *charge-transfer* amplifier and is one of the two major T/H amplifier architectures. Notice that in the formula of the closed-loop time constant, there is a coefficient of 1/2 of the GBW term. This is due to the feedback factor of the charge-transfer topology, which is similar to that of a continuous-time inverting amplifier (see *Question 17, Feedback factors of inverting amplifiers and unity-gain buffers*).

Track-and-hold (T/H) circuits, also known as sample-and-hold (S/H) circuits, are essential front-end components in ADCs.

Short Answer to Question 29

(a) What is the relationship between the output voltage and the input voltage? **(b)** What is the DC gain requirement for the op-amp if the gain error of the T/H needs to be less than 0.1%? **(c)** What is the requirement for the op-amp gain-bandwidth product if the output settling error needs to be less than 0.1% of its steady-state value?

(a) The relationship between the output voltage and input voltage:

$$v_o = v_{op} - v_{on} = v_i = v_{ip} - v_{in}$$

(b) The DC gain requirement for the op-amp if the gain error of the T/H needs to be less than 0.1%: $A(0) > 1998$

(c) The requirement for the op-amp gain-bandwidth product if the output settling error needs to be less than 0.1% of its steady-state value: GBW > 446 MHz.

Question 30

For the fully differential track-and-hold (T/H) circuit in Question 29, we have assumed that the input capacitance of the op-amp is zero. In practice, however, the input capacitance of the op-amp might not be insignificant in comparison to the sampling and feedback capacitance. What would be the impact of this parasitic capacitance, C_p, as shown in Figure Q30a, on the DC gain and closed-loop time constant of the T/H? Assume $A(0) \gg 1$ and the resistance of the switches is negligible.

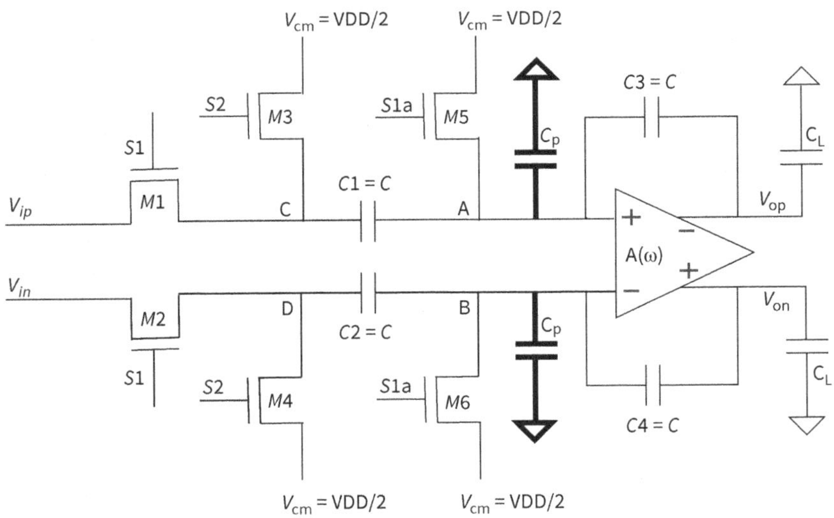

Figure Q30a

Intuitive and Quantitative Analysis: First, let us determine whether the relationship between the output voltage and input voltage is affected by the input parasitic capacitance of the amplifier. In the tracking phase, both sides of parasitic capacitance, C_p, are AC grounds, and there is no signal charge stored at the capacitance. Notice that the DC offset charge due to the V_{cm} and ground difference is cancelled at the hold phase due to differential operation. In the hold phase, nodes A and B become virtual grounds ($A(0) \gg 1$), and the signal charges at the sampling capacitors (C_1 and C_2) are transferred to the feedback capacitors (C_3 and C_4). The total charge at node A and B is conserved and the output voltage equals the last sampled input voltage on the sampling capacitor, C_1 and C_2. The input capacitance of the amplifier has little impact on the DC transfer function and the DC gain of the T/H.

What about the dynamic characteristic of the T/H with the input parasitic capacitance? We now need to update the transfer function with the presence of the input parasitic capacitance C_p. Figure Q30b is the simplified single-ended representative of the T/H circuit at hold phase with the complex impedance of all capacitors, including the parasitic input capacitance of the amplifier.

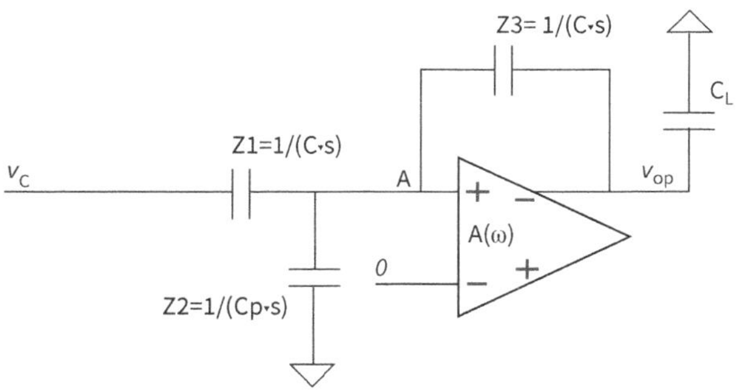

Figure Q30b

Applying KCL, we have the following system of equations:

$$\begin{cases} \dfrac{v_C - v_A}{Z_1} = \dfrac{v_A - v_{\text{op}}}{Z_3} + \dfrac{v_A}{Z_2} \\ v_A \cdot \dfrac{A(0)}{\dfrac{1}{2 \cdot \pi \cdot f_{-3\text{dB}}} \cdot s + 1} = -v_{\text{op}} \\ Z_1 = Z_3 = \dfrac{1}{C \cdot s} \\ Z_2 = 1/(C_p \cdot s) \\ A(0) \cdot f_{-3\text{dB}} = \text{GBW} \end{cases}$$

where f_{-3dB} is the -3 dB frequency of the first-order op-amp. $A(0)$ and GBW are the DC gain and gain-bandwidth product of the op-amp, respectively. Solving the system of equations and expanding to a fully differential configuration lead to:

$$H(s) = \frac{v_{op} - v_{on}}{v_C - v_D} = \frac{-1}{\frac{2 + \frac{C_p}{C}}{2 \cdot \pi \cdot \text{GBW}} \cdot s + 1 + \frac{2 + \frac{C_p}{C}}{A(0)}} \approx \frac{-1}{\frac{2 + \frac{C_p}{C}}{2 \cdot \pi \cdot \text{GBW}} \cdot s + 1}$$

for $A(0) \gg 1$. The DC gain of the closed-loop switched-capacitor amplifier at hold phase is still approximately -1, but the closed-loop time constant, τ_{cl}, changes to:

$$\tau_{\text{cl}} = \frac{2 + \frac{C_p}{C}}{2 \cdot \pi \cdot \text{GBW}}$$

As a sanity check, if $C_p = 0$, the closed-loop time constant becomes:

$$\tau_{\text{cl}:Cp=0} = \frac{2}{2 \cdot \pi \cdot \text{GBW}} = \frac{1}{2 \cdot \pi \cdot \frac{\text{GBW}}{2}}$$

This is the same as the result from Question 29.

We observe that the closed-loop time constant increases in the presence of the input parasitic capacitance. For example, when the input capacitance is equal to the sampling capacitance, or $C_p = C$, the closed-loop time constant becomes 50% larger compared to the T/H with no op-amp input capacitance. This is one of the reasons that in high-speed switched-capacitor applications, amplifiers with low input parasitic capacitance are preferred.

Note that we have assumed the resistance of the switches is negligible in the analysis. In applications where the time constant associated with the switch resistance is significant, we also need to take switch resistance into account. In the tracking phase, the resistance of the sampling switch affects the time constant for charging/discharging the sampling capacitor, as discussed in Question 26. At the hold phase, the switch resistance of M3 and M4 causes the closed-loop amplifier to become a higher-order system, changing its settling time and behavior, resulting in overshoot and ringing under certain conditions. These are beyond the complexity that most interviews would involve, but they are factors that designers need to consider in real-world design tasks.

Short Answer to Question 30

What would be the impact of this parasitic capacitance, C_p, as shown in Figure Q30a, on the DC gain and closed-loop time constant of the T/H?

The transfer function of the T/H in the hold phase with C_p is:

$$H(s) = \frac{v_{\text{op}} - v_{\text{on}}}{v_C - v_D} = \frac{-1}{\dfrac{2+\dfrac{C_p}{C}}{2\cdot\pi\cdot\text{GBW}}\cdot s + 1 + \dfrac{2+\dfrac{C_p}{C}}{A(0)}} \approx \frac{-1}{\dfrac{2+\dfrac{C_p}{C}}{2\cdot\pi\cdot\text{GBW}}\cdot s + 1}$$

for $A(0) \gg 1$. The DC gain of the closed-loop switched-capacitor amplifier in the hold phase is still approximately −1, but the closed-loop time constant increases, resulting in slower settling.

Question 31

For the fully differential charge-transfer track-and-hold (T/H) circuit in Question 29 (circuit diagram redrawn in Figure Q31a), what is the total output noise voltage (RMS) of the T/H? Assume only the thermal noise of the switch resistance is of concern, and the op-amp is noiseless. Further assume the resistance of switches M5 and M6 is zero so that they do not contribute to the total thermal noise of the circuit.

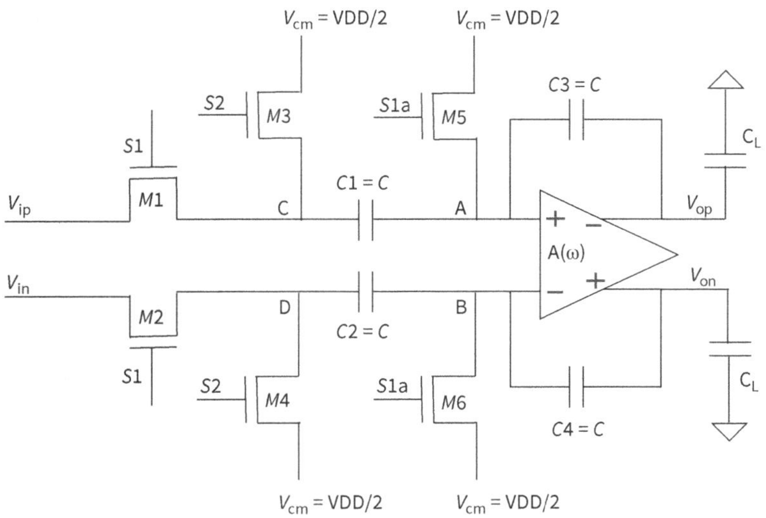

Figure Q31a

Intuitive and Quantitative Analysis: $\sqrt{KT/C}$ is the answer I most frequently hear; it is not entirely wrong, but it is incomplete for this circuit. One of the important features of switched-capacitor circuits is that the operation is usually divided into two or more phases, and charge is transferred between or among phases. A common mistake in switched-capacitor noise analysis is to consider only one of the phases. In the charge-transfer T/H circuit in Figure Q31a, we need to consider noise generated in both tracking phase and hold phase.

Let us start with the tracking phase and focus on noise sources and how they contribute to the noise charge at nodes A and B, which will be transferred to the feedback capacitors C_3 and C_4 and affect the output voltage during the hold phase. Figure Q31b is a simplified single-ended version of the circuit at the tracking phase for the noise analysis, with the thermal noise of the switch modeled as a voltage noise source v_1 in series with the switch resistance R_1. The input V_{ip} is modeled as a voltage source v_i with AC voltage of 0 V in the noise analysis. For simplicity, we use ideal switches to represent the switch functions and use the gate control signal names (e.g., S1 and S2) to identify the switches.

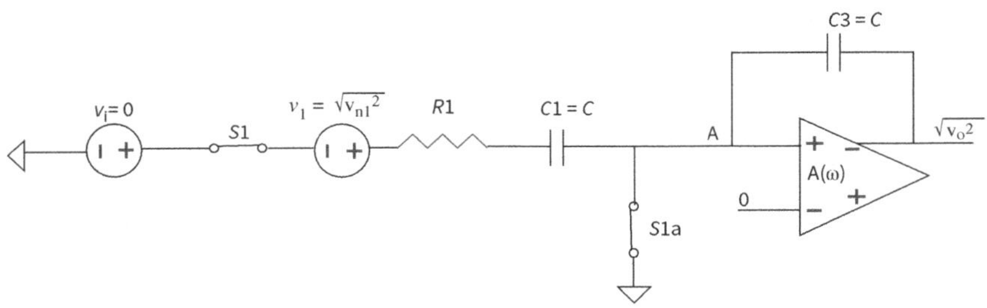

Figure Q31b

Before we dive into noise analysis, we need to clarify a basic concept. Switched-capacitor circuits are sampling systems – continuous-time signals and noises are sampled and become discrete-time analog signals and noises. For example, the final sampled input voltage corresponds to the instantaneous voltage across the sampling capacitor C_1 at the moment switch S1a is disabled. A natural question is: "What is the impact of the sample system on noises such as thermal noise of the sampling switch?"

The resistance of the sampling switches, such as M1, generates thermal noise. The power spectral density (PSD) of thermal noise, or thermal noise spectral density (NSD), is white, i.e., approximately flat over the relevant frequency range. In the tracking phase, the switch resistance and sampling capacitor form a first-order low-pass filter and the PSD of the thermal noise rolls off at high frequencies, as shown in Figure Q31c. When the noise is sampled, noise power with frequencies exceeding half the sampling frequency (1/2 fs, or Nyquist frequency) aliases back to the frequencies below the Nyquist frequency. This means even though sampling alters the noise frequencies through aliasing, the total noise power, and the noise RMS value, is not changed.

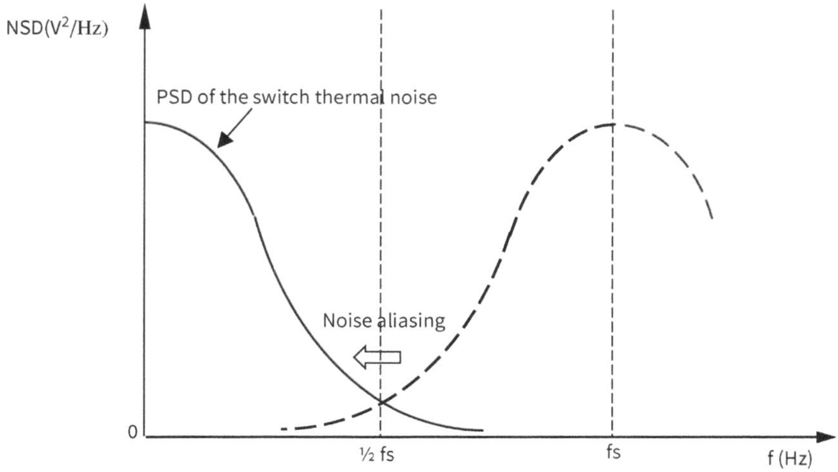

Figure Q31c

Now let us calculate the RMS value of the noise charge on the sampling capacitor C_1 during the tracking phase. The resistance of the switch S1 (R_1) generates thermal noise charge which is stored on the sampling capacitor C_1. The resistance also forms a first-order low-pass filter in conjunction with the sampling capacitor which limits the noise power at high frequencies. As a result, the total noise power and noise RMS depends only on the capacitance value – this brings us the KT/C noise (see *Appendix 4, Section 6, Circuit noise fundamentals*).

The thermal noise voltage (RMS) on the sampling capacitor C_1 during the tracking phase is:

$$v_{n:c1:T} = \sqrt{\frac{K \cdot T}{C}}$$

where K is Boltzmann's constant, T is absolute temperature (in K).

The noise charge (RMS) stored on the sampling capacitor C_1 during the tracking phase is:

$$Q_{n:c1:T} = C \cdot v_{n:c1:T} = \sqrt{K \cdot T \cdot C}$$

Note that both sides of the feedback capacitor, C_3, are DC voltage (or AC ground), and there is no noise charge on it during the tracking phase (under the assumption that the switch S1a is of zero resistance, and the amplifier is noise-free).

In the hold phase, switches M3 (or simply notated as switch S2) is turned on, and the op-amp is configured in negative feedback, operating as an inverting amplifier. Figure Q31d illustrates the simplified single-ended version of the circuit at the hold phase for noise analysis. Similar to the tracking phase, the thermal noise of the switch is modeled as a voltage noise source v_2 in series with the switch resistance R_2, and V_{cm} is modeled as a voltage source v_{cm} with AC voltage of 0 V in the noise analysis.

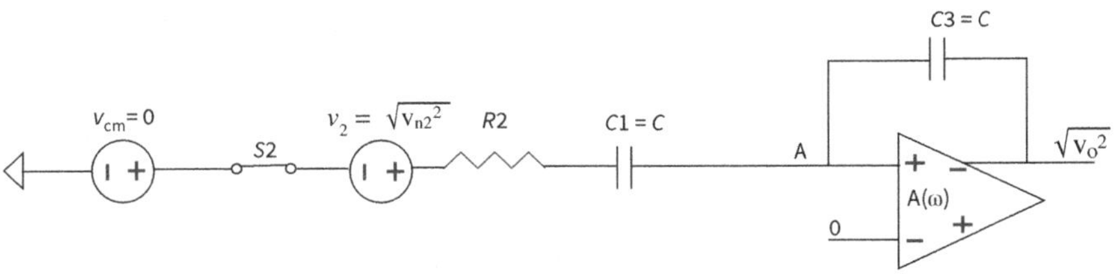

Figure Q31d

During the hold phase, node A is regulated by the closed-loop op-amp and becomes a virtual (AC) ground (within the closed-loop bandwidth of the amplifier). Because the sampling capacitor C_1 already has noise charge on it from the tracking phase, to simplify the analysis, we will calculate the output noise due to each one of the two input voltage sources (v_{cm} and v_2) independently and add them together by applying the superposition theorem (see *Appendix 2, Section 3, The superposition theorem*).

With voltage source v_{cm} acting alone, and noticing charge is conserved at node A, we have the following equation:

$$Q_{n:c1:T} = \sqrt{K \cdot T \cdot C} = C \cdot v_{o:n1}$$

where $v_{o:n1}$ is the output noise voltage (RMS) by v_{cm}.

This means the noise charge on $C1$ from the tracking phase is transferred to the feedback capacitor C_3. The output noise voltage (RMS) by v_{cm} is:

$$v_{o:n1} = \sqrt{\frac{K \cdot T}{C}}$$

The thermal noise source v_2 charges the sampling capacitor C_1 through switch resistance R_2 during the hold phase, and the noise voltage (RMS) on C_1 during the hold phase is:

$$v_{n:c1:H} = \sqrt{\frac{K \cdot T}{C}}$$

Because the op-amp is configured as an inverting amplifier with a gain of -1 ($C_1 = C_3 = C$), the output noise voltage (RMS) by v_2 is:

$$v_{o:n2} = v_{n:c1:H} = \sqrt{\frac{K \cdot T}{C}}$$

Applying the superposition theorem, and noticing noise can only add up by power, we obtain the total output noise power:

$$v_{o:n}^2 = v_{o:n1}^2 + v_{o:n2}^2 = 2 \cdot \frac{K \cdot T}{C}$$

The noise RMS at the amplifier output is:

$$v_{o:n}(\text{RMS}) = \sqrt{2} \cdot \sqrt{\frac{K \cdot T}{C}}$$

Do not forget that this is derived from the single-ended version of the circuit. As the noise power from each side of the differential topology adds together, the total output noise power is doubled, and the RMS noise of the differential voltage output is:

$$v_{\text{out}:n}(\text{RMS}) = 2 \cdot \sqrt{\frac{K \cdot T}{C}}$$

where C is the capacitance of the sampling capacitors.

This analysis demonstrates that both the tracking and hold phases contribute noise to the output signal. The final output noise power is therefore the sum of the noise powers contributed by all phases. To simplify the current problem, the op-amp's noise is excluded from the analysis, despite it typically dominating the overall output noise in practical switched-capacitor circuits. Consequently, the assumption that noise power in a switched-capacitor circuit is strictly limited to KT/C is technically flawed. Such an oversimplification may lead to a significant underestimation of the total noise power present in the system.

Regarding noise in fully differential architectures, we have observed that the total output noise power is twice that of a single-ended implementation. However, the differential topology also doubles the peak-to-peak input signal swing. Because this results in a fourfold increase in signal power against a twofold increase in noise power, the signal-to-noise ratio improves by a factor of $\sqrt{2}$, or approximately 3 dB.

Short Answer to Question 31

What is the total output noise voltage (RMS) of the T/H? Assume only thermal noise of the switch resistance is of concern, and the op-amp is free of noise.

Noise contributions from both the tracking and hold phases must be considered. The total output noise (RMS) is:

$$v_{\text{out}:n}(\text{RMS}) = 2 \cdot \sqrt{\frac{K \cdot T}{C}}$$

where C is the capacitance of the sampling capacitors.

Question 32

Figure Q32a is a switched-capacitor common-mode feedback (CMFB) circuit for stabilizing the common-mode output voltage of the fully differential amplifier as shown in Figure Q32b. V_{op} and V_{on} are the differential output voltages of the amplifier, and V_{cm} is the reference (target) output common-mode voltage. The CMFB circuit exhibits full symmetry between its left and right halves, where $C_{1A} = C_{1B} = C_1$, and $C_{2A} = C_{2B} = C_2$. S1 and S2 are nonoverlapping switch control (clock) signals. In phase 1, transistors with gates connected to S1 (designated as S1 switches) are turned on. In phase 2, transistors with gates connected to S2 (designated as S2 switches) are turned on. The amplifier is designed in such a way so that when V_{cmfb} is at V_b, the output common-mode voltage is centered around the desired value of V_{cm}. **(a)** Explain how the CMFB circuit works to stabilize the output common-mode. **(b)** What is the relationship between V_{cmfb} and V_{op}/V_{on}? **(c)** What are the advantages of the switched-capacitor CMFB circuit over the continuous-time CMFB control (e.g., as described in Question 15)?

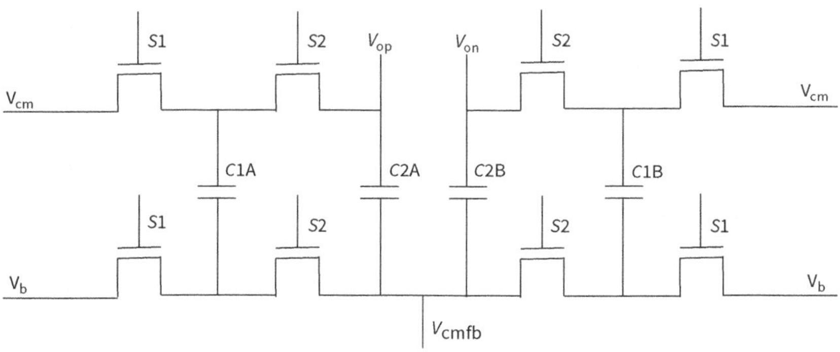

Figure Q32a

Intuitive Analysis: The switched-capacitor CMFB circuit in Figure Q32a might look daunting with eight switches and four capacitors involved. Less so if we focus on "what essential functions CMFB circuits perform?" Importantly, CMFB circuits need to sense the actual output common-mode voltage. Notice that in Figure Q32a, V_{cmfb} is the average of V_{op} and V_{on}, obtained through the two identical capacitors (C_{2A} and C_{2B}). If the common-mode voltage changes, say, goes up, V_{cmfb} would also go up which is used to adjust the "knob," namely the tail current of M0a, to bring the output common-mode voltage down. We cannot, however,

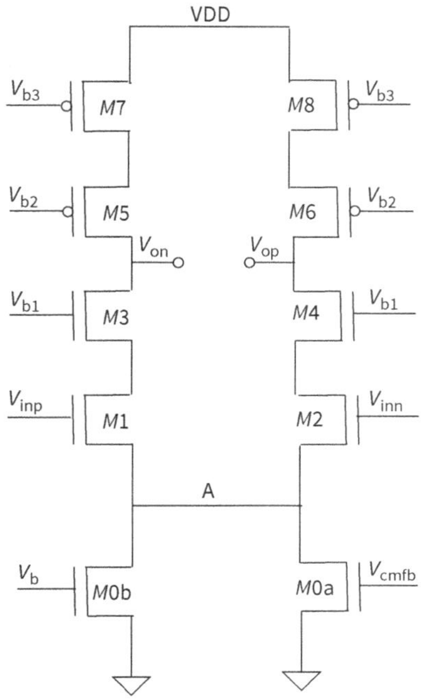

Figure Q32b

simply connect V_{cmfb} to the gate of M0a, as the initial charges on the C_{2A} and C_{2B} are uncontrolled and the V_{cmfb} could be too high or too low to provide a proper bias voltage to M0a (i.e., the capacitor averaging circuit only works for AC signals, not DC levels). When the output common-mode voltage is at the target, V_{cm}, the gate voltage for M0a should be around V_b. This tells us that the control voltage should shift down by the amount of $V_{cm} - V_b$. This "offset" charge is stored on C_{1A} and C_{1B} in phase 1 when all S1 switches are turned on. In phase 2, the charges on C_{1A} and C_{1B} are shared with C_{2A} and C_{2B} to level-shift the V_{cmfb}. With the help of the negative-feedback loop, the output common-mode voltage stabilizes around the V_{cm} after a certain settling time. Note that the offset charge on C_{2A} and C_{2B} could drift or change over time due to leakage and noise coupling, and periodic refreshment is needed even after the common-mode voltage is stabilized.

To analyze the relationship between V_{cmfb} and V_{op}/V_{on}, we need to realize that unlike continuous-time CMFB circuits, the switched-capacitor CMFB circuit operates in a discrete-time manner. During phase 1, the capacitors, C_{1A} and C_{1B}, are charged to the offset voltage, $V_{cm} - V_b$. The charge at node V_{cmfb} is:

$$Q_1 = C_1 \cdot (V_b - V_{cm})$$

During phase 2, the charges on C_{1A} and C_{1B} are shared with C_{2A} and C_{2B}, and charge is conserved at node V_{cmfb}:

$$\begin{cases} Q_1 + C_2 \cdot \left[V_{\text{cmfb}}(n-1) - V_{\text{op}}\right] = (C_1 + C_2) \cdot \left[V_{\text{cmfb}}(n) - V_{\text{op}}\right] \\ Q_1 + C_2 \cdot \left[V_{\text{cmfb}}(n-1) - V_{\text{on}}\right] = (C_1 + C_2) \cdot \left[V_{\text{cmfb}}(n) - V_{\text{on}}\right] \end{cases}$$

where $V_{\text{cmfb}}(n-1)$ is the common-mode control voltage at the previous clock cycle (before the start of phase 2 of the current clock cycle), and $V_{\text{cmfb}}(n)$ is the common-mode control voltage at the current clock cycle (at the end of phase 2). The CMFB loop typically operates much more slowly than the clock; therefore, the output common-mode voltage is assumed to be stationary during a single clock cycle.

Simplify the system of equations and solve for V_{cmfb}:

$$\left(1 + \frac{C_2}{C_1}\right) \cdot V_{\text{cmfb}}(n) - \frac{C_2}{C_1} \cdot V_{\text{cmfb}}(n-1) = \frac{V_{\text{op}} + V_{\text{on}}}{2} - (V_{\text{cm}} - V_b)$$

This is a (discrete) *difference equation* that can be solved using the *z-transform*. Without being involved in a *discrete-time z-domain analysis*, we can obtain some intuition by investigating what happens when the CMFB loop reaches its steady state. At the steady state, the common-mode control voltage, V_{cmfb}, also stabilizes and ceases to change. This means: $V_{\text{cmfb}}(n) = V_{\text{cmfb}}(n-1)$. The relationship between V_{cmfb} and $V_{\text{op}}/V_{\text{on}}$ now becomes:

$$V_{\text{cmfb}}(n) = \frac{V_{\text{op}} + V_{\text{on}}}{2} - (V_{\text{cm}} - V_b)$$

The first term at the right side of the equation, $(V_{\text{op}} + V_{\text{on}})/2$, is the actual output common-mode voltage of the amplifier. This suggests that when the V_{cmfb} stabilizes around V_b, the output common-mode voltage is around the target value of V_{cm}.

The switched-capacitor CMFB circuit realizes the averaging and level shifting functions through a simple and fast switched-capacitor network and enables adjusting the output common-mode voltage without having to implement an error amplifier (e.g., the error amplifier in Question 15). The removal of the error amplifier lowers the open-loop gain, reduces the number of poles, and helps to improve the stability of the common-mode control loop, which often is a challenging task in continuous-time high-gain differential amplifiers. Switched-capacitor CMFB needs a clock to operate, which is readily available in switched-capacitor systems.

Short Answer to Question 32

(a) Explain how the CMFB circuit works to stabilize the output common-mode. **(b)** What is the relationship between V_{cmfb} and $V_{\text{op}}/V_{\text{on}}$? **(c)** What are the advantages of the switched-capacitor CMFB circuit over the continuous-time CMFB control (e.g., as described in Question 15)?

(a) Capacitors C_{2A} and C_{2B} function as a voltage divider to sense the output common-mode voltage. This voltage is used to tune the bias voltage of the tail transistor to adjust the output common-mode voltage. This forms a negative-feedback loop to stabilize the common-mode voltage. Capacitors C_{1A} and C_{1B} maintain the required offset charge and act as level shifters, translating the control voltage to the appropriate bias level V_b to achieve the target output common-mode voltage V_{cm}.

(b) The relationship between V_{cmfb} and V_{op}/V_{on} is:

$$\left(1+\frac{C_2}{C_1}\right)\cdot V_{cmfb}(n)-\frac{C_2}{C_1}\cdot V_{cmfb}(n-1)=\frac{V_{op}+V_{on}}{2}-(V_{cm}-V_b)$$

When the CMFB loop reaches its steady state, this becomes:

$$V_{cmfb}(n)=\frac{V_{op}+V_{on}}{2}-(V_{cm}-V_b)$$

(c) Advantages: the switched-capacitor CMFB circuit realizes the averaging and level shifting functions through a simple and fast switched-capacitor network and regulates the output common-mode voltage without an error amplifier. The removal of the error amplifier lowers the open-loop gain, reduces the number of poles, and helps to improve the stability of the common-mode control loop.

Question 33

A CMOS latch comparator is shown in Figure Q33a. The latch has a two-phase operation controlled by the clock signal *strobe*. When the *strobe* is high, the latch is in regeneration phase, with its outputs latched into one of two stable states depending on the input signal. If the *strobe* goes low, the latch enters reset phase, with both outputs pulled high. Assume the load capacitance at the outputs is C_L. **(a)** What is the (DC) gain of the latch in regeneration phase? **(b)** Please draw a diagram showing how the outputs would change over time with a fixed DC input (step response) after the *strobe* goes high. **(c)** What is the advantage of using a latch instead of an amplifier for comparator design?

Intuitive Analysis: When the *strobe* is high, the latch comparator is in regeneration phase with its positive-feedback circuit formed by the cross-coupled load transistors M3 and M4. To understand the transient response of the circuit in

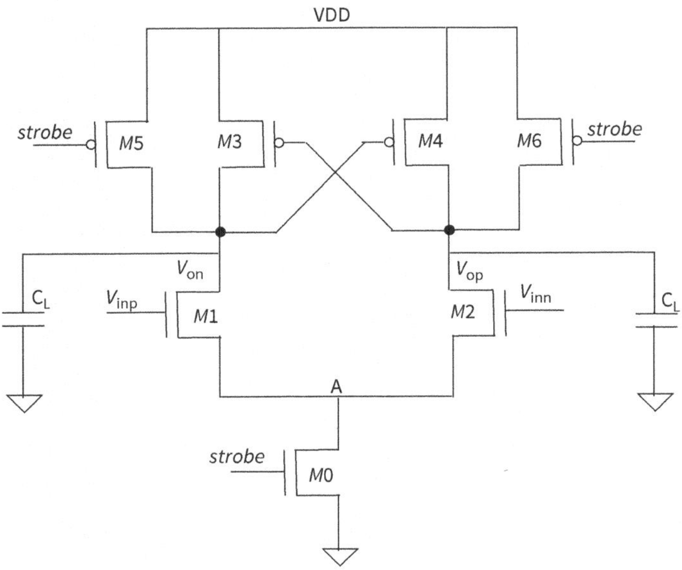

Figure Q33a

regeneration phase, we need to know the (small-signal) transfer function of the circuit first. The analysis of Question 6, *Impedance of a cross-coupled load* demonstrates that the impedance of a drain-to-gate cross-coupled transistor pair is $-1/g_m$ (negative impedance due to the positive feedback). Since the circuit in regeneration mode functions as a common-source amplifier comprising input transistors M1 and M2, a load impedance of $-1/g_m$, and a load capacitance C_L, the s-domain transfer function can be expressed as:

$$H(s) = \frac{v_{op} - v_{on}}{v_{inp} - v_{inn}} = \frac{-\dfrac{g_{m1}}{g_{m3}}}{\dfrac{-C_L}{g_{m3}} \cdot s + 1} = \frac{\dfrac{g_{m1}}{g_{m3}}}{\dfrac{C_L}{g_{m3}} \cdot s - 1}$$

where g_{m1} and g_{m3} are the transconductance of M1 and M3.

This is a standard transfer function of a first-order positive-feedback system, also known as an *unstable first-order system.* Without diving into the control system theories, we can think of this as a first-order amplifier (or a stable *first-order system*) BUT with a *negative* time constant:

$$\tau = -\frac{C_L}{g_{m3}}$$

For a (stable) first-order amplifier with a positive time constant, the step response is:

$$v_o(t) = v_{\text{op}} - v_{\text{on}} = \frac{g_{m1}}{g_{m3}} \cdot \left(v_{\text{inp}} - v_{\text{inn}}\right) \cdot \left(1 - e^{-\frac{t}{\tau}}\right)$$

where τ is the time constant and is greater than zero.

The exponential function in the equation converges to its final, steady-state value, which is given by the DC gain (g_{m1}/g_{m3}) multiplied by the input.

In an unstable first-order positive-feedback system, the equivalent time constant is negative, and the step response becomes:

$$v_o(t) = v_{\text{op}} - v_{\text{on}} = \frac{g_{m1}}{g_{m3}} \cdot \left(v_{\text{inp}} - v_{\text{inn}}\right) \cdot \left(e^{+\frac{t}{|\tau|}} - 1\right)$$

Now the exponential function diverges, meaning that the output will grow exponentially without an upper limit (mathematically). Consequently, the DC gain is infinite.

The (small-signal) step response of the circuit in Figure Q33a is sketched in Figure Q33b. The step response of an amplifier with diode-connected PMOS as load is also plotted to serve as a comparison.

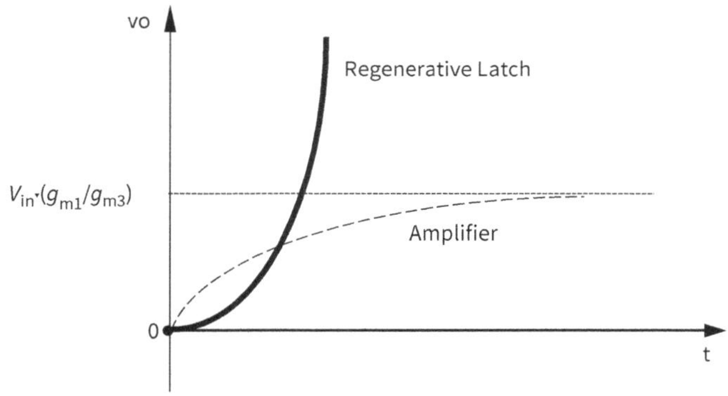

Figure Q33b

The regenerative latch has a much higher DC gain and responds to its input much faster than an amplifier. For example, at the time equal to 3 time constants, the output of a first-order amplifier settles to 95% to its final value, but the output of a regenerative latch already reaches 20× of this value (Assuming the output signal is small and not restricted by transistor nonlinearity and supply rails)! Fast resolution (decision) and high gain are the main advantages of a regenerative latch comparator. Comparators usually need to resolve a small input into a sufficiently large output to drive the digital logic in a short period of time. For amplifiers, this stands for a large gain and bandwidth, which means a large power consumption even if feasible.

Please note that Figure Q33b is based on results from the small-signal analysis. In reality, the transient response of the latch is a highly nonlinear process. At the beginning of the regeneration phase, both outputs are pre-charged to VDD and transistors M3/M4 are turned off. The two load capacitors (C_L) start to provide current to the common-source amplifier formed by M1/M2 (and the tail transistor M0, also referred to as the *pulse current source*) while being discharged. The input difference $(V_{ip} - V_{in})$ is amplified by the differential pair during this stage. Once V_{op} and V_{on} drop to below VDD − V_{THP}, where V_{THP} is the threshold voltage of the PMOS transistors, M3/M4 are turned on, enabling the positive-feedback loop. The separation of V_{op} and V_{on} is accelerated. Eventually, one of the outputs is pulled to ground potential, and the other output reaches a high voltage level below VDD. Figure Q33c illustrates the large-signal settling process of the latch outputs during the regeneration phase.

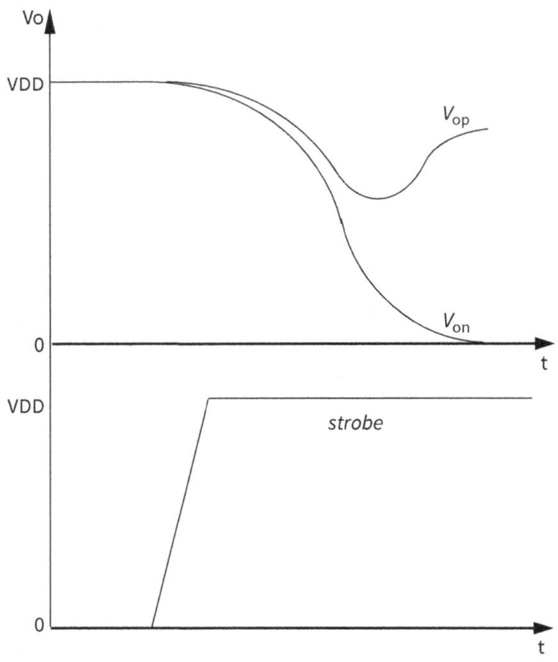

Figure Q33c

The positive feedback during the regeneration phase makes the latch highly sensitive to the imbalances, such as mismatch of the input differential pair, imbalances in load and parasitic capacitances, timing skews, and interferences. Because of the high sensitivity to these factors, the regenerative latch often presents larger equivalent input-referred offset than an open-loop amplifier. Preamplifiers are usually necessary to be placed before the latch to attenuate the large latch offset for precision circuits.

Quantitative Analysis: First let us quantitatively derive the impedance of node V_{on} in Figure Q33a and verify that it is a negative resistance. The small-signal model of the topology in Figure Q33a is shown in Figure Q33d. As the circuit is symmetrical between the two differential inputs, we only draw v_{inp} (left) side of the circuit in our small-signal model. A test voltage source v_{Test} is connected to node V_{on} to use as the stimulus to measure the node impedance. Also note that for the purpose of analyzing the impedance on node V_{on}, the input v_{inp} and v_{inn} are shorted to the AC ground (node A). Importantly, the control voltage for the dependent current source (VCCS) of M3, V_C, equals v_{op}, which is in the same magnitude of v_{on}, but of opposite polarity, i.e., $V_C = v_{op} = -v_{on}$.

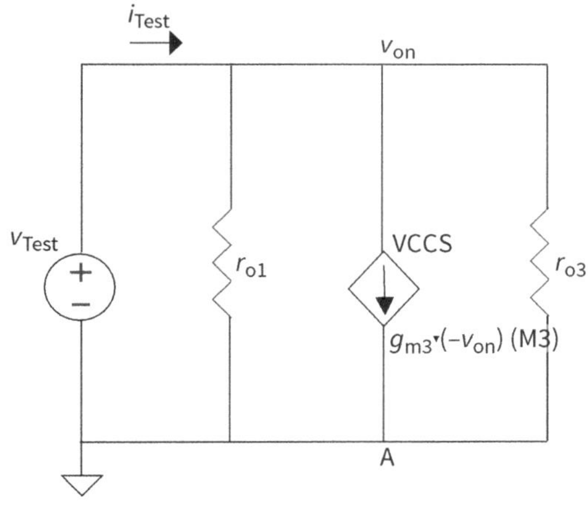

Figure Q33d

The i_{Test} is equal to the sum of all branch currents:

$$i_{Test} = \frac{v_{Test}}{r_{o1}} + \frac{v_{Test}}{r_{o3}} - g_{m3} \cdot v_{Test}$$

The impedance of the node V_{on}, rout, therefore is:

$$r_{out} = \frac{1}{\frac{1}{r_{o1}} + \frac{1}{r_{o3}} - g_{m3}} \approx \frac{-1}{g_{m3}}$$

Next, we derive the step response of an unstable first-order system, to support our argument in the intuitive analysis. A standard unstable first-order system has a transfer function as shown below:

$$H(s) = \frac{1}{\tau \cdot s - 1}$$

where τ is the time constant of the system and is a positive number in the standard form.

We can also treat the equivalent time constant as a negative number to map it to a standard stable first-order system like what we did in the intuitive analysis:

$$H(s) = \frac{-1}{(-\tau) \cdot s + 1}$$

With a unit step input (i.e., the input is defined as a constant value of 1 for all $t \geq 0$; the Laplace transform of the unit step input is $1/s$), the output of the regenerative latch is:

$$v_o(s) = H(s) \cdot \frac{1}{s} = \frac{1}{\tau \cdot s - 1} \cdot \frac{1}{s} = \frac{\tau}{\tau \cdot s - 1} - \frac{1}{s}$$

Using the inverse Laplace transform, we obtain the time-domain step response of the regenerative latch:

$$v_o(t) = \mathcal{L}^{-1}\left\{\frac{\tau}{\tau \cdot s - 1} - \frac{1}{s}\right\} = \mathcal{L}^{-1}\left\{\frac{1}{s - \frac{1}{\tau}} - \frac{1}{s}\right\} = e^{\frac{t}{\tau}} - 1$$

Assuming $V_o(0) = 0$ (initial condition).

The step response of an unstable first-order system is a diverging exponential function.

Regenerative latch comparators are widely used in ADCs, sense amplifiers, and other analog and mixed-signal circuits. The drawbacks of the circuit in Figure Q33a include static current in regeneration phase and non-rail-to-rail output levels. A modified version is shown in Figure Q33e, where a cross-coupled NMOS pair is added to the latch core. Interested readers are encouraged to analyze the topology and tell the difference between the two.

Short Answer to Question 33

(a) What is the (DC) gain of the latch in regeneration mode? **(b)** Please draw a diagram showing how the outputs would change over time with a fixed DC input (step response) after the *strobe* goes high. **(c)** What is the advantage of using a latch instead of an amplifier for comparator design?

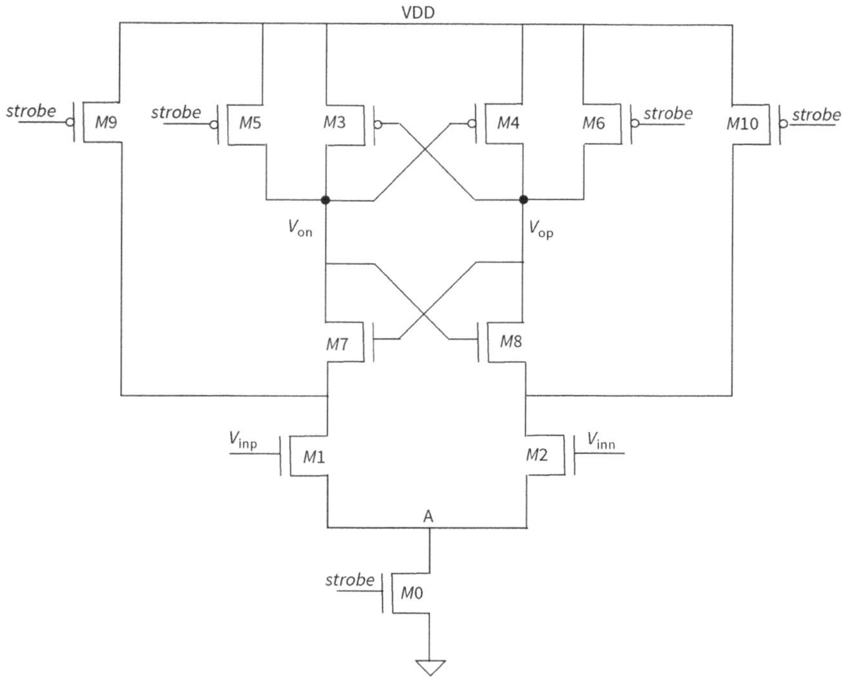

Figure Q33e

(a) The DC gain of the latch in regeneration mode is infinite due to positive feedback.

(b) Please refer to Figure Q33c.

(c) High gain and fast decision speed are the primary advantages of a regenerative latch comparator over an open-loop amplifier.

Question 34

A differential switched-capacitor comparator is shown in Figure Q34a. It consists of a preamplifier (preamp) and a latch. V_{inp}/V_{inn} are differential input signals, and V_{refp}/V_{refn} are differential reference voltages to which the inputs are compared. S1/S1a and S2 are nonoverlapping clocks, and *strobe* is the control signal of the latch (*strobe* = high: latch in regeneration phase). Figure Q34b shows the timing relationship of the control signals. C_p is the capacitance at nodes A and B

(the input capacitance of the preamp and other parasitics). C_L is the load capacitance at preamp outputs (input capacitance of the latch and other parasitics). Assume the latch has an input-referred offset voltage of 50 mV (a value of $\pm 3\sigma$, where σ is the standard deviation of the offset voltage) and the offset voltage of the preamp is negligible. The $\pm 3\sigma$ criterion for the offset voltage applies throughout this problem unless otherwise specified. The input-referred offset voltage of the comparator should be less than 5 mV when clocking at 100 MHz. What is the gain-bandwidth product requirement for the preamp to fulfill the design target?

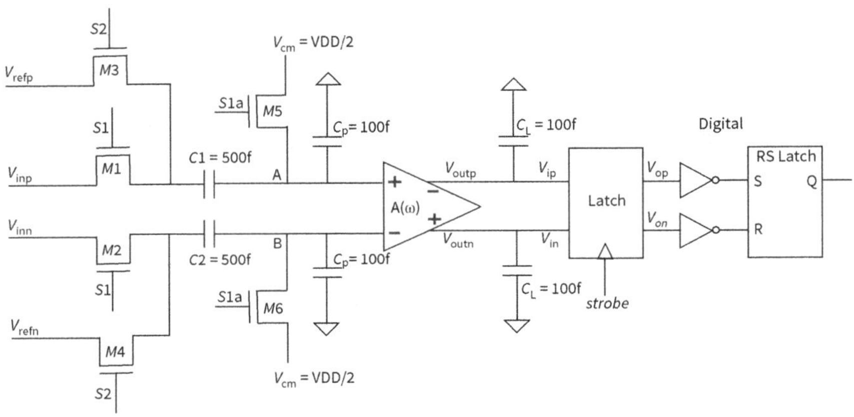

Figure Q34a

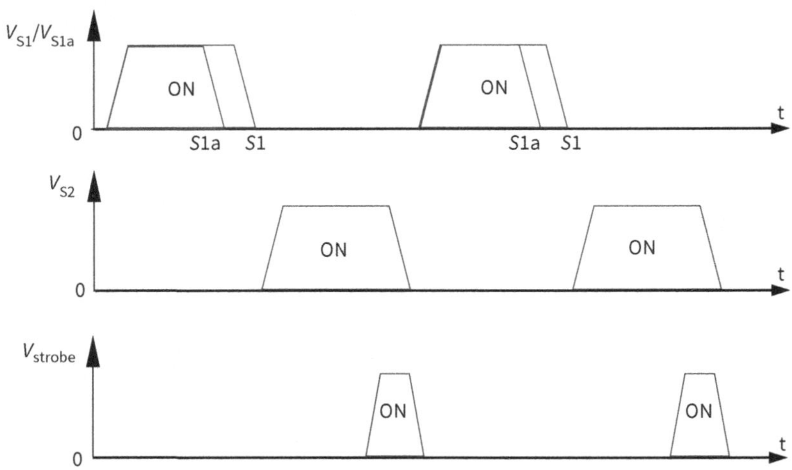

Figure Q34b

Intuitive and Quantitative Analysis: While the topology of the switched-capacitor preamplifier resembles the track-and-hold (T/H) circuits in Question 29, a fundamental distinction exists: the preamplifier operates in open-loop, whereas the T/H circuit utilizes closed-loop operation. This is because the preamplifier has no stringent precision requirements; its primary function is to provide sufficient gain for the latch.

When S1/S1a are high (the sampling phase), the differential inputs are sampled by the sampling capacitors C_1 and C_2. The falling edge of S1a is ahead of that of S1 to prevent signal-dependent charge injection, as discussed in the analysis of Question 28. Due to the circuit's symmetry, we will utilize half-circuit analysis. The charge stored at node A at the end of sampling phase is:

$$Q_1 = \left(0 - V_{\text{inp}}\right) \cdot C_1 = -V_{\text{inp}} \cdot C_1$$

Note that V_{cm} is a fixed DC voltage and its influence on the differential output is cancelled out due to circuit symmetry. In our analysis, we treat V_{cm} as AC ground. Also, there is no charge on the parasitic capacitance C_p because it is shorted to AC ground by M5 during sampling.

In the next phase (the comparison phase), S2 becomes high, reference voltage V_{refp} is applied to the input side of the sampling capacitor C_1, and node A is disconnected from V_{cm}. Applying charge conservation at node A (an isolated node), we have the following equation:

$$Q_1 = -V_{\text{inp}} \cdot C_1 = \left(V_A - V_{\text{refp}}\right) \cdot C_1 + V_A \cdot C_p$$

This yields:

$$V_A = \frac{-C_1}{C_1 + C_p} \cdot \left(V_{\text{inp}} - V_{\text{refp}}\right)$$

$$V_{\text{outp}} = -V_A \cdot A(\omega) = \frac{C_1}{C_1 + C_p} \cdot A(\omega) \cdot \left(V_{\text{inp}} - V_{\text{refp}}\right)$$

The final differential output of the preamp is:

$$V_{\text{out}} = V_{\text{outp}} - V_{\text{outn}} = \frac{C_1}{C_1 + C_p} \cdot A(\omega) \cdot (V_i - V_{\text{ref}});$$

where $V_i = V_{\text{inp}} - V_{\text{inn}}; V_{\text{ref}} = V_{\text{refp}} - V_{\text{refn}}$

The parasitic capacitance, C_p, attenuates the gain of the preamp by charge sharing. Another way to explain this attenuation is that the sampling capacitor C_1

and parasitic capacitance C_p form a voltage divider between V_{refp} and AC ground in the comparison phase.

Now let us calculate the gain-bandwidth product requirement for the preamp. To maintain an input-referred offset voltage of less than 5 mV, the total gain before the latch must satisfy:

$$G = \frac{C_1}{C_1 + C_p} \cdot A(\omega) > \frac{50\text{ mV}}{5\text{ mV}}$$

which yields:

$$A(\omega) > \frac{10}{\frac{5}{6}} = 12$$

Assume the time period for the comparison phase before *strobe* is applied is about half of the clock period, i.e., 5 ns. If we allow 3 time constants for the preamp to settle to this equivalent gain (12×), the time constant and the −3 dB bandwidth of the amplifier (assuming first-order system) are:

$$\tau = \frac{5\text{nS}}{3} = 1.67\text{ ns};\ f_{-3\text{dB}} = \frac{1}{2 \cdot \pi \cdot \tau} \approx 96\text{ MHz}$$

As the first-order system reaches 95% of its target value at a settling time of 3 time constants, the DC gain of the preamp should be:

$$A(0) > \frac{12}{0.95} = 12.6$$

And the gain-bandwidth product (GBW) of the amplifier is:

$$\text{GBW} > 12.6 \cdot 96\text{ MHz} = 1.21\text{ GHz}$$

In practice, we may need to reserve more margin on the GBW due to factors such as settling time losses caused by the gaps of the nonoverlapping clocks, and other possible timing arrangement between the preamp and the latch.

Demands on the preamp's gain-bandwidth product increase as the system requires either a higher operating clock rate or a smaller input-referred offset. If a large DC gain and bandwidth are needed, multiple preamp stages can sometimes lower the power consumptions (see *Question 19, Power optimization of multistage open-loop amplifiers*).

How do we determine the input offset requirement in a real system? Consider ADCs as an example. For the comparators of an n-bit flash ADC, the minimum distinguishable input is typically one LSB (least significant bit):

$$1\text{ LSB} = \frac{V_{\text{FS}}}{2^n}$$

V_{FS} is the full-scale analog input range of the ADC, and n is the resolution of the ADC. Keep in mind that the polarity and magnitude of the comparator offset in a flash ADC is usually random. This means that the input-referred offset of the comparators of a flash ADC should be less than $\pm 1/2$ LSB.

Short Answer to Question 34

What is the gain-bandwidth product requirement for the preamp to fulfill the design target?

The required preamp DC gain and −3 dB bandwidth are: 12.6 V/V and 96 MHz, respectively. The gain-bandwidth product requirement is therefore 1.21 GHz.

Question 35

The same comparator as shown in Figure Q34a of Question 34 is used in this question. The input-referred offset of the preamp is now 5 mV (a value of 3σ, where σ is the standard deviation of the offset voltage), instead of offset-free as assumed in Question 34. The 3σ criterion for the offset voltage applies throughout this problem unless otherwise specified. **(a)** What is the input-referred offset of the comparator? **(b)** Figure Q35a shows a new preamp circuit with an offset-cancellation scheme (auto-zero). The timing of S1/S1a/S2 is the same as in Question 34. What is the input-referred offset of the preamp with the offset cancellation, and what is the total input-referred offset of the comparator?

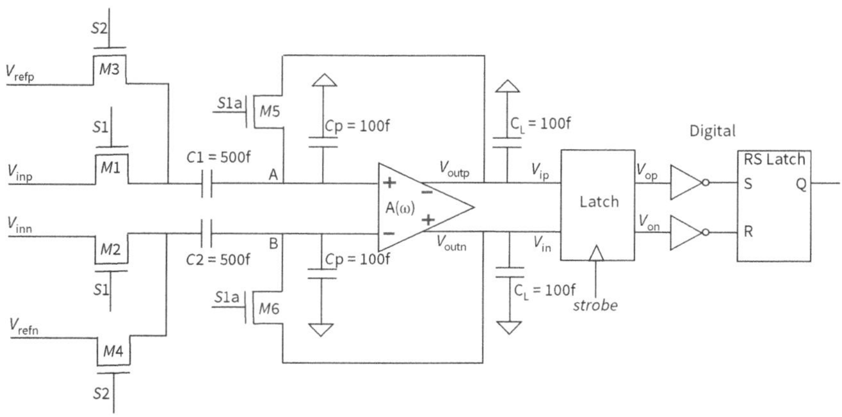

Figure Q35a

Intuitive and Quantitative Analysis: As we have discussed in Question 34, with the preamp's effective gain of 10 and the latch's offset of 50mV, the input-referred offset of the comparator is 5 mV if the preamp itself is offset-free. If the preamp has an offset, since there is no gain stage before it, its input-referred offset contributes directly to the total input-referred offset of the comparator without attenuation. Recall that in Question 34, we determined the relationship between the comparator input voltage V_{inp} and preamp input voltage V_A:

$$V_A = \frac{-C_1}{C_1 + C_p} \cdot \left(V_{\text{inp}} - V_{\text{refp}}\right)$$

Therefore, the (magnitude) contribution of the preamp input-referred offset voltage at node A (v_{OSA}) to the total equivalent input offset of the comparator (v_{OS}) is expressed as:

$$v_{\text{OS}} = \frac{C_1 + C_p}{C_1} \cdot v_{\text{OSA}} = \frac{6}{5} \cdot 5\ \text{mV} = 6\ \text{mV}$$

Intuitively, this is because the input signal is attenuated by the parasitic capacitance $\left(C_p\right)$, and therefore the offset of the preamp appears larger at the comparator input.

One common mistake is to assume that the preamp's 6 mV equivalent input-referred offset adds arithmetically to the 5mV equivalent input-referred offset caused by the latch, and the total input-referred offset of the comparator becomes 11mV.

It is important to realize that the offset of a comparator is made up of two parts: random offset, and systematic offset. Random offset is caused by unpredictable factors such as device random mismatch during the manufacture, while systematic offset is caused by deterministic factors such as mismatch due to a layout pattern or a process gradient. For a well-designed circuit (including physical layout), systematic offset can be minimized. The random offset, however, is governed by manufacturing variations and other unpredictable factors, and often dominates the comparator's offset.

The random offset is like random noise (it can be regarded as a form of "DC noise"): its value is often specified using statistics such as the *standard deviation*, similar to the RMS (root-mean-square) value of the noise. Because of its random and independent nature, the offsets of different parts of the circuit should be combined in a root-sum-square manner, similar to adding uncorrelated noise sources together:

$$v_{\text{OS}} = \sqrt{v_{\text{OS1}}^2 + v_{\text{OS2}}^2 + \ldots + v_{\text{OSN}}^2}$$

Going back to part (a), the total input-referred offset is:

$$v_{OS} = \sqrt{6^2 + 5^2} \approx 7.81(\text{mV})$$

As there is no gain stage before the preamp, the only way to meet the design target of the total input-referred offset (5 mV) is to lower the input-referred offset of the preamp itself. An auto-zero circuit, as shown in Figure Q35a, is often used for this purpose. Let us first analyze the mechanism of its operation.

At the sampling phase, transistors M1/M2 and M5/M6 are turned on. The amplifier is in close-loop operation with negative feedback. The inputs of the amplifier are therefore virtual (AC) ground, and the V_{inp}/V_{inn} charge the sampling capacitors against this virtual ground. Figure Q35b illustrates the equivalent single-ended circuit at the sampling phase.

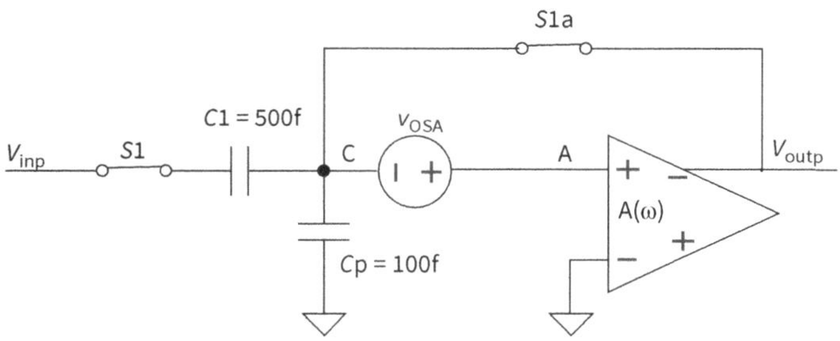

Figure Q35b

In Figure Q35b, v_{OSA} represents the input-referred offset voltage of the amplifier. Because $V_A = 0$ (virtual ground), $V_C = -v_{OSA}$. The charge stored at node C at the end of sampling phase is:

$$Q_1 = \left(V_C - V_{inp}\right) \cdot C_1 + V_C \cdot C_P = -\left(V_{inp} + v_{OSA}\right) \cdot C_1 - v_{OSA} \cdot C_P$$

During the comparison phase, M1/M2 and M5/M6 are turned off, and M3/M4 are turned on. The amplifier is in open-loop mode. Figure Q35c illustrates the equivalent single-ended circuit in the comparison phase.

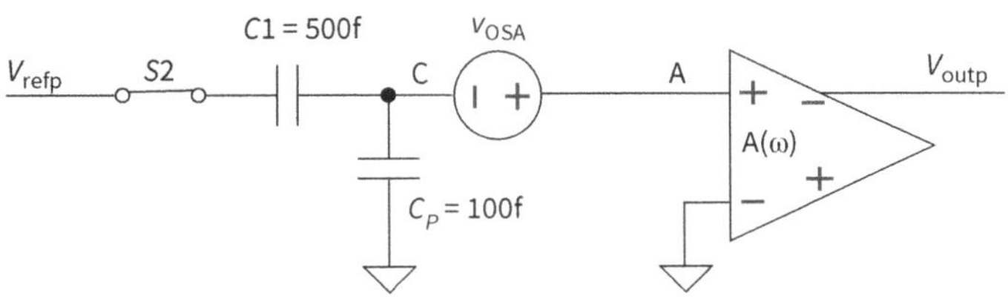

Figure Q35c

Charge at node C in the comparison phase is:

$$Q_2 = \left(V_C - V_{\text{refp}}\right) \cdot C_1 + V_C \cdot C_P$$

Applying charge conservation at isolated node C, we obtain the following equation:

$$Q_1 = -\left(V_{\text{inp}} + v_{\text{OSA}}\right) \cdot C_1 - v_{\text{OSA}} \cdot C_P = Q_2 = \left(V_C - V_{\text{refp}}\right) \cdot C_1 + V_C \cdot C_P$$

Because $V_C = V_A - v_{\text{OSA}}$, this yields:

$$V_A = -\left(V_{\text{inp}} - V_{\text{refp}}\right) \cdot \frac{C_1}{C_1 + C_P}$$

$$V_{\text{outp}} = -V_A \cdot A(\omega) = \frac{C_1}{C_1 + C_p} \cdot A(\omega) \cdot \left(V_{\text{inp}} - V_{\text{refp}}\right)$$

Extending the result to the fully differential circuit, the transfer of the preamp is:

$$V_{\text{out}} = V_{\text{outp}} - V_{\text{outn}} = \frac{C_1}{C_1 + C_P} \cdot A(\omega) \cdot (V_i - V_{\text{ref}})$$

where $V_i = V_{\text{inp}} - V_{\text{inn}}; V_{\text{ref}} = V_{\text{refp}} - V_{\text{refn}}$

The offset voltage of the preamp is therefore cancelled by the two-phase switched-capacitor operation. The final input-referred offset of the comparator is the input-referred offset of the latch (attenuated by the preamp), which is 5 mV.

Note that the above analysis assumes infinite gain of the preamp so that in the sampling phase, node A is a perfect virtual ground. In reality, with a finite amplifier open-loop gain of $A(\omega)$, node A deviates from virtual ground in the sampling phase:

$$V_A = -\frac{V_{\text{outp}}}{A(\omega)} \neq 0$$

This means an inaccurate offset charge of the amplifier stored on the sampling capacitor, and subsequently an imperfect offset cancellation in the comparison phase. The uncancelled (residual) comparator input-referred offset voltage is:

$$V_{\text{OS:err}} = \frac{C_1 + C_p}{C_1} \cdot \frac{1}{1 + A(\omega)} \cdot v_{\text{OSA}}$$

For a preamp with a DC open-loop gain of 12.6 and bandwidth supporting a settling of three time constants (see *Question 34, Comparator – preamplifier*), this would leave a residual input-referred offset of:

$$V_{\text{OS:err}} = \frac{500+100}{500} \cdot \frac{1}{1+12.6 \cdot 0.95} \cdot v_{\text{OSA}} = 1.2 \cdot 0.077 \cdot 5 = 0.46\ (\text{mV})$$

And the total input-referred offset of the comparator is:

$$V_{\text{OS}} = \sqrt{0.46^2 + 5^2} \approx 5.02\ (\text{mV})$$

The auto-zero scheme in Figure Q35b is also known as input offset storage (IOS). There are other preamp offset-cancellation topologies such as output offset storage (OOS), an auto-zeroing mechanism that measures the amplifier's offset in an open-loop state and cancels it at the output. Interested readers are encouraged to explore these topics further, as offset cancellation is a key technique in switched-capacitor circuits that not only mitigates DC offset but also suppresses low-frequency flicker noise.

Short Answer to Question 35

(a) What is the input-referred offset of the comparator? **(b)** Figure Q35a shows a new preamp circuit with an offset-cancellation scheme (auto-zero). What is the input-referred offset of the preamp with the offset cancellation, and what is the total input-referred offset of the comparator?

(a) The input-referred offset voltage (without offset cancellation) of the comparator is:

$$V_{\text{OS}} = \sqrt{6^2 + 5^2} \approx 7.81(\text{mV})$$

(b) The input-referred offset of the preamp with offset cancellation is:

$$V_{\text{OS:err}} = \frac{C_1 + C_p}{C_1} \cdot \frac{1}{1+A(\omega)} \cdot v_{\text{OSA}} = 0.46(\text{mV})$$

The total input-referred offset of the comparator is:

$$V_{\text{OS}} = \sqrt{0.46^2 + 5^2} \approx 5.02(\text{mV})$$

Question 36

A switched-inductor circuit is shown in Figure Q36a. The input (V_{in}) is driven by an ideal voltage source. Switches S1 and S2 are controlled by complementary clock signals with equal ON and OFF periods (50% duty cycle), as shown in Figure Q36b. Assume initially $(t = 0)$ the voltage across and the current in the inductor L

are zero. Also assume that the ON and OFF periods are sufficiently long compared with the circuit time constant. **(a)** Sketch the voltage waveform at the output (V_{out}) of the switched-inductor circuit. **(b)** What happens if the OFF period of S1 becomes significantly shorter than its ON period, as shown in Figure Q36c?

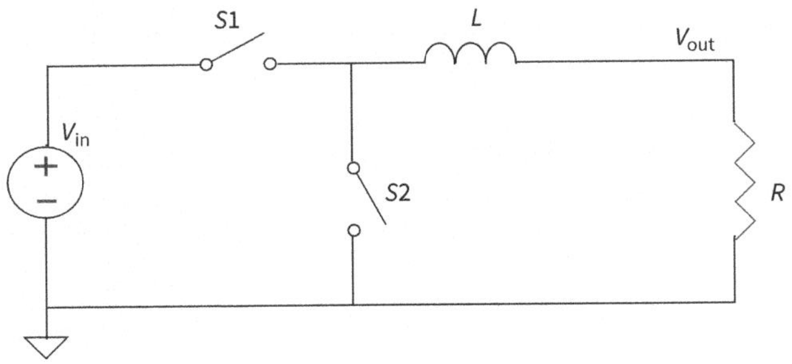

Figure Q36a

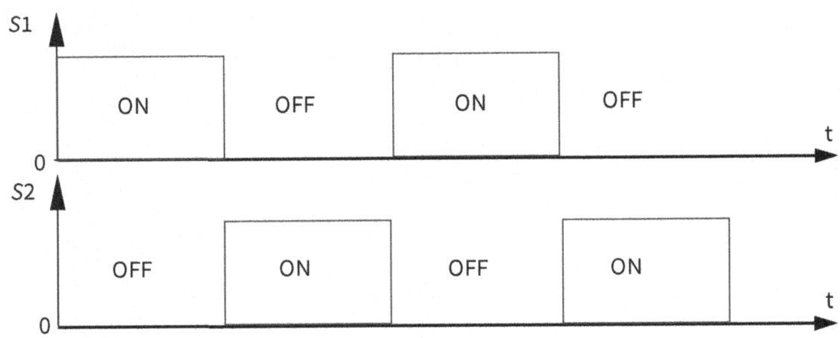

Figure Q36b

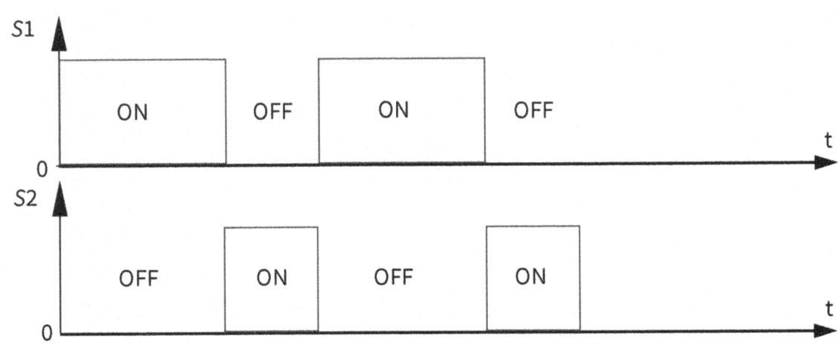

Figure Q36c

Intuitive Analysis: A switched-inductor circuit question in a chapter of switched-capacitor circuits? Yes, and for good reasons. Inductor-related circuits are seldom discussed in many circuit design courses. Because of this, a lot of students feel that they are less familiar with and prepared for questions regarding the operations of inductors. In real applications, there are cases where the circuit systems involve inductors as key functional components (e.g., RF) or parasitic components which cannot be ignored (especially in high-speed circuits). Switched-inductor circuits also find wide applications in power management systems and ICs (e.g., buck and boost converters). The chances of encountering a switched-inductor or other inductor-related questions in a technical interview are not unlikely.

Although inductors and capacitors exhibit distinct circuit behaviors, their current-voltage (*I-V*) relationships and associated analysis methods are remarkably similar. The *I-V* relationship for a capacitor is:

$$i = C \cdot \frac{dv}{dt}; \; v = \frac{1}{C} \cdot \int i \cdot dt$$

These equations reveal the most essential feature of a capacitor: the voltage across a capacitor cannot change instantaneously. A large current will result if the capacitor is forced to experience a quick voltage change.

The *I-V* relationship equations of an inductor is in the same mathematical formats as that of a capacitor, with position of *V* and *I* swapped in the equations:

$$v = L \cdot \frac{di}{dt}; i = \frac{1}{L} \cdot \int v \cdot dt$$

This means the current through an inductor cannot change instantaneously. A large voltage will result if the inductor is forced to experience a quick current change. Note that the polarity of this voltage *opposes* the direction of the current change (*Lenz's Law* – this is why a negative sign is sometimes included in the *I-V* equations of an inductor).

Now let us analyze the circuit in Figure Q36a. At $t = 0$, switch S1 is turned on, the voltage source V_{in} forces a current flowing through the inductor L and resistor R. This intended sudden current change induces a voltage jump across the inductor, V_L, with a polarity opposing the current change I_L, as illustrated in Figure Q36d. The current in the inductor is zero at the beginning as it cannot change instantaneously, which means the voltage of the inductor, V_L, is the same as V_{in} and there is no voltage drop on the resistor at time zero.

The current in the inductor increases after $t = 0$, with its induced voltage decreasing over time. Voltage drops more on the resistor, R, and the output voltage (V_{out}) increases over time.

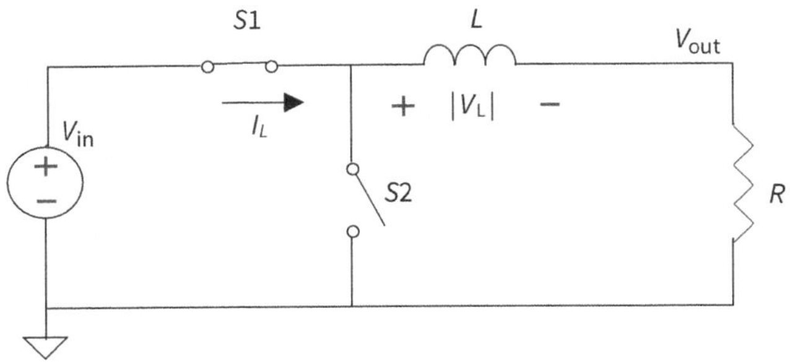

Figure Q36d

Given sufficient settling time, the induced voltage across the inductor (V_L) approaches zero, and all V_{in} drops to the resistor R, with final steady-state current $I_L = V_{in}/R$.

During the second phase, switch S1 is turned off and switch S2 is turned on. Now the circuit forces the current to become zero as there is no supply voltage in the loop. This instantaneously induces a voltage, V_L, across the L, with the polarity opposing this intended change of current, as illustrated in Figure Q36e. At the very beginning, $V_L = V_{in}$, and the current flowing through the inductor, L, keeps the final current of the previous phase (V_{in}/R).

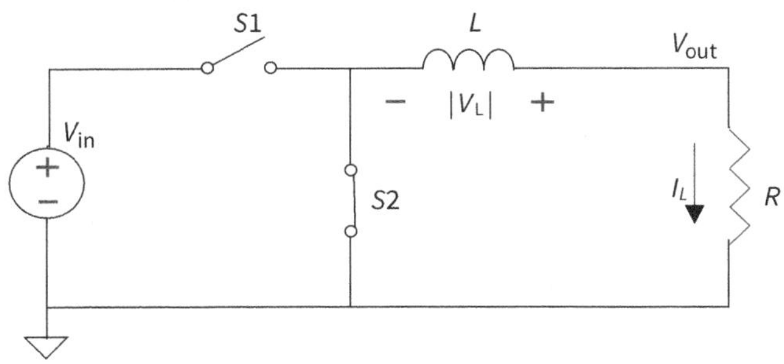

Figure Q36e

The current in the inductor decreases over time after the initial moment of the second phase, as well as the induced voltage across the inductor. Eventually the V_L approaches zero and so does the I_L and V_{out}. The waveforms of output voltage V_{out} as well as the voltage and current of the inductor L, V_L and I_L, are shown in Figure Q36f.

If the OFF period of S1 is significantly shorter than its ON period, the current of the inductor may not reach its final steady state, i.e., zero, before the switch S2 is turned off and S1 is turned on again. The voltage source V_{in} forces the inductor current to go up. The inductor responds to this intention of the current change with an opposing induced voltage, keeping the current unchanged at the moment of the circuit switching, and then goes up to its steady state (V_{in}/R). Figure Q36g illustrates the waveform of the output voltage V_{out} in this scenario.

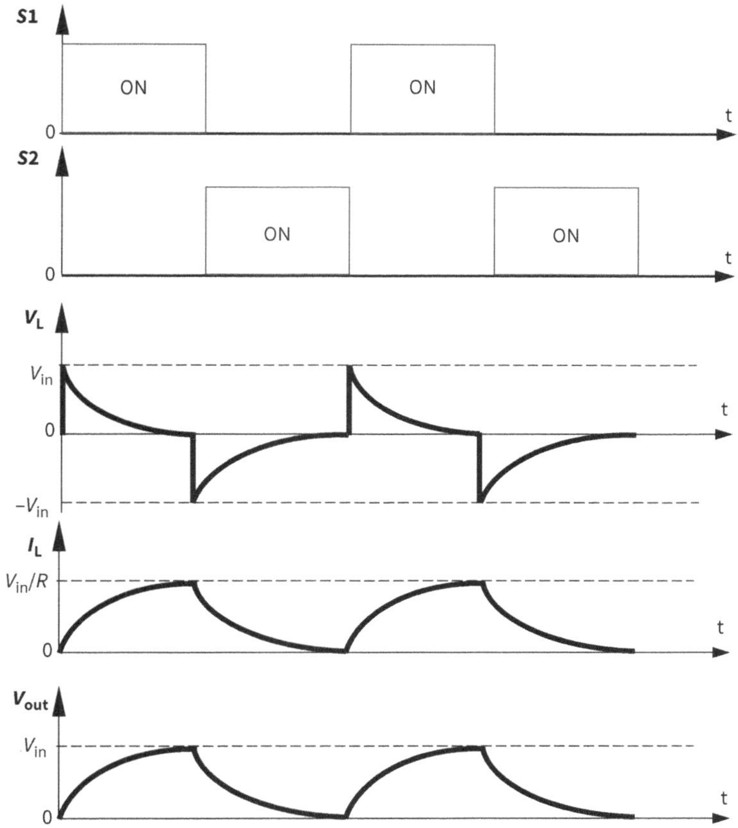

Figure Q36f

The average output voltage is lower than the input voltage for the switched-inductor circuit in Figure Q36a. The average of the V_{out} can be adjusted by changing the duty cycle of the S1/S2 clocks. The larger the duty cycle of the ON period of the switch S1, the higher the average output voltage. This type of switched-inductor circuit is widely used in *step down (buck) DC/DC converters*, converting DC supply voltage from high levels to lower levels (e.g., 12V to 5V). In practice, a capacitor, *C*, is added in parallel with the load resistor *R*, as shown in

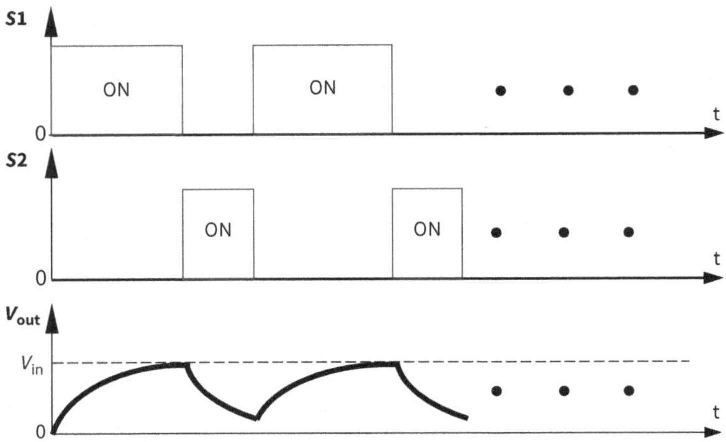

Figure Q36g

Figure Q36h, to reduce the ripple on the output voltage. Similar to the inductor, the capacitor also stores electrical energy when the switch S1 is turned on, and supplies current, along with the inductor, to the load when the switch S1 is turned off and switch S2 is turned on. If the capacitor is sufficiently large, it could keep the V_{out}, and therefore the V_L, almost constant, and the current in the inductor (I_L) ramps up/down linearly. Figure Q36i illustrates the waveforms of the circuit in Figure Q36h with a large capacitor.

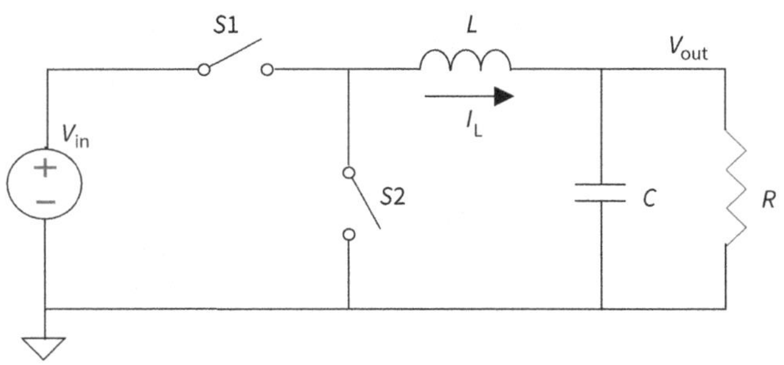

Figure Q36h

Quantitative Analysis: Let us do some quantitative analysis on the switched-inductor circuits in Figure Q36a (without the load capacitor) and Figure Q36h (with the load capacitor), to verify the waveforms that we have drawn through the intuitive analysis. A convenient way to analyze the dynamics of the circuit is to utilize the complex frequency (s-domain) transfer function.

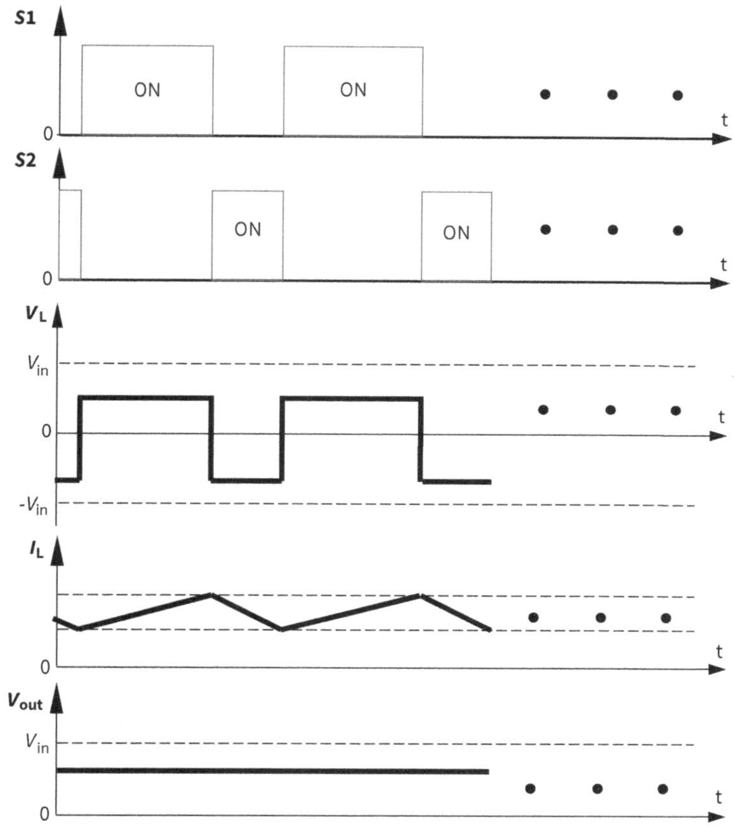

Figure Q36i

The circuit in Figure Q36a can be viewed as a voltage divider between L and R. Noting that the s-domain impedance of the inductor is $L \cdot s$, the voltage transfer function of the circuit can be written as:

$$H(s) = \frac{V_{out}(s)}{V_{in}(s)} = \frac{1}{\frac{L}{R} \cdot s + 1}$$

This is a first-order system, with the time constant τ equal to:

$$\tau = \frac{L}{R}$$

This confirms the exponential settling behavior of V_{out} in Figure Q36f. The larger the inductor L, the slower the transient response.

For the circuit in Figure Q36h, the current in the inductor, I_L, can be written as:

$$I_L(s) = \frac{V_{in}(s)}{L \cdot s + \frac{1}{C \cdot s + \frac{1}{R}}} = \frac{R \cdot C \cdot s + 1}{L \cdot R \cdot C \cdot s^2 + L \cdot s + R} \cdot V_{in}(s)$$

When C is sufficiently large, the equation can be approximated as:

$$I_L(s) = \frac{R \cdot C \cdot s + 1}{L \cdot R \cdot C \cdot s^2 + L \cdot s + R} \cdot V_{in}(s) \approx \frac{R \cdot C \cdot s}{L \cdot R \cdot C \cdot s^2} \cdot V_{in}(s) = \frac{1}{L \cdot s} \cdot V_{in}(s)$$

Now it behaves like an integrator ($1/s$), and the current ramps up or down linearly when V_{in} is a step function. Intuitively, when C is large enough, the impedance of the R and C in parallel becomes negligible – there is no AC voltage across L (meaning V_L and V_{out} are constant), and the current flowing through L changes linearly (recall $i = \frac{1}{L} \cdot \int v \cdot dt$).

Short Answer to Question 36

(a) Sketch the voltage waveform at the output (V_{out}) of the switched-inductor circuit. **(b)** What happens if the OFF period of S1 becomes significantly shorter than its ON period, as shown in Figure Q36c?

(a) Please refer to Figure Q36f.
(b) If the OFF period of S1 becomes significantly shorter than its ON period, the average value of V_{out} will increase. Please refer to Figure Q36g.

General References

1 Allen PE and Holberg DR. *CMOS Analog Circuit Design*. 3rd ed. New York (NY): Oxford University Press; 2012.
2 Baker RJ. *CMOS: Mixed-Signal Circuit Design*. 2nd ed. Hoboken (NJ): Wiley-IEEE Press; 2009.
3 Baker RJ. *CMOS: Circuit Design, Layout, and Simulation*. 3rd ed. Hoboken (NJ): Wiley/IEEE Press; 2010.
4 Carusone TC, Johns DA and Martin KW. *Analog Integrated Circuit Design*. 2nd ed. Hoboken (NJ): Wiley; 2012.
5 Fischer, M.C. *Comparators*. New York (NY): Springer; 2017.

6 Gray PR, Hurst PJ, Lewis SH and Meyer RG. *Analysis and Design of Analog Integrated Circuits.* 6th ed. Hoboken (NJ): Wiley; 2024.

7 Gregorian, R. *Introduction to CMOS OP-AMPs and Comparators.* New York (NY): John Wiley & Sons; 1999.

8 Pelgrom MJM. *Analog-to-Digital Conversion.* 4th ed. [place unknown]: Springer; 2022.

9 Quinn PJ and van Roermund AHM. Switched-Capacitor Techniques for High-Accuracy Filter and ADC Design. New York (NY): Springer; 2007.

10 Razavi B. *Analysis and Design of Data Converters.* Cambridge (UK): Cambridge University Press; 2025.

11 Razavi B. *Design of Analog CMOS Integrated Circuits.* 2nd ed. New York (NY): McGraw-Hill Education; 2017.

12 Sansen WMC. *Analog Design Essentials.* New York (NY): Springer; 2006.

13 Wicht B. *Design of Power Management Integrated Circuits.* Hoboken (NJ): Wiley-IEEE Press; 2024.

5

Digital Circuits from an Analog Perspective

Today's digital system design is largely automated – complex logic and state machines are synthesized by EDA tools, and layout implementations are automatically placed and routed. Do designers still need to know anything about digital logic at the transistor level? The answer is still *yes*.

First of all, not all digital circuits are automatically synthesized. Many manually designed and implemented digital circuits exist inside analog and mixed-signal systems for multiple purposes: more compact design for local logic and analog-to-digital interfaces, optimized performance, especially in high-speed or high-precision applications, or simply the absence of digital standard-cell libraries for I/O or other larger geometry transistors.

Even for digital designers who primarily use automatic digital design flows, a good understanding of how logic gates work at the transistor level is beneficial: a transistor-level-informed designer can set up the design constraints and margins more effectively in the EDA environment to avoid potential timing and signal integrity issues resulting from the lack of certain physical details in most digital design flows. It also helps with digital system debugging and identifying digital problems of analog nature.

These reasons explain why questions regarding transistor-level digital circuits are asked in technical interviews from time to time. In this chapter, we have arranged seven questions to test your "digital circuit sense." We start with inverter-based ring oscillators in Question 37. Question 38 investigates inverter sizing for speed and clock duty cycle optimization. Buffer design for driving long RC lines is studied in Question 39. Question 40 explores estimation of inverter current and power consumption. The next two questions (Questions 41 and 42) are on D flip-flops and clock dividers. Question 43 discusses the digital phase-frequency detector and its application in phase-locked loops (PLLs). The last question (Question 44) examines metastability in digital systems. You will likely find that, beneath the surface, digital and analog circuits share many common traits.

Question 37

To make a single-ended ring oscillator using CMOS inverters as shown in Figure Q37a, would one inverter work? How about three inverters? What about two?

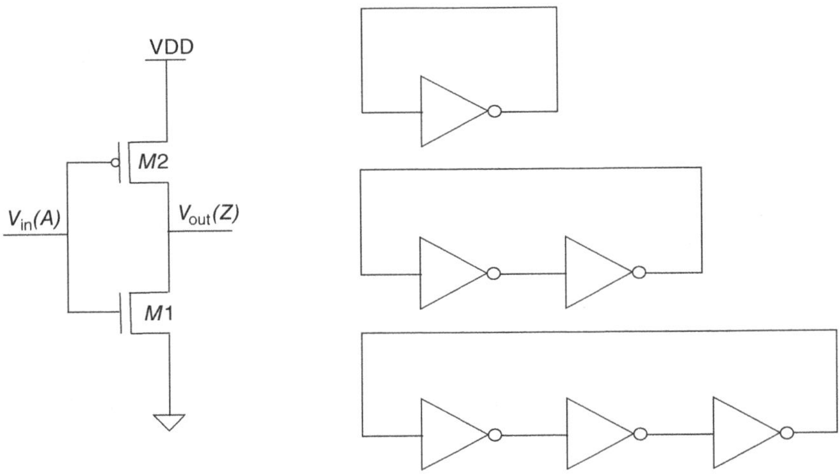

Figure Q37a

Intuitive Analysis: A single-ended CMOS ring oscillator requires an odd number of inverting stages to sustain oscillation. But will one inverter work? Let us first analyze it in the time domain.

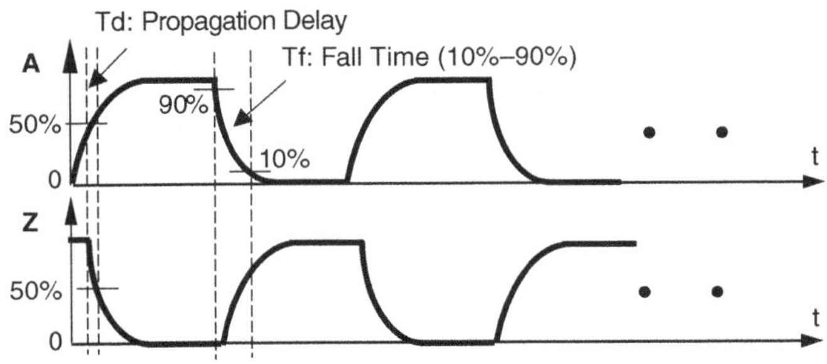

Figure Q37b

Figure Q37b illustrates the timing relationship between the input and output voltage of the inverter. The propagation delay (measured at the switching point of the input and output of the inverter, usually around 50% of the signal swing) of a CMOS inverter is (see *Appendix 4, Section 7, Basic digital circuits*):

$$T_d \approx 0.7 \cdot R \cdot C$$

where R is the effective resistance of PMOS or NMOS (assuming PMOS and NMOS are balanced, i.e., with equal effective resistance), and C is the load capacitance of the inverter.

The rise/fall times (10%–90%) of the inverter are:

$$T_r = T_f \approx 2.2 \cdot R \cdot C$$

This means that if a single inverter is looped back like that in Figure Q37a, shortly after the input starts transitioning from low to high (or from high to low), the output starts to change from high to low (or from low to high) and is fed back to the input. This is negative feedback; it weakens the transition of the input signal, and oscillation cannot be sustained.

From another angle, connecting the output of an inverter to its input forms negative feedback due to the inversion (180° phase shift). The question is, is this negative-feedback loop stable? From a small-signal point of view, a CMOS inverter is a common-source amplifier with a single pole at its output (inverters are used as amplifiers in ultra-low-cost electronic devices!). Its phase shift approaches 90° at high frequencies, and therefore the loop with negative feedback is guaranteed stable. So, it is not possible for the loop to oscillate due to the phase delay if there is only one inverter in the loop.

This analysis can be extended to a ring oscillator with three inverters in the loop (or, more generally, any ring oscillator with an *odd* number of inverters greater than one). Obviously, it is still connected as negative feedback, but now there are three amplifiers with nearly identical pole frequencies. With an overall phase shift approaching 270° and a −60 dB/decade roll-off rate, the loop is unstable. In the time domain, the propagation delay is now longer, separating the transition edges of the input and the inverted output and therefore avoiding their interactions (of cancelling each other).

Finally, what about a loop with two inverters? Now it is connected as positive feedback. This has the same structure as the latch of a comparator (see *Question 33, Comparator – regenerative latch*), and the behavior is that of a latch. The output would quickly lock to a stable state.

As a final remark, keep in mind that although, around the middle of the input range, an inverter behaves like an amplifier, for most of the voltage swing, it is a highly nonlinear circuit. For example, in a digital circuit, when the input of the inverter goes from ground to VDD, the NMOS transistor transitions from cutoff to saturation and then to the triode (linear) region, with drain current ramping up to peak and dropping back to zero.

Short Answer to Question 37

Would one inverter work? How about three inverters? And two?

One inverter would not work, as it has only a single pole and 90° of phase shift – the negative-feedback loop is stable. Three inverters would work because the negative-feedback loop with three identical single-pole amplifiers is unstable. Two inverters form a latch and will lock up instead of oscillating.

Question 38

Figure Q38a shows a CMOS inverter (M1/M2) with a fanout of 1, i.e., driving another inverter (M3/M4) of the same size. Assume NMOS and PMOS both use the minimum length, L_{min}, of the process technology. W_p and W_n are the widths of the PMOS and NMOS, respectively. Further assume that $K_{pn} = 3 \cdot K_{pp}$, where K_{pn} and K_{pp} are the transconductance parameters of NMOS and PMOS transistors, respectively. **(a)** How should we choose the ratio between W_n and W_p so that the propagation delays of the inverter M1/M2 is minimal? **(b)** How to choose the ratio between W_n and W_p so that the duty cycle distortion of the inverter M1/M2 is minimal?

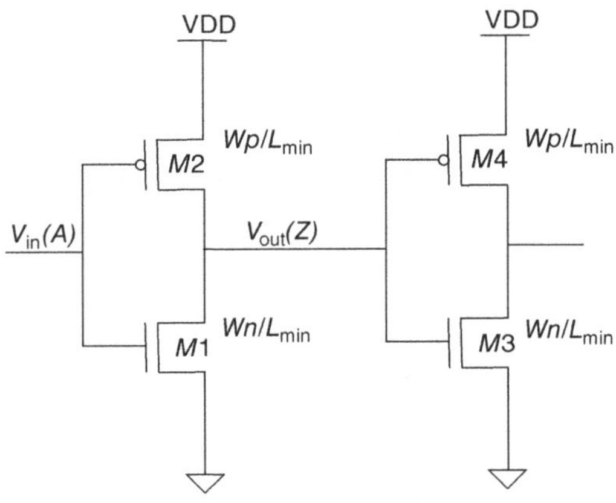

Figure Q38a

Intuitive and Quantitative Analysis: The propagation delays of an inverter depends on the transistor on-resistance and the load capacitance. In digital logic circuits, the load is typically the inputs of other inverters or logic gates. The propagation delay when an inverter output goes from low to high is (see *Appendix 4, Section 7, Basic digital circuits*):

$$T_{\mathrm{dLH}} \approx 0.7 \cdot R_p \cdot C_L$$

where R_p is the equivalent on-resistance of the PMOS, and C_L is the load capacitance.

Similarly, the propagation delay when an inverter output goes from high to low is:

$$T_{\mathrm{dHL}} \approx 0.7 \cdot R_n \cdot C_L$$

where R_n is the equivalent on-resistance of the NMOS.

The effective switching resistance of the NMOS and PMOS of an inverter is actually not easy to model, due to the highly nonlinear process during the low-to-high, or high-to-low transitions. One estimation method is to use the DC resistance of a transistor when both its V_{GS} (or V_{SG}) and V_{DS} (or V_{SD}) are equal to the VDD to represent the average resistance during the transition (see *Appendix 4, Section 7, Basic digital circuits*):

$$R_n = \frac{\mathrm{VDD}}{\frac{K_{pn} \cdot W_n}{2 \cdot L_{\min}} \cdot \left(\mathrm{VDD} - V_{\mathrm{thn}}\right)^2}$$

$$R_p = \frac{\mathrm{VDD}}{\frac{K_{pp} \cdot W_p}{2 \cdot L_{\min}} \cdot \left(\mathrm{VDD} - V_{\mathrm{thp}}\right)^2}$$

where K_{pn} and K_{pp} are transconductance parameters of the NMOS and PMOS transistors, respectively. V_{thn} and V_{thp} are the threshold voltage of NMOS and PMOS transistors, respectively.

The load capacitance is the sum of the output capacitance of the first-stage inverter and the input capacitance of the second-stage inverter, which are both roughly equal to the total gate capacitance of the NMOS and PMOS transistors for a quick estimation (see *Appendix 4, Section 7, Basic digital circuits*). That is:

$$C_L = 2 \cdot C'_{\mathrm{ox}} \cdot \left(W_p + W_n\right) \cdot L_{\min}$$

where W_p and W_n are the widths of the PMOS and NMOS transistors, respectively.

Note that this load capacitance is just a very rough approximation. Both the NMOS and PMOS transistors transition through different operating regions (cutoff, saturation, and triode) during the switching, and the parasitic capacitances (C_{gd}, C_{gs}, and C_{db}) vary substantially.

To minimize the propagation delays, we need to make the sum of the low-to-high and high-to-low propagation delays, T_p, a minimum value. Assuming $V_{\mathrm{thn}} = V_{\mathrm{thp}} = V_{\mathrm{th}}$, the following equation can be written:

$$T_p = 0.7 \cdot \left[\frac{\mathrm{VDD}}{\frac{K_{pn} \cdot W_n}{2 \cdot L_{\min}} \cdot (\mathrm{VDD} - V_{\mathrm{th}})^2} + \frac{\mathrm{VDD}}{\frac{K_{pp} \cdot W_p}{2 \cdot L_{\min}} \cdot (\mathrm{VDD} - V_{\mathrm{th}})^2} \right] \cdot 2 \cdot C'_{\mathrm{ox}} \cdot (W_p + W_n) \cdot L_{\min}$$

Let $W_p = n \cdot W_n$ and $K_{pp} = \alpha \cdot K_{pn}$. This yields:

$$T_p = \frac{2.8 \cdot C'_{\mathrm{ox}} \cdot L_{\min}^2 \cdot \mathrm{VDD}}{K_{pn} \cdot (\mathrm{VDD} - V_{\mathrm{th}})^2} \cdot \frac{(n+1) \cdot (1 + \alpha \cdot n)}{\alpha \cdot n}$$

The minimum total propagation delay, T_p, occurs at:

$$\frac{dT_p(n)}{dn} = \frac{2.8 \cdot C'_{\mathrm{ox}} \cdot L_{\min}^2 \cdot \mathrm{VDD}}{K_{pn} \cdot (\mathrm{VDD} - V_{\mathrm{th}})^2} \cdot \frac{\alpha \cdot n^2 - 1}{\alpha \cdot n^2} = 0$$

This yields:

$$n = \sqrt{\frac{1}{\alpha}}$$

As $\alpha = 1/3 (K_{pn} = 3 \cdot K_{pp})$, $n \approx 1.7$.

So size $W_p = 1.7 \cdot W_n$ for minimizing the total propagation delay. Note that this ratio is smaller than the ratio for the *β-matched* NMOS and PMOS transistors, or $W_p = 3 \cdot W_n$, at which the device transconductance parameters β (defined as $K_p \cdot W/L$) of NMOS and PMOS transistors match each other. Intuitively, this is because NMOS transistors offer large transconductance with small transistor size and therefore small time constant, and the ratio shifts towards NMOS transistors from the balance point (*β-matched* point) in favor of speed.

Note that this optimization equation is coarse as some elements used in the estimation, including on-resistance and gate capacitance, are far from accurate, and the threshold voltages of NMOS and PMOS transistors are not precisely matched.

To minimize duty cycle distortions (DCDs), we require that the rise time and fall time be equal. Unmatched rise and fall times result in DCDs, as illustrated in Figure Q38b.

This means the on-resistance of the NMOS and PMOS should match each other. That is:

$$\frac{\mathrm{VDD}}{\frac{K_{pn} \cdot W_n}{2 \cdot L_{\min}} \cdot (\mathrm{VDD} - V_{\mathrm{th}})^2} = \frac{\mathrm{VDD}}{\frac{K_{pp} \cdot W_p}{2 \cdot L_{\min}} \cdot (\mathrm{VDD} - V_{\mathrm{th}})^2}$$

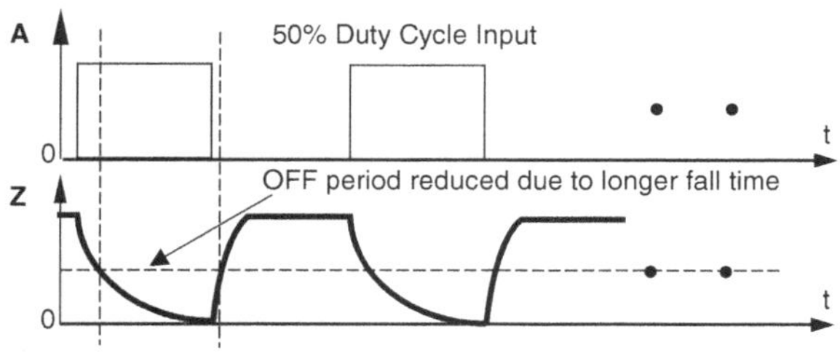

Figure Q38b

Or simply:

$$\frac{W_p}{W_n} = \frac{K_{pn}}{K_{pp}} = 3$$

Which is the β-matched condition.

It is interesting to know that the β-matched NMOS and PMOS transistors in an inverter not only make the rise and fall times matched to each other, but also keep the switching point, V_{sp}, in the middle of the inverter voltage swing (VDD/2). The switching point is defined as the input voltage at which the output equals the input. Around this switching point, both NMOS and PMOS transistors work in the saturation region and the inverter behaves like an amplifier.

In practice, inverters and other logic gates are often grouped into two categories based on their design goals. Sizing optimization for standard logic gates typically focuses on achieving minimal propagation delays, whereas optimization for clock buffers/drivers prioritizes minimizing DCDs.

Short Answer to Question 38

(a) How should we choose the ratio between W_n and W_p so that the propagation delays of the inverter M1/M2 is minimal? **(b)** How to choose the ratio between W_n and W_p so that the duty cycle distortion (DCD) of the inverter M1/M2 is minimal?

(a) For minimal propagation delays, the transistor width ratio (n) between the PMOS and NMOS transistors should be:

$$n = \frac{W_p}{W_n} = \sqrt{\frac{1}{\alpha}}$$

where α is the transconductance parameter ratio between PMOS and NMOS $(\alpha = K_{pp}/K_{pn})$.

As $\alpha = 1/3(K_{pn} = 3 \cdot K_{pp})$, $n = 1.7$, or $W_p = 1.7 \cdot W_n$.

(b) For minimal DCD, the transistor width ratio (n) between PMOS and NMOS transistors should be:

$$n = \frac{1}{\alpha}$$

As $\alpha = 1/3(K_{pn} = 3 \cdot K_{pp})$, $n = 3$, or $W_p = 3 \cdot W_n$.

Question 39

A CMOS buffer, comprising two cascaded inverters, drives a long interconnect (metal line) before reaching its terminal load (Figure Q39a). The (lump-sum) parasitic capacitance and resistance of the routing are 250 fF and 300 Ω, respectively. The load capacitance is 100 fF. The system requires that both the rise/fall times of V_{out} and the propagation delays from V_{in} to V_{out} not exceed 100 ps. Minimum DCD is desired. Design the W/L for transistors of each inverter to achieve this target. Assume the input and output parasitic capacitances of the inverters can be ignored, and the effective on-resistances of NMOS and PMOS transistors are:

$$R_n = 5000 \cdot \frac{L}{W}\ (\Omega);\ R_p = 10000 \cdot \frac{L}{W}\ (\Omega)$$

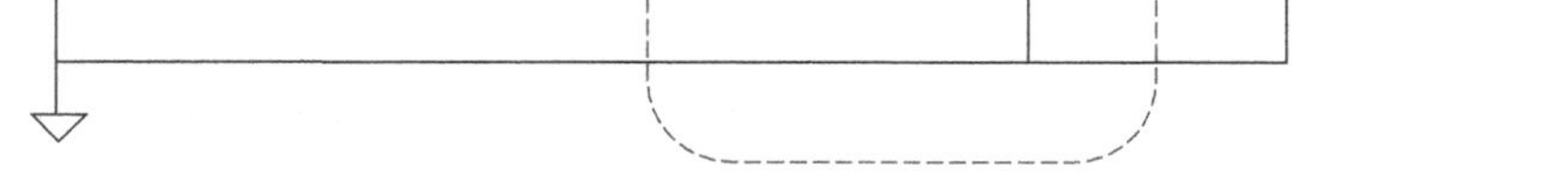

Figure Q39a

Intuitive and Quantitative Analysis: First of all, for long routing lines, we need to check their time constant first. With R_L of 300 Ω and C_L of 250 fF, the time constant of the routing itself is:

$$\tau_{\text{routing}} = R_L \cdot C_L = 300 \cdot 250f = 75\ \text{ps}$$

The propagation delay and rise/fall times just considering the routing are:

$$T_{d:\text{routing}} = 0.7 \cdot 75\ \text{pS} = 52.5\ \text{ps}$$

$$T_{rf:\text{routing}} = 2.2 \cdot 75\ \text{pS} = 165\ \text{ps}$$

The rise/fall times by the routing itself (165 ps) are longer than the design target (100 ps), meaning that even if the buffer has zero output resistance and load capacitance is zero, the target cannot be met.

Because the routing parasitic dominates the time constant, we need to break it up for faster transitions. If we break the routing line at the middle of the line, then the time constant of each routing section would drop to a quarter (25%) of the original value. This can help achieve faster rise/fall times at the load and, at the same time, relax the propagation delay constraints. Apparently, the on-resistances of the inverters must be accounted for, as they contribute to the cumulative time constant of the circuit. We need to perform some calculations to verify whether this would work.

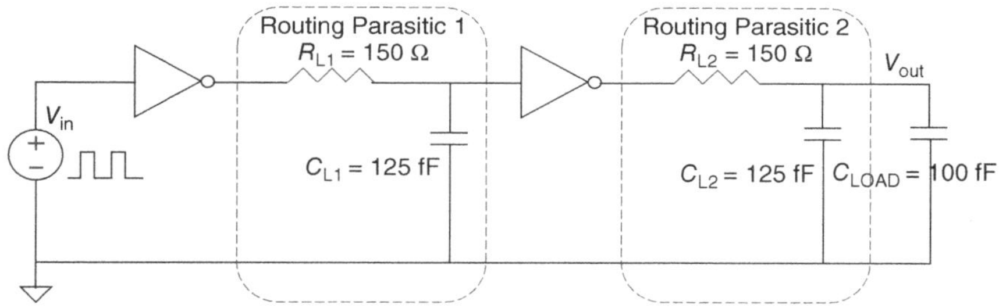

Figure Q39b

Figure Q39b shows the new arrangement: each inverter now drives one half of the routing line. The time constant of each section of the routing alone is reduced to:

$$\tau_{\text{routing1}} = R_{L1} \cdot C_{L1} = R_{L2} \cdot C_{L2} = 150 \cdot 125f = 18.75\ \text{ps}$$

To keep duty cycle distortion low, we need to make the on-resistances of the NMOS and PMOS transistors equal: $R_{n1} = R_{p1}$. The propagation delay of the first driver stage is:

$$T_{p:\text{stage1}} = 0.7 \cdot (R_{n1} + R_{L1}) \cdot C_{L1}$$

If we assign half of the target propagation delay (50 ps) to this stage, we can calculate the allowed on-resistance of the inverter:

$$R_{n1} = R_{p1} \leq 420\ \Omega$$

The corresponding W/L ratios of NMOS and PMOS transistors are:

$$\frac{W_{n1}}{L_{n1}} \geq \frac{5000}{420} \approx 12; \frac{W_{p1}}{L_{p1}} \geq \frac{10000}{420} \approx 24$$

The rise/fall times of this stage are:

$$T_{rf:\text{stage1}} = T_{r1} = T_{f1} = 2.2 \cdot (R_{n1} + R_{L1}) \cdot C_{L1} \approx 157 \text{ ps}$$

Note that these are the rise/fall times at the output of the first stage, NOT the rise/fall times at the load which must be less than 100 ps.

The rise/fall times of the second driver stage are (again making the switching resistance of NMOS and PMOS equal to minimize duty cycle distortion):

$$T_{rf:\text{stage2}} = 2.2 \cdot (R_{L2} + R_{n2}) \cdot (C_{L2} + C_{\text{LOAD}})$$

With the upper limit of rise/fall set to 100 ps, this yields:

$$R_{n2} = R_{p2} = 52\ \Omega$$

The corresponding W/L ratios of NMOS and PMOS are:

$$\frac{W_{n2}}{L_{n2}} \geq \frac{5000}{52} \approx 96; \frac{W_{p2}}{L_{p2}} \geq \frac{10000}{52} \approx 192$$

Checking the propagation delay of the second stage yields:

$$T_{p:\text{stage2}} = 0.7 \cdot (R_{L2} + R_{n2}) \cdot (C_{L2} + C_{\text{LOAD}}) \approx 31.8 \text{ ps}$$

The total propagation delay is:

$$T_{p:\text{total}} = 50 \text{ ps} + 31.8 \text{ ps} = 81.8 \text{ ps} < 100 \text{ ps}$$

And the rise/fall times of the V_{out} (second stage) are:

$$T_{rf:\text{stage2}} = 99.9 \text{ ps} < 100 \text{ ps}$$

As we have some margin on the propagation delay, we may be able to move the split point farther toward the load while still meeting the propagation delay target and reducing the size (and therefore the switching current consumption) of the second-stage inverter. On the other hand, as the parasitic input and output capacitances of the inverters are ignored in the above analysis, we run the risk of underestimating the time constant (especially for large size inverters and buffers). By accurately modeling the drivers, interconnects, and loads, a mathematically optimized driving strategy can be derived.

We have used the lump-sum resistance and capacitance for calculating the time constant of the routing line; this could result in errors due to the distributed RC effect. Calculations based on the distributed RC line model can be taken for more accuracy if needed. Also, in high-speed circuits, if the signal wavelength is shorter than the length of the physical routing lines, *reflection* might happen and we need to treat the routing lines as *transmission lines*. These topics are beyond the scope of this book but interested readers are encouraged to explore more, especially for readers who work in high-speed circuit design and applications.

Short Answer to Question 39

Design the *W*/*L* for transistors of each inverter to achieve the target.

Because the time constant of the metal routing alone exceeds the rise/fall-time requirements, the interconnect must be partitioned into multiple segments driven by sequential stages. One potential configuration involves partitioning the interconnect into two segments of equal length. To optimize performance, the two inverter stages are sized as follows: a first stage with PMOS = 24/1 and NMOS = 12/1, and a second stage with PMOS = 192/1 and NMOS = 96/1 (where transistor lengths are normalized to unity).

Question 40

A CMOS inverter drives a load capacitance C_L as shown in Figure Q40a. The input is a clock signal with rise time T_r, fall time T_f and period T. **(a)** If the input clock frequency goes up, what happens to the average power consumption of the inverter? **(b)** What happens to the power consumption if the input rise/fall times increase?

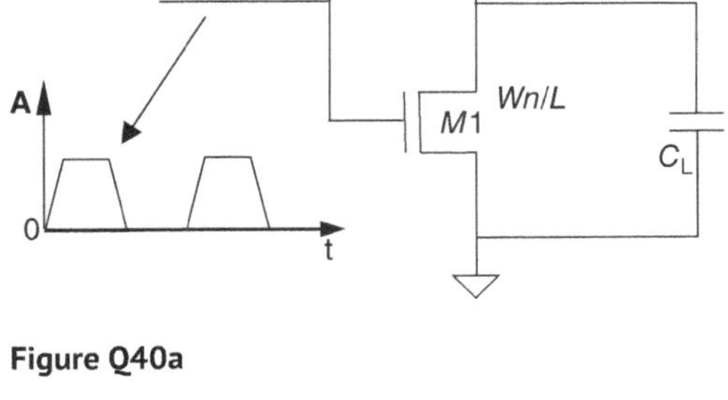

Figure Q40a

Intuitive and Quantitative Analysis: To estimate the power consumption of an inverter, we need to understand how current flows when the inverter is switching. Ideally, an inverter can be modeled as a pull-up switch (the PMOS transistor) and a pull-down switch (the NMOS transistor), as shown in Figure Q40b and Figure Q40c.

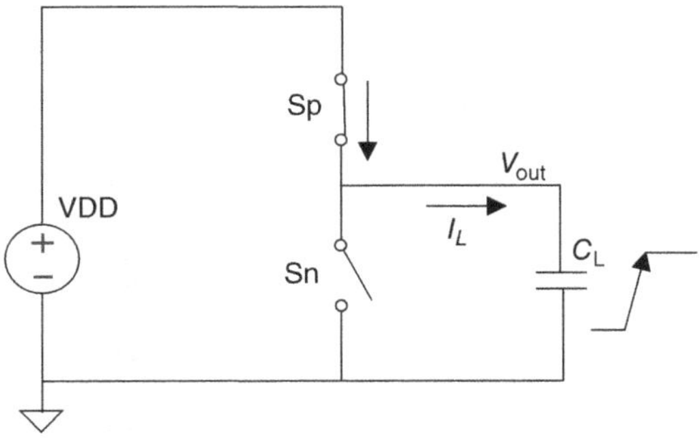

Figure Q40b

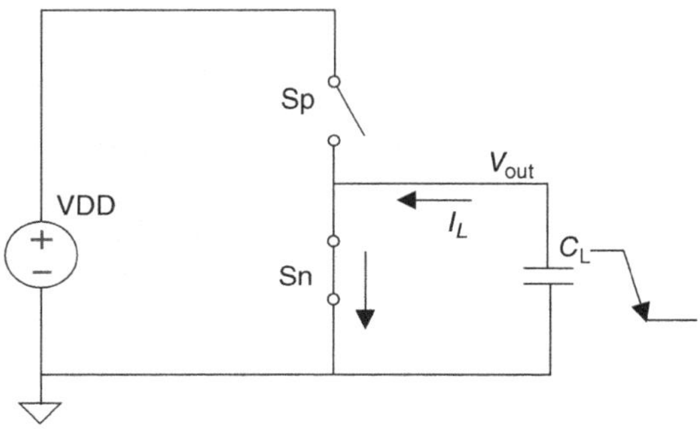

Figure Q40c

When the input changes from high to low (VDD to ground), the pull-up switch Sp (PMOS) is closed, and pull-down switch Sn (NMOS) is open, as shown in Figure Q40b. The load capacitor is charged up towards VDD with the current (or charge) supplied by the power supply. Charging completes when V_{out} reaches VDD, and the current becomes zero.

The energy transferred from the power supply during this process is:

$$E_{\text{pull-up}} = \text{VDD} \cdot Q = \text{VDD} \cdot C_L \cdot \text{VDD} = \text{VDD}^2 \cdot C_L$$

where Q is the total charge transferred to the load capacitor C_L. Keep in mind that the change in potential energy (ΔE) equals the change in potential (VDD) multiplied by the total charge involved (Q).

When the input changes from low to high (ground to VDD), the pull-down switch Sn (NMOS) is closed, and pull-up switch Sp (PMOS) is open, as shown in Figure Q40c. The load capacitor is discharged towards ground. The discharging completes when V_{out} reaches ground. Note that the discharging current (or charge) is supplied by the load capacitor, and no current is drawn from the power supply during the pull-down process.

The average power consumption of the inverter for a complete clock cycle due to switching is:

$$P_s = \frac{E_{\text{pullup}}}{T} = \frac{\text{VDD}^2 \cdot C_L}{T} = \text{VDD}^2 \cdot C_L \cdot f$$

where T and f are the clock period and clock frequency, respectively.

The power consumption of the inverter, therefore, goes up with increasing clock frequency. The power consumption also goes up with larger load capacitance and higher VDD. As a matter of fact, this equation is widely used in quick power consumption estimation of synchronous digital circuits in general:

$$P = \text{VDD}^2 \cdot C \cdot f \cdot \alpha$$

where α is the switching density of the digital circuit under estimation (not all gates in a digital circuit switching at the same clock edge). The equation highlights why scaling to smaller process nodes significantly reduces power consumption; the concurrent reduction in supply voltage (VDD) and parasitic capacitance (C) enables large-scale integrated systems to operate with greater power efficiency.

The question in part (b) may appear contradictory, as the above analysis suggests that an inverter's power consumption is independent of input rise and fall times. This conclusion, however, stems from the idealized pull-up/pull-down switch model, which assumes instantaneous input transitions and immediate transistor switching. In practice, there is a finite interval during both low-to-high and high-to-low transitions where both the NMOS and PMOS transistors are simultaneously conductive. As illustrated in Figure Q40d, this overlap creates a direct current path from VDD to ground, resulting in short-circuit power dissipation.

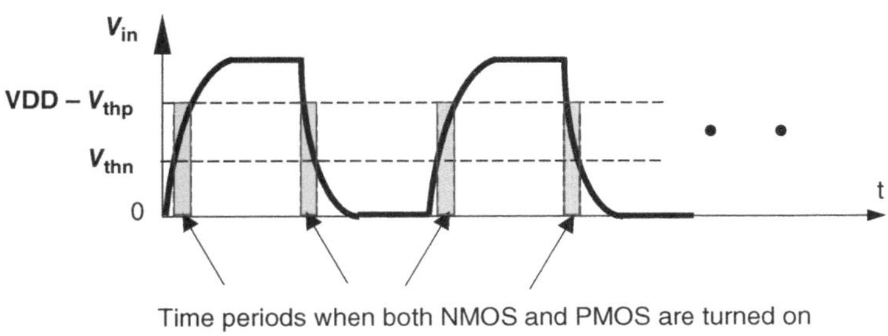

Figure Q40d

Both the NMOS and PMOS transistors are turned on when the input voltage falls into the range of $(V_{\text{thn}}, \text{VDD} - V_{\text{thp}})$. In this time period, the inverter behaves like an amplifier and there is a direct conduction path via the two transistors from VDD to ground. This short-circuit current is distinct from the charging and discharging currents associated with the load capacitance. Notably, this current draw occurs during both low-to-high and high-to-low transitions, as illustrated by the current flow diagrams in Figures Q40e and Q40f.

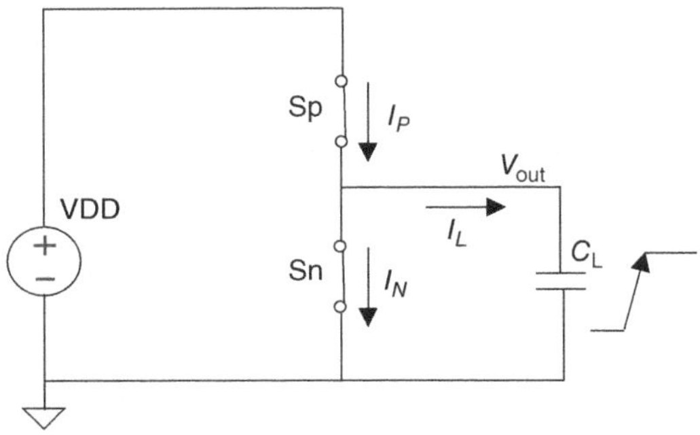

Figure Q40e

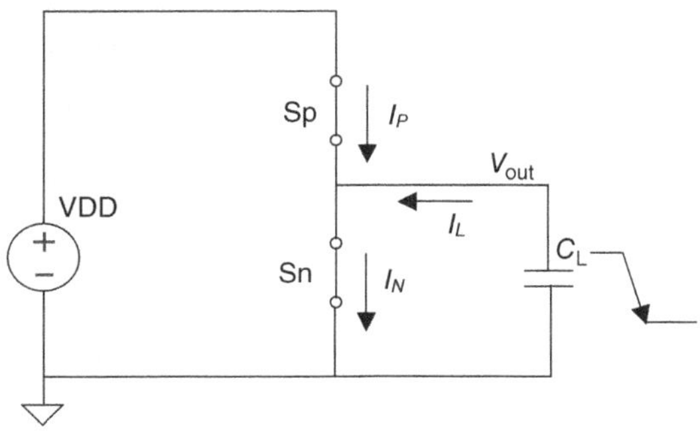

Figure Q40f

The magnitude of the current when both NMOS and PMOS are turned on depends on the transconductance of both transistors, the input voltage, and the output voltage, and varies during the transition. The energy transferred from the power supply through the short-circuit path (E_{on}) is:

$$E_{\text{on}} = \text{VDD} \cdot Q_{\text{on}} = \text{VDD} \cdot \int_{V_{\text{out:L}}}^{V_{\text{out:H}}} I_N(t) \cdot dt + \text{VDD} \cdot \int_{\text{Vout:H}}^{\text{Vout:L}} I_P(t) \cdot dt$$

And the average power consumption (P_{on}) is:

$$P_{\text{on}} = \frac{E_{\text{on}}}{T} = \frac{\text{VDD}}{T}\left[\int_{V_{\text{out:L}}}^{V_{\text{out:H}}} I_N(t)\cdot dt + \int_{V_{\text{out:H}}}^{V_{\text{out:L}}} I_P(t)\cdot dt\right]$$

where Q_{on} is the total charge transferred directly from VDD to ground when both transistors are on. I_N and I_P are the current of NMOS and PMOS, respectively. Notice that we use I_N for output low-to-high transition and I_P for output high-to-low transition, to avoid double counting the charging/discharging current on which the power consumption has been estimated previously.

When the input rise/fall times increase, the time interval during which both transistors are on becomes longer, more current is drawn from the power supply, and power consumption increases. Thus, slower transitions do not imply lower power consumption for inverters (and other logic gates).

The total power consumption of the inverter therefore is:

$$P_{\text{total}} = P_S + P_{\text{on}} = \text{VDD}^2\cdot C_L\cdot f + \frac{\text{VDD}}{T}\left[\int_{V_{\text{out:L}}}^{V_{\text{out:H}}} I_N(t)\cdot dt + \int_{V_{\text{out:H}}}^{V_{\text{out:L}}} I_P(t)\cdot dt\right]$$

This equation also applies to synchronous digital circuits. In a digital system with large internal parasitic capacitance (e.g., large layout area and long routing lines), the power consumption associated with capacitance charging, P_S, may dominate. Yet power consumption due to short-circuit conduction, P_{on}, might dominate in small-scale digital circuits with relatively large transistor sizes.

Short Answer to Question 40

(a) If the input clock frequency goes up, what happens to the average power consumption of the inverter? **(b)** What happens to the power consumption if the input rise/fall times increase?

(a) The average power consumption of the inverter for a complete clock cycle due to switching is:

$$P_S = \frac{E_{\text{pullup}}}{T} = \frac{\text{VDD}^2\cdot C_L}{T} = \text{VDD}^2\cdot C_L\cdot f$$

The power consumption of the inverter, therefore, goes up as clock frequency increases.

(b) When the input rise/fall times increase, the time interval during which both transistors are on becomes longer, more current is drawn from the power supply, and power consumption increases.

Question 41

The schematic of an edge-triggered D flip-flop (DFF) is shown in Figure Q41a. clk and $\overline{\text{clk}}$ are complementary clock signals. Explain what determines the setup time and hold time in this DFF circuit. Would slowing the clock help with setup-time violations? Would slowing the clock help with hold-time violations?

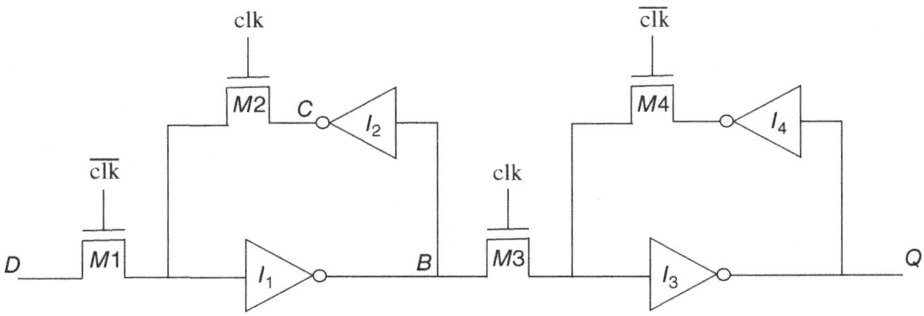

Figure Q41a

Intuitive Analysis: To explain what determines the setup and hold times, we need to understand how an edge-triggered DFF works. Figure Q41b shows the timing diagram of the key signals of the circuit.

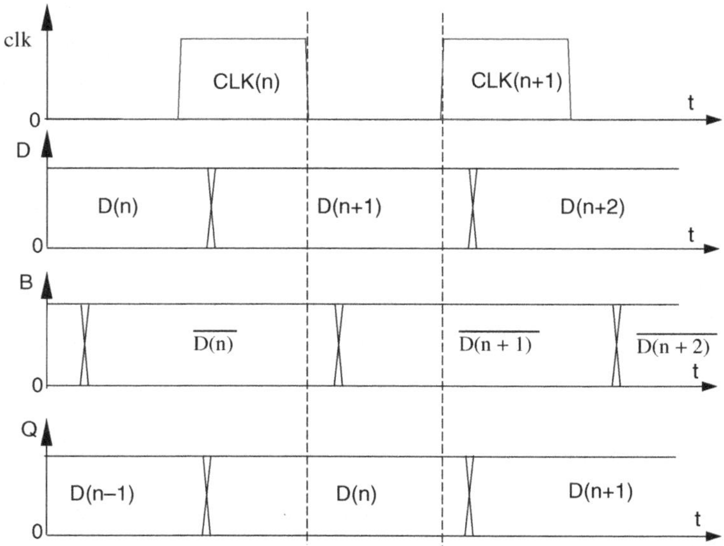

Figure Q41b

Let us start with the falling edge of clk(n). After this falling edge, the input data $D(n+1)$ is connected to inverter I_1 (the output of inverter I_2 is disconnected), and node B is driven to $\overline{D}(n+1)$ after the inverter's propagation delay. In the meantime, inverters I_3 and I_4 form a latch, disconnected from the output of I_1, and maintain the previous data $D(n)$.

At the rising edge the next clock cycle, clk$(n+1)$, I_1 and I_2 form a latch, are disconnected from the input D, and the output, $\overline{D}(n+1)$, is latched at node B. At the same time, I_4 is disconnected and I_3 is connected to node B, driving the output of the DFF to the value $D(n+1)$. The data will be latched at the falling edge of clk $(n+1)$ and remains latched until the rising edge of the subsequent clk$(n+2)$.

The edge-triggered DFF circuit in Figure Q41a is often referred to as a *master-slave* DFF. It consists of two level-sensitive latches: the master latch (I_1 and I_2) and the slave latch (I_3 and I_4).

The setup time of a DFF (T_S) is defined as the minimum duration the data input D must be ready and remain stable before the clock's rising edge to ensure the value is reliably captured by the master latch. In the timing diagram in Figure Q41b, the input data is ready well before the rising edge of the clock. How close can we push this to the point when the clock signal is applied?

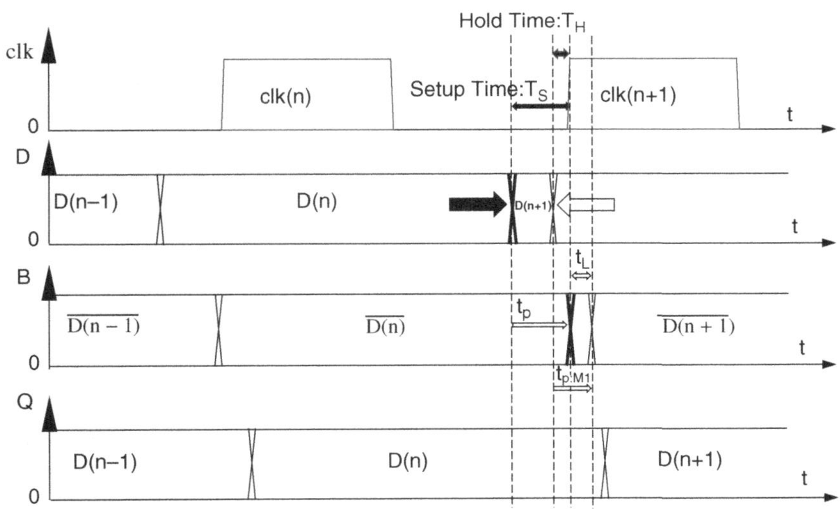

Figure Q41c

As shown in Figure Q41c, the critical timing path for the setup time is the propagation delay, t_p, from the input (D) to the output of I_2 (node C). This includes the propagation delay of the passing transistor M1, inverter I_1, and inverter I_2. The input signal must be ready at least this amount of time before the rising edge of the clock to be latched into the master latch reliably. Therefore, the setup-time requirement is:

$$T_S \geq t_p = t_{p:I1} + t_{p:I2} + t_{p:M1}$$

where $t_{p:M1}$, $t_{p:I1}$, and $t_{p:I2}$ are the propagation delays of the passing gate M1, inverters I_1, and I_2, respectively.

The hold time T_H is defined as the time interval for which the input data must be kept unchanged (valid) after the rising edge of the clock signal (clk). This requirement is due to the fact that latches usually need a time, t_L, to resolve to the stable state after the clock is applied, during which a change at the input could disrupt or adversely affect the outcome. Because of the positive-feedback mechanism, latches settle quickly and t_L is usually small. In fact, at the rising edge of the clock, the passing transistor M1 is turning off and disconnecting the master latch from the input. This, however, cannot happen instantaneously, and it is safe to hold the input signal steady until the completion of the latch-up process. From Figure Q41c, we conclude that the hold time for the DFF in Figure Q41a is:

$$T_H \geq t_L - t_{p:M1}$$

The hold time T_H can be negative when $t_{p:M1}$ is longer than t_L.

Note that the circuit diagram in Figure Q41a is a simplified version of *D* flip-flops for illustration purposes. In practical implementations, internal clock buffers are often employed. These buffers cause a delay in the actual rising edge of the latch clock, which results in a positive hold time. Furthermore, transmission gates are often used as switches instead of passing transistors to ensure full rail-to-rail voltage levels for both logic low and high. Due to inverter delays, the clock signals clk and $\overline{\text{clk}}$ are often not perfectly complementary. Some *D* flip-flops use physically weaker inverters in the feedback paths of the latches (e.g., typically configured with a longer channel length for the transistors) to prevent race conditions during the latching process. The actual setup and hold times must be determined by analyzing the real circuit topology.

Would slowing the clock help with setup-time violations? Yes. A slower clock, or longer clock cycle, means more time margin for the input at *D* to be ready before the clock is applied. A slower clock, however, would not help with hold-time violations as it also means the input data has to hold longer. Because of this, using a slower clock to test whether the timing related errors of a digital circuit are reduced or eliminated is a good indication of whether the problems are caused by the setup-time violations.

Short Answer to Question 41

Explain what determines the setup time and hold time in this DFF circuit. Would slowing the clock help with setup-time violations? Would slowing the clock help with hold-time violations?

The propagation delay, t_p, from the input (D) to the output of I_2 (node C) determines the setup time (T_S). This includes the propagation delay of the passing transistor M1 $(t_{p:M1})$, inverter I_1 $(t_{p:I1})$, and inverter I_2 $(t_{p:I2})$. The setup-time requirement is:

$$T_S \geq t_p = t_{p:I1} + t_{p:I2} + t_{p:M1}$$

The hold time T_H is determined by the propagation delay of the passing transistor M1 $(t_{p:M1})$ and the resolution time of the master latch (t_L). The requirement is:

$$T_H \geq t_L - t_{p:M1}$$

Slowing the clock frequency would alleviate setup-time violations, but it would not help the hold-time violations.

Question 42

A rising-edge-triggered DFF is connected in a feedback loop as shown in Figure Q42a. After the power supply VDD ramps up to its stable level, the clock signal, CLK1, starts to toggle. Please draw the waveform of the DFF output CLK2.

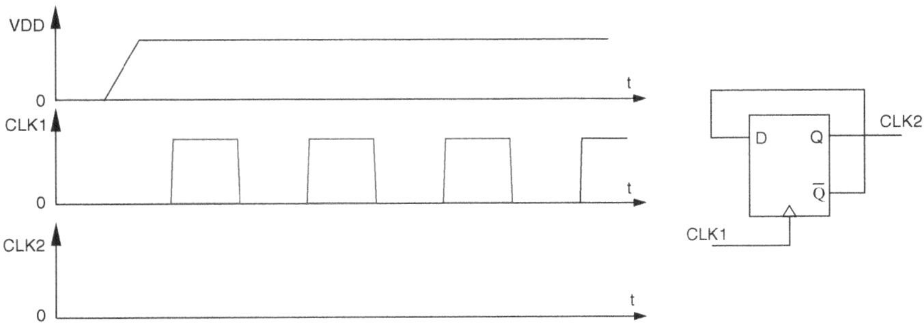

Figure Q42a

Intuitive Analysis: The circuit in Figure Q42a is well known as a divide-by-two clock divider, and the analysis of its output and input relationship is straightforward, except one thing: the initial state of the DFF. One perspective assumes that after power-up, the initial voltages of the DFF outputs (Q) and its complementary $(\overline{Q})$ are both zero. Because $\overline{Q}$ is connected to the input D, when the first rising edge of the CLK1 comes, the output Q is latched at logic low. The output changes to logic high at the next rising edge of the CLK1, as the $\overline{Q}$ is logic high after the previous rising edge of the input clock. The output Q, or CLK2, therefore is a clock signal with half the frequency of the input clock (CLK1) and a 180° phase lag (in terms of the divided-down frequency), as shown in Figure Q42b.

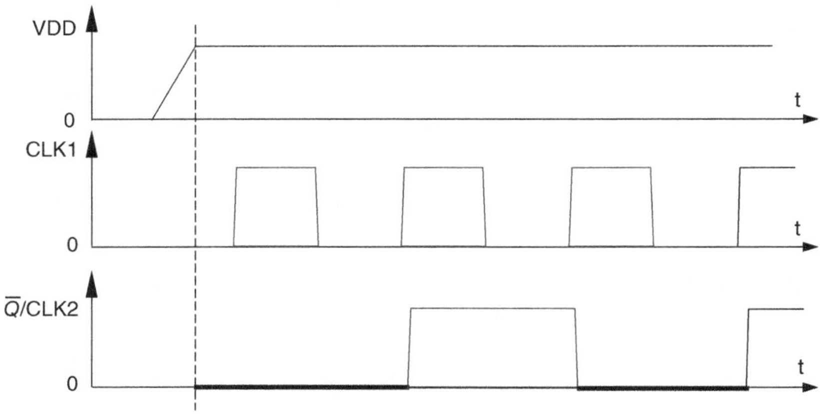

Figure Q42b

Alternatively, it can be argued that, after power-up, if the initial voltage of Q is zero, then $\overline{Q}$ must be logic high as the latch would force Q and $\overline{Q}$ to be complementary to each other. As such, the output signal, CLK2, is a divided clock with half the frequency of CLK1 and zero phase lag, as illustrated in Figure Q42c.

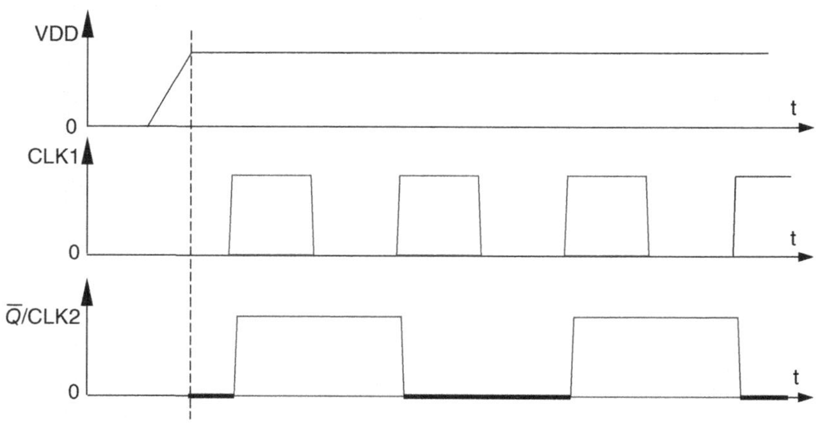

Figure Q42c

Which output waveform is correct? Unfortunately, neither is entirely correct. However, the second argument has a valid and important point: after power-up, a latch will resolve into a definite state due to its positive feedback. Let us take a closer look at what would happen when the power supply voltage is stable after the initial ramp up. According to the time sequence in Figure Q42a, the input clock (CLK1) remains low for some time after the power reaches the target value. This means the first stage of the DFF (see *Question 41, Master/slave D flip-flops and setup/hold time*) works as an inverter, and the second stage forms a latch (slave latch) with its input floating during this period of time. The positive feedback of the slave latch would force its outputs to settle to a definite state (i.e., Q = High or Q = Low) quickly once adequate supply voltage is reached. Which one of the two possible states it would resolve into depends on how imbalances build up in the positive-feedback loop of the slave latch and is not of a deterministic nature. The DFF might end up with latching into either one of the two states after the power-up.

This means the divide-by-two clock divider in Figure Q42a has *two* possible output phases – the timing sequences in Figure Q42b and Figure Q42c are both probable. While this might not be a problem in some systems and applications, it is impermissible when a fixed output phase is required. To fix this problem, a DFF with asynchronous or synchronous *set* (i.e., set Q = High) or *reset* (set Q = Low, sometimes called clear) could be used so that the output of the DFF is set/reset to a known state after power-up but before the operation starts. This is also the key reason that a reset is usually asserted before the operation of digital systems to ensure all state machines start from a known state. Interested readers are encouraged to explore how to add set/reset functions to the master-slave DFF in Question 41.

Short Answer to Question 42

Please draw the waveform of the output of the DFF, CLK2.

The divide-by-two clock divider in Figure Q42a has two possible output phases; please refer to the timing sequences in Figure Q42b and Figure Q42c.

Question 43

Figure Q43a shows a DFF-based phase-frequency detector (PFD), along with the subsequent charge pump, loop filter, and voltage-controlled oscillator (VCO), together used as the forward path in a phase-locked loop (PLL). Assume that the phase of the input clock and feedback clock are ϕ_{in} and ϕ_{fb}, respectively. Also assume that the DFFs are triggered by the rising edge of the clock. **(a)** Sketch the relationship between the change in the charge pump output voltage ΔV_{CP} and the phase difference between the input clock and the feedback clock, $\Delta\phi = \phi_{in} - \phi_{fb}$. Ignore the resistor R of the loop filter. **(b)** What limits the smallest phase difference the PFD can effectively detect? **(c)** What is the function of the resistor R in the loop filter?

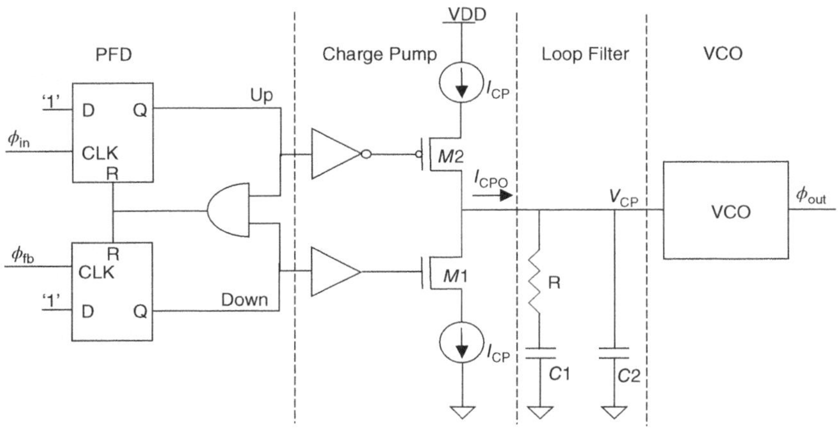

Figure Q43a

Intuitive Analysis: A good way to understand the function of the PFD is to analyze its outputs at a few different input phase relationships.

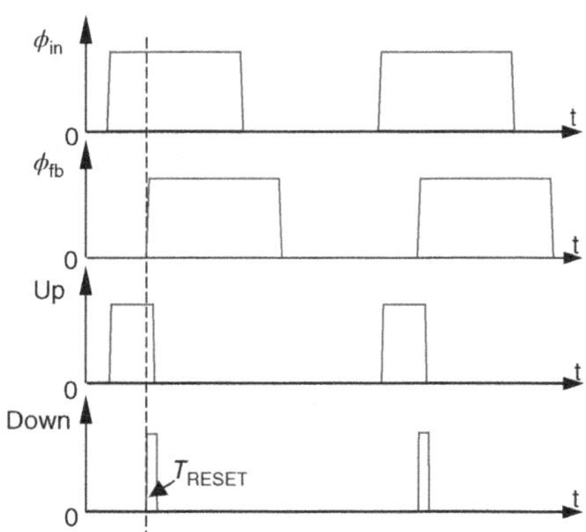

Figure Q43b

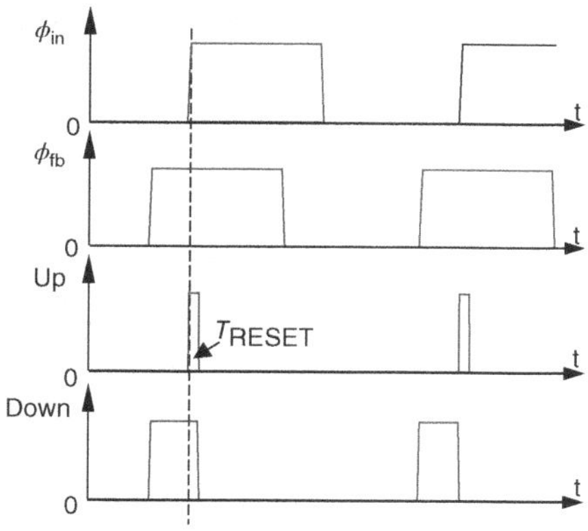

Figure Q43c

Figure Q43b illustrates the timing diagram when the phase of the input clock ϕ_{in} leads that of the feedback clock ϕ_{fb}. After the two DFFs in the PFD reset, the first (or leading) clock rising edge, in this case the rising edge of the input clock, triggers the corresponding DFF and sets the signal "Up" to high. The "Up" stays at high until the rising edge of the lagging clock, i.e., the feedback clock, sets the signal "Down" to high and subsequently resets (clears) the two DFF outputs to low. Similarly, the signal "Down" is set to high for the duration which the feedback clock leads the input clock, as seen in Figure Q43c. As both DFFs are triggered by only the rising edge of the clock, the above conclusion is valid for the entire period of the clocks.

The propagation delay of the PFD's reset logic, T_{RESET}, is usually very short so that the DFF output pulse of the lagging clock is rather narrow. As it takes some time for the gates of the CMOS switches to accumulate sufficient amount of charge, or gate voltage, to turn on and conduct current, the narrow pulses at the output of the lagging DFF might not be able to produce any current flowing to the filter capacitor.

Mathematically, the pulse width of the Up/Down signals for the leading clock is directly proportional to the phase difference of the two clocks. In the charge pump, a constant current, I_{CP}, charges or discharges the filter capacitor for the duration of the corresponding pulse width. The relationship between the change of the charge pump output voltage, ΔV_{CP}, and the phase difference between the input clock and the feedback clock, $\Delta\phi = \phi_{in} - \phi_{fb}$, can be written as:

$$\Delta V_{CP} = \frac{I_{CP} \cdot \Delta\phi \cdot T_{clock}}{2 \cdot \pi \cdot (C_1 + C_2)}$$

where T_{clock} is the clock period which is measured by the time between the leading edges of the clocks. Note that the phase difference, $\Delta\phi$, has the unit of rad. Also note that we ignore the loop filter resistor R, and the charge pump current I_{CPO} directly charges or discharges a capacitance of $C_1 + C_2$.

The relationship between the ΔV_{CP} and $\Delta\phi$ is drawn in Figure Q43d.

Because the output of the PFD depends on the timing of the rising edges of the clocks, the PFD is also sensitive to the clock frequencies. PLLs based on a PFD are not sensitive to input clock duty cycle and will not lock onto harmonics of the input clock.

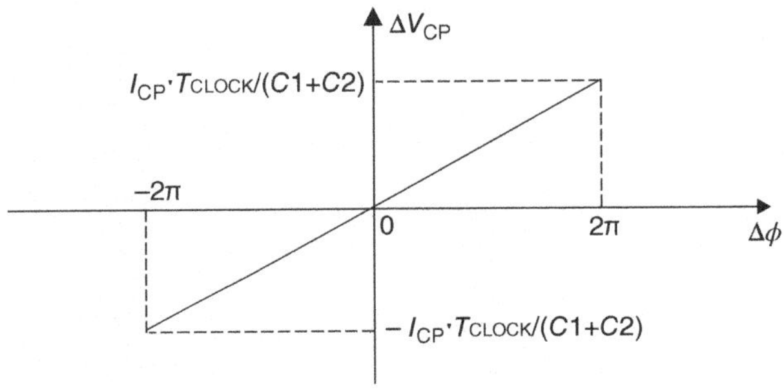

Figure Q43d

What is the smallest phase difference that a PFD can effectively detect? When the phase difference between the input clock and feedback clock becomes smaller, the corresponding Up/Down pulses become narrower. Eventually, the duration of the pulse is less than the time required for the gate voltage of the NMOS and PMOS switches to reach the level needed for them to turn on and conduct current. From this point, the output current of the charge pump becomes zero and no longer responds to changes in phase (in real circuits there will be a region of reduced (softened) gain before the gain drops to zero). This introduces a "dead zone" to the transfer function of the PFD/charge pump (Figure Q43e). The dead zone is a type of nonlinearity and can degrade system performance. In a PLL with phase detector dead zone, clock jitter increases and the loop might experience instability problems in some applications.

Question in (c) asks about the function of the resistor R in the loop filter. We just analyze the circuit without R – the charge pump charges the two capacitors, C_1 and C_2, directly, which forms an *integrator*. Connecting R in series with C_1 introduces a zero, which brings in a phase lead to the system. Why does adding a zero help? To answer this question, we need to consider the closed-loop operation of the PLL.

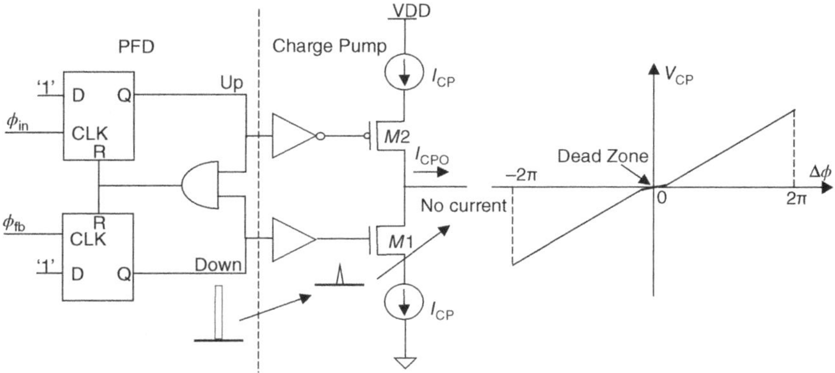

Figure Q43e

Figure Q43f illustrates the complete control loops of the PLL. The clock generated by the VCO is fed back to the PFD, forming a negative-feedback loop for both clock phase and frequency. In some applications a clock divider (divided by *N*) is used to divide down the VCO clock before it is fed back to the PFD so that a clock with higher frequency than the reference frequency can be generated (*N* times higher).

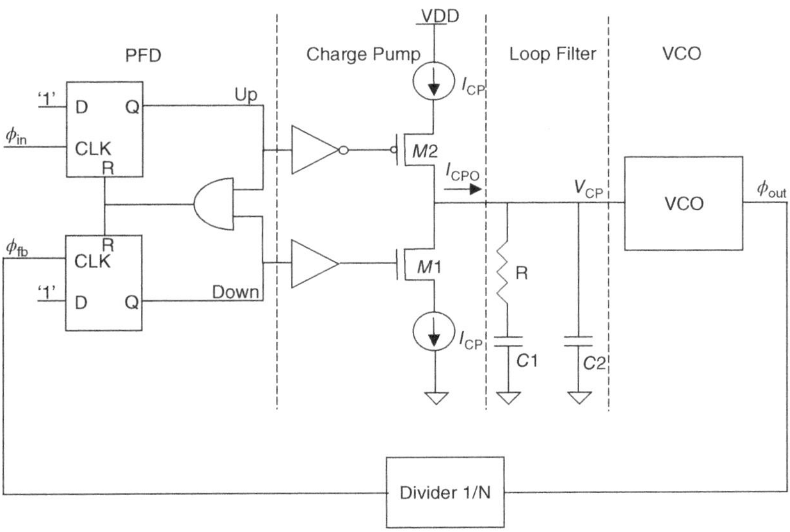

Figure Q43f

Notice that the VCO's output frequency is proportional to its input voltage, which is the output voltage (V_{CP}) of the loop filter. As phase is the variable for the transfer function of the system, we need to know the relationship between the phase and frequency of the VCO's output clock. Mathematically phase is the time integral of frequency:

$$\phi_{out} = \int \omega_{out} \cdot dt$$

Now there are two integrators in the loop: the loop filter and the VCO. Integrators have large DC gain but they cannot respond to fast input change. Having two integrators together worsens the situation – any changes at the phase detector would experience a large delay until it reflects at the VCO output (phase). This excessive delay can cause instability of the loop as the corrections always come in late.

The zero provided by R introduces a phase lead and reduces the overall delay of the system. Intuitively, when there are fast variations in phase, the charge pump current drives the resistor directly (recall the voltage across a capacitor cannot change instantaneously), resulting in a fast voltage change at the input of the VCO. The reduction of the excessive loop delay can improve system stability.

In the next section, we explore the PLL's dynamics further by deriving the closed-loop transfer function and applying control system theory.

Quantitative Analysis: To better understand the dynamics of the PLL (Figure Q43f), we need to quantitatively study the negative-feedback loop through its transfer function. Readers without a control system background can refer to *Appendix 3. Circuit Analysis in Complex Frequency Domain (s-domain)* and gain some fundamental knowledge of feedback system analysis.

We start with the loop filter without R. The transfer function of the PFD and charge pump, $G_{pc}(s)$, is:

$$G_{pc}(s) = \frac{I_{CPO}(s)}{\Delta\phi(s)} = \frac{I_{CP}}{2 \cdot \pi} = K_{pc}$$

Note that we approximate the charge pump output current as a continuous and constant current with a value proportional to the phase difference between the two clocks. K_{pc} is referred to as the gain of the PFD.

Now the transfer function of the loop filter, $G_f(s)$, is:

$$G_f(s) = \frac{V_{CP}(s)}{I_{CPO}(s)} = \frac{1}{(C_1 + C_2) \cdot s}$$

The VCO has the following transfer function:

$$G_{VCO}(s) = \frac{\phi_{out}(s)}{V_{CP}(s)} = \frac{K_{VCO}}{s}$$

where K_{VCO} is the gain of the VCO.

The open-loop transfer function of the PLL therefore is:

$$G(s) = \frac{\phi_{out}(s)}{\Delta\phi(s)} = G_{pc}(s) \cdot G_f(s) \cdot G_{VCO}(s) = \frac{K_{pc} \cdot K_{VCO}}{(C_1 + C_2) \cdot s^2}$$

This is a *Type 2* system (system having two integrators in its open-loop transfer function). The phase shift at DC is −180°. The closed-loop transfer function is:

$$H(s) = \frac{\phi_{out}(s)}{\phi_{in}(s)} = \frac{\frac{K_{pc} \cdot K_{VCO}}{C_1 + C_2}}{s^2 + \frac{1}{N \cdot (C_1 + C_2)} \cdot K_{pc} \cdot K_{VCO}}$$

The characteristic equation contains a depressed polynomial and has a pair of pure imaginary roots. In fact, this is a second-order system with the damping ratio (ζ) of zero (see *Appendix 3, Section 4, Second-order systems*). The system would exhibit sustained oscillation due to both of its closed-loop poles (roots of the characteristic equation) being on the imaginary axis.

Adding a zero can move the roots of the closed-loop system out of the imaginary axis and towards the left-hand plane (roots in the left-hand plane represent a stable system with a converged impulse response). With R in series with C_1, the transfer function of the loop filter becomes:

$$G_f(s) = \frac{V_{CP}(s)}{I_{CPO}(s)} = \frac{R \cdot C_1 \cdot s + 1}{s \cdot (R \cdot C_1 \cdot C_2 \cdot s + C_1 + C_2)}$$

And the open-loop transfer function of the PLL is:

$$G(s) = \frac{\phi_{out}(s)}{\Delta\phi(s)} = \frac{K_{pc} \cdot K_{VCO} \cdot (R \cdot C_1 \cdot s + 1)}{s^2 \cdot (R \cdot C_1 \cdot C_2 \cdot s + C_1 + C_2)}$$

It is still a Type 2 system, but with an added left-hand-plane pole and left-hand-plane zero. The closed-loop transfer function now becomes:

$$H(s) = \frac{N \cdot K_{pc} \cdot K_{VCO} \cdot (R \cdot C_1 \cdot s + 1)}{N \cdot R \cdot C_1 \cdot C_2 \cdot s^3 + N \cdot (C_1 + C_2) \cdot s^2 + K_{pc} \cdot K_{VCO} \cdot R \cdot C_1 \cdot s + K_{pc} \cdot K_{VCO}}$$

This is a third-order system and the stability of the system can be achieved by carefully designing the parameters of the components (R, C, and the gains) so that all closed-loop poles reside at the left-hand plane.

In practice, C_1 is usually significantly larger than C_2. In fact, C_1 and R can produce the needed zero for the loop stability. The purpose of C_2 is to dampen frequency jumps caused by R (i.e., effect of the zero). When $C_1 \gg C_2$, the closed-loop transfer function of the PLL can be simplified to the following:

$$H(s) = \frac{\frac{K_{pc} \cdot K_{VCO}}{C_1} \cdot (R \cdot C_1 \cdot s + 1)}{s^2 + \frac{1}{N} \cdot K_{pc} \cdot K_{VCO} \cdot R \cdot s + \frac{K_{pc} \cdot K_{VCO}}{N \cdot C_1}}$$

The natural frequency and damping ratio of the system is:

$$\omega_n = \sqrt{\frac{K_{pc} \cdot K_{VCO}}{N \cdot C_1}}$$

$$\zeta = \frac{R \cdot \sqrt{\frac{C_1 \cdot K_{pc} \cdot K_{VCO}}{N}}}{2} = \frac{\omega_n}{2} \cdot R \cdot C_1$$

For a second-order system, overshoot and ringing in step response exist when the damping ratio, ζ, is smaller than 1, and if it is too small, longer lock in time and excessive output clock jitter would occur (see *Appendix 3, Section 4, Second-order systems*).

Please note that the small-signal transfer function of the PLL has its limitations – for example, it ignores the nonlinearity that exists in the components of the PLL, such as PFD, charge pump, and VCO, which could cause performance degradation and stability concerns.

PLL is a good example on how digital circuits and concepts are used in analog design methodologies and mixed-signal systems. PLL and related circuits are widely used in the field of frequency synthesis, data recovery and signal synchronization, and SERDES (Serializer-Deserializer).

Short Answer to Question 43

(a) Sketch the relationship between the change in the charge pump output voltage ΔV_{CP} and the phase difference between the input clock and the feedback clock, $\Delta\phi = \phi_{in} - \phi_{fb}$. Ignore the resistor R of the loop filter. **(b)** What limits the smallest phase difference the PFD can effectively detect? **(c)** What is the function of the resistor R in the loop filter?

(a) The relationship between the change of the charge pump output voltage, ΔV_{CP}, and the phase difference between the input clock and the feedback clock, $\Delta\phi = \phi_{in} - \phi_{fb}$, can be written as:

$$\Delta V_{CP} = \frac{I_{CP} \cdot \Delta\phi \cdot T_{clock}}{2 \cdot \pi \cdot (C_1 + C_2)}$$

Please refer to Figure Q43d for the sketch.

(b) The minimum PFD output pulse duration required for the charge pump switch to fully turn on and conduct current.

(c) The loop filter resistor R introduces a phase lead (zero) that reduces the system's overall phase delay, thereby enhancing the loop stability.

Question 44

The outputs of three comparators, A, B, and C, are connected to a combinational logic circuit, whose outputs, Y_1, Y_2, and Y_3, are clocked into three DFFs, as shown in Figure Q44a. The comparators are strobed (latched) by the rising edge of the clock signal, clk1. A short time after a certain rising edge of clk1, both comparator outputs B and C are resolved to logic high. Comparator output A should resolve to logic low, but instead it enters a metastable state. List all possible logic states of the signals Y_1, Y_2, and Y_3 due to the metastability of comparator output A.

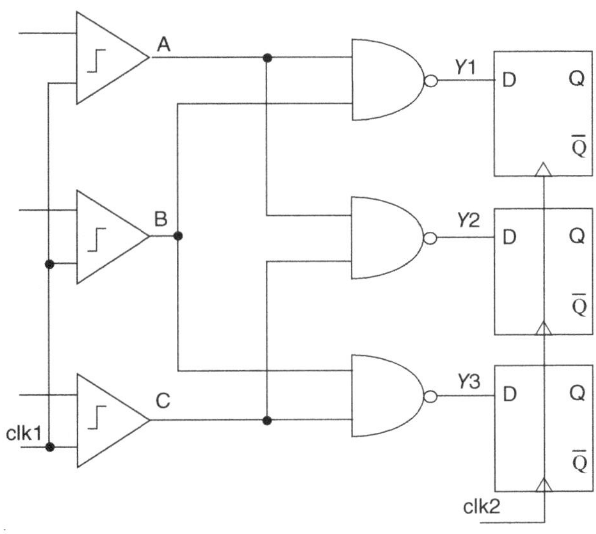

Figure Q44a

Intuitive Analysis: To answer this question, we need to understand what metastability stands for. Metastability refers to an unresolved output state, which is neither logic high, nor logic low, at the output of a comparator, a latch, or a flip-flop. Some of the possible causes of metastability include inadequate net input voltage for a comparator, insufficient latch time, and input transition too close to the clock edge of a DFF.

It is important to realize that a metastable output could be interpreted as different logic states by the subsequent digital circuits. Let us take inverters as an example. CMOS inverters typically exhibit significant variations in their switching thresholds (V_{SW}) due to process-induced mismatches, causing them to deviate from the ideal design point. This means an inverter might interpret a metastable input as a logic high or logic low, or *unknown* (denoted here as *logic X*) if the input is too close to the V_{SW}. This also means that it is possible for one inverter to interpret a metastable input as logic low, and another inverter, connected to the same metastable input, to interpret it as logic high, or logic X. Same applies to other logic gates as well.

Now let us return to the circuit in Figure Q44a, the comparator output A is connected to two NAND gates (let us identify the logic gates by their output name, i.e., Y_1 and Y_2, for simplicity). Each NAND gate can interpret signal A as three possible states (logic low, logic high, and logic X) independently, as illustrated in Figure Q44b. As such, there are nine possible combinations of input logic states for the two NAND gates (Y_1 and Y_2).

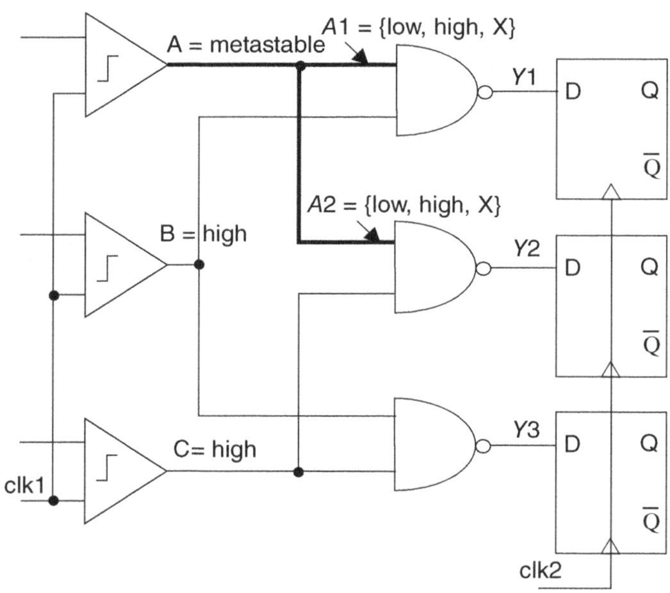

Figure Q44b

The rest of the analysis is truth table based: comparator output A would only affect the output Y_1 and Y_2, and we only list this part of the true table, as shown in Table Q44a.

Table Q44a

Y_1	Y_2	Y_3	A_1	A_2	B	C	note
1	1	0	0	0	1	1	0 = logic low
1	0		0	1			1 = logic high
1	X		0	X			X = logic unknown
0	1		1	0			
0	0		1	1			
0	X		1	X			
X	1		X	0			
X	0		X	1			
X	X		X	X			

So there could be nine (9) different logic combinations for Y_1, Y_2, and Y_3.

Note that the metastable state is not a steady state – it evolves over time and eventually resolves into a known logic state (low or high) if given sufficient time. In the circuit of Q44a, delaying the rising edge of clk2 could give time for Y_1 and Y_2 to resolve into a known logic state. For comparators, increasing the gain of the preamp can lower the probability of metastability occurring. Adding more latch stages can help logic signals resolve and lower the risk of metastability in digital circuits. Also keep in mind that circuits with metastable states are vulnerable to noise, and the outputs can resolve into an incorrect logic state under disturbances. In certain applications, forcing a known but potentially incorrect logic outcome is less harmful than having an unknown logic state.

Short Answer to Question 44

List all possible logic states of the signals Y_1, Y_2, and Y_3 due to the metastability of the comparator output A.

Please refer to Table Q44a for all possible logic states. Note that a metastable output can be interpreted as different logic states by subsequent digital circuits.

General References

1 Baker RJ. *CMOS: Circuit Design, Layout, and Simulation*. 3rd ed. Hoboken (NJ): Wiley/IEEE Press; 2010.

2 Kang SM and Baran JD. *CMOS Digital Integrated Circuits: Analysis & Design*. 4th ed. New York (NY): McGraw-Hill; 2019.

3 Rabaey JM, Chandrakasan A and Nikolic B. *Digital Integrated Circuits: a Design Perspective*. 2nd ed. Upper Saddle River (NJ): Prentice Hall; 2003.

4 Razavi B. *Design of CMOS Phase-Locked Loops: From Circuit Level to Architecture Level*. Cambridge (UK): Cambridge University Press; 2020.

5 Weste NHE and Harris DM. *CMOS VLSI Design: A Circuits and Systems Perspective*. 4th ed. Boston (MA): Pearson; 2011.

6

Circuit Layout Techniques

In a circuit design technical interview, when candidates are asked a layout question, some might think, "Maybe I didn't do very well in answering the circuit design questions." On the contrary, this might actually be a good sign. Why? Because layout is an integral and important part of circuit design, and not many students or even professional designers are considered well trained in this field.

As a matter of fact, the higher the performance requirements are, the more important the layout becomes. In cutting-edge high-speed or high-precision circuits, layout is critical. For example, parasitic capacitances and other impedances, if not well controlled in layout, can severely degrade the bandwidth and signal integrity of a high-speed circuit system. In high-precision circuits, overlooked noise coupling between routings, or a suboptimal layout for matching, can degrade the overall noise and accuracy performance. Good habits and intuition in layout are of utmost importance – when something is wrong with a circuit, it is often difficult and costly to find the underlying flaws in layout.

In this chapter, we assemble six questions to prepare candidates with some fundamental layout concepts and considerations. The topics range from transistor-level layout concerns – parasitics (Question 45), matching (Question 46), and substrate/well contacts and latch-up (Question 48), to interconnectivity issues such as noise shielding (Question 49) and top-level matters such as decoupling (Question 47) and power/ground arrangement (Question 50). After all, an investment in learning layout techniques is highly rewarding – not only for doing well in interviews but also for one's long-term career as an accomplished circuit designer.

Question 45

Propose a layout design for an NMOS transistor with a *W*/*L* ratio of 100/1. The transistor is used in a high-speed circuit and requires an overall compact layout in both the *x* and *y* directions to minimize (transistor and routing) parasitic capacitance. The source of the transistor is grounded (Figure Q45a).

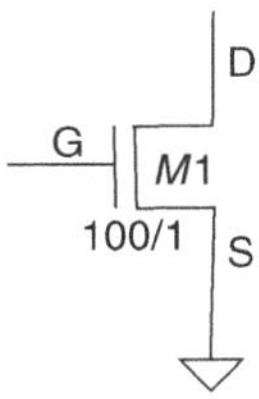

Figure Q45a

Intuitive Analysis: A straightforward layout of a 100/1 *W*/*L* transistor would result in a tall (or long) and skinny shape, which is not compact overall and is not convenient for routing. From a geometry point of view, a square-shaped transistor has a minimum overall layout dimension (*x* and *y* directions). Splitting a transistor with a large *W*/*L* ratio into multiple transistors in parallel with smaller widths can help achieve this goal. Figure Q45b shows an example with two identical NMOS transistors connected in parallel, forming a transistor with doubled width:

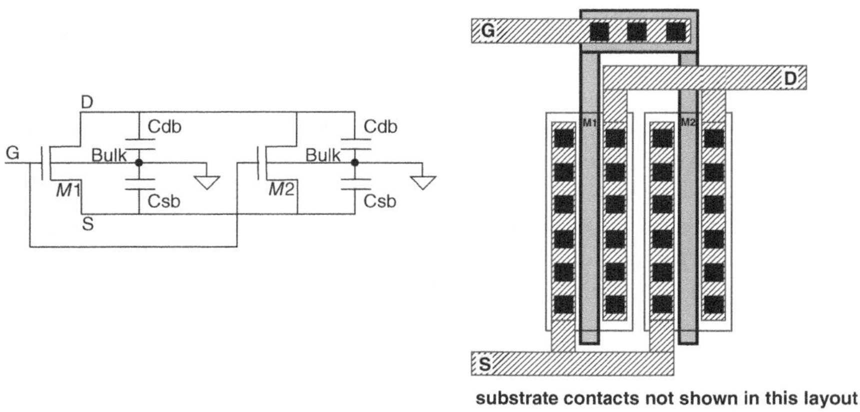

Figure Q45b

The parasitic capacitances of CMOS transistors that are affected by the layout include the depletion capacitances of the drain and source, C_{db} and

C_{sb}, respectively. For large-size transistors, the depletion capacitance could be significant and impact the speed of the circuit.

Drain and source sharing is an effective method to make large-size transistors more compact and reduce the drain and source depletion capacitance. Assume that five transistors in parallel, each with a size of 20/1, would make the layout shape close to a square (one needs to check the layout design rules of the process technology to determine the proper splitting configuration). Figure Q45c illustrates the concept of drain and source sharing.

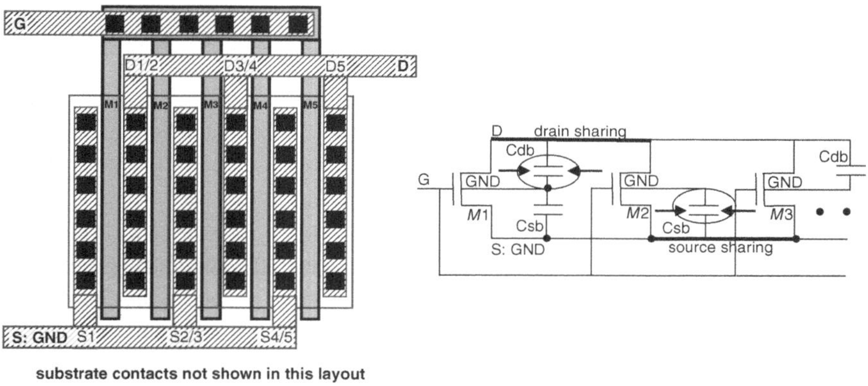

Figure Q45c

As shown in Figure Q45c, transistors M1 and M2 share a common drain diffusion, while transistors M2 and M3 share a common source diffusion, continuing this pattern for subsequent devices. In this way, about half of the total parasitic depletion capacitance of the drain and source is eliminated, and the total layout area is reduced significantly compared to the non-sharing layout in Figure Q45b.

Note that in the above example, we split the transistor into an odd number of parts (also referred to as an odd number of *fingers*, which is equal to the number of poly gate branches). The numbers of source and drain branches are equal, and so are the total depletion capacitances of the drain and source. If we split the transistor into an even number of fingers, such as four fingers, the numbers of source and drain branches would differ by one, and the parasitic capacitances of the drain and source would no longer be balanced. This is illustrated in Figure Q45d. We would either have three drain branches and two source branches, or vice versa. The optimal design choice depends on the circuit topology and design concerns. Very often, the sources of NMOS transistors are connected to ground (or AC ground). If this is the case, it is preferable to let the source have more parasitic capacitance as this capacitance experiences minimal charge redistribution.

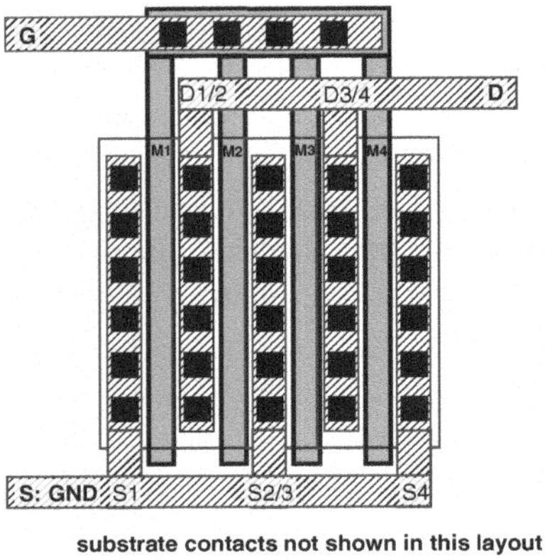

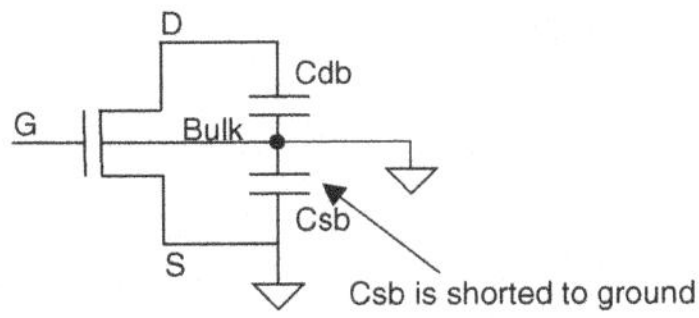

Figure Q45d

Short Answer to Question 45

Propose a layout design for an NMOS transistor with a *W*/*L* ratio of 100/1. The transistor is used in a high-speed circuit and requires an overall compact layout in both the *x* and *y* directions to minimize (transistor and routing) parasitic capacitance.

Please refer to Figure Q45c for the layout design. A multi-finger structure with drain and source sharing is used to minimize the overall layout size and parasitic capacitance.

Question 46

Sketch a layout design for the current mirror shown in Figure Q46a. Transistors A and B have the same size (*W* and *L*) and each has four fingers (nf = 4). The physical design should help with the matching between A and B, and keep the layout compact to reduce parasitic capacitance. For this latter purpose, A and B are required to share diffusion (drain/source), if possible.

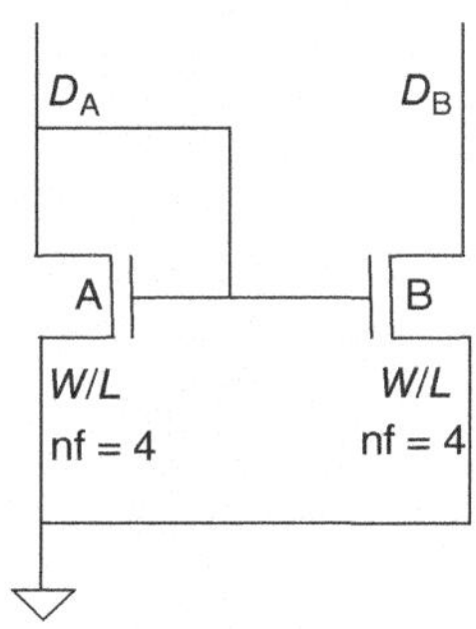

Figure Q46a

Intuitive Analysis: Matching is one of the most important design concerns for circuits like current mirrors and differential pairs. There are two major layout techniques to improve matching: *interdigitation* and *common-centroid*. Interdigitation improves matching among transistors in one layout dimension, and common-centroid enhances matching along two layout dimensions. While it is true that a common-centroid layout could result in better overall matching, especially for transistors with a large layout size, it has the burden of extra layout area and complex metal routings, which add parasitic capacitance, compared to its counterpart. For relatively small-size transistors, interdigitation is more area-efficient with less parasitic capacitance. Because compactness and less parasitic capacitance are important in addition to matching in this question, it is reasonable to choose an interdigitation design.

The most thorough interdigitation arrangement for two matching transistors (A and B) is the fully alternating pattern: ABAB. Drain and source (diffusion) sharing is required in this design to reduce parasitic capacitance. The question is: can we do the two things at the same time? A quick sketch can help us have a clearer view of the feasibility.

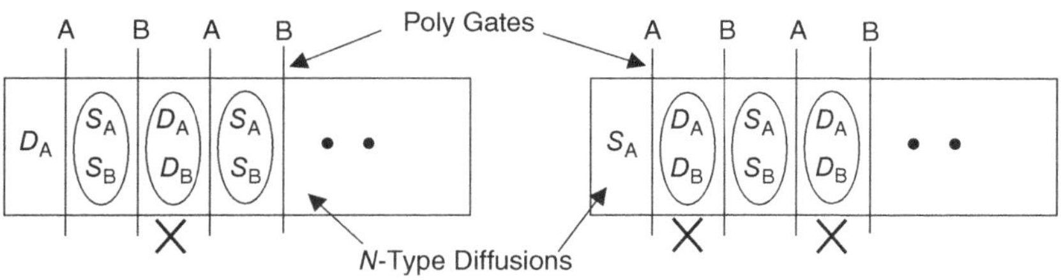

Figure Q46b

Notice that in the current mirror in Figure Q46a, the sources of the two transistors are connected (both tied to ground), but the drains have to be separated. Two possible drain and source sharing arrangements are sketched in Figure Q46b, and the results suggest that a fully alternating pattern (ABAB) does not work for drain/source sharing.

The problem with the arrangements in Figure Q46b is that a fully alternating pattern results in both drain and source sharing between the two transistors, but drain sharing is not allowed. We can make a slight adjustment by grouping two fingers of the same transistor together, as shown in Figure Q46c, thereby avoiding drain sharing.

For the layout planning of the transistor pair in Figure Q46a, this means the interdigitation pattern should be AABBAABB or ABBAABBA. Figure Q46d illustrates a layout example of using the former arrangement. Note that in a typical

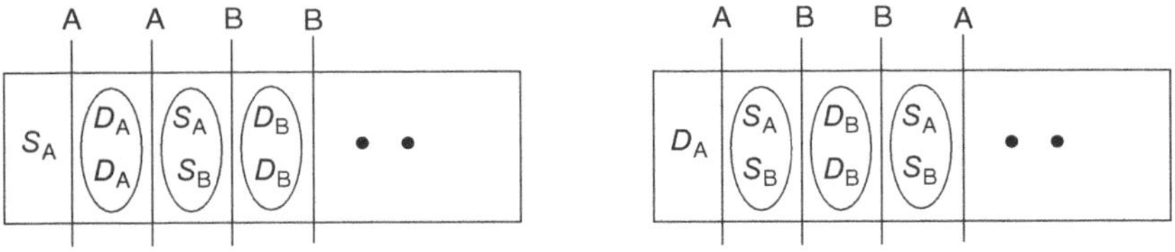

Figure Q46c

technical interview, a sketch of the layout planning, such as those in Figure Q46b and Figure Q46c, is an efficient way to present layout ideas, and usually is considered good enough for interview discussion purposes.

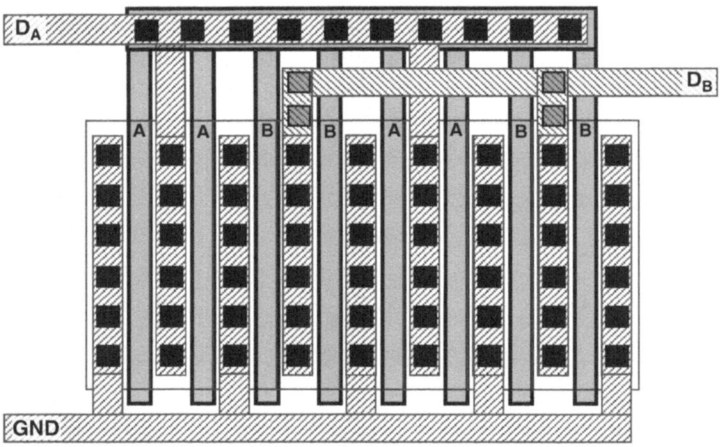

Figure Q46d

A matched transistor pair (NMOS or PMOS) is widely used in many analog circuits, such as current mirrors, amplifier differential input pairs, matched load pairs, and so on. Interested readers are encouraged to explore further into the layout implementations for matched transistor pairs, for example, how to properly add dummy transistors (non-functional devices placed at the ends of an array to ensure the active transistors experience identical etching and photolithography conditions, thereby improving matching and symmetry), substrate contacts, etc.

Short Answer to Question 46

Sketch a layout design for the current mirror shown in Figure Q46a (transistors A and B are required to share diffusion).

Please refer to Figure Q46c for the layout sketch. Both patterns (AABBAABB and ABBAABBA) allow the two transistors to share diffusion.

Question 47

An NMOS decoupling capacitor with 500 fF capacitance needs to be implemented. Assume the gate oxide capacitance density (C'_{ox}) is 5fF/µm^2, a total gate area ($W \cdot L$) of 100 µm^2 is required. Three transistor layout arrangements are shown in Figure Q47a. Examine the relative strengths and weaknesses of each approach.

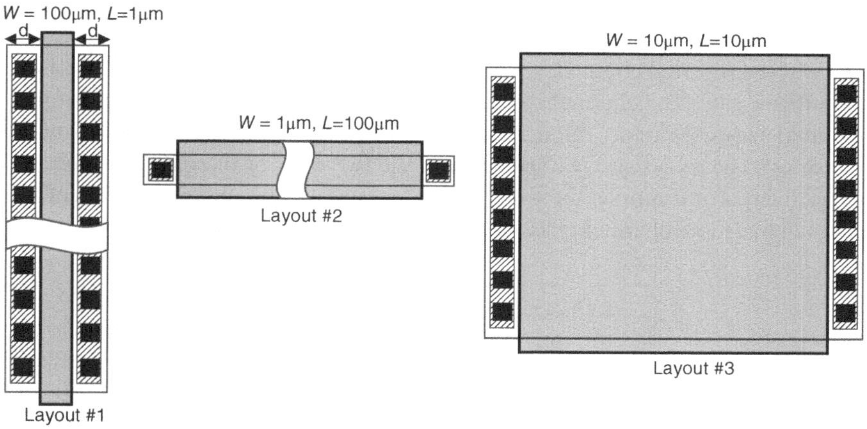

Figure Q47a (illustration only, size not to scale)

Intuitive Analysis: MOSFET decoupling capacitors utilize the gate capacitance of NMOS or PMOS transistors with strongly inverted channels. Like other types of capacitors, the *capacitance density* is defined as the effective capacitance per unit area of the implementation. Because some area of the capacitor is allocated for drain and source contacts, substrate contacts, and metal routing, the efficiency of the area usage for capacitance is always less than 100%.

Using Layout #1 as an example, assume the length of the drain/source contact area (plus all necessary clearance) is d, as illustrated in Figure Q47a. The capacitance density α is:

$$\alpha \approx \frac{W \cdot L}{(2 \cdot d + L) \cdot W} \cdot C'_{ox} = \frac{L}{2 \cdot d + L} \cdot C'_{ox} = \frac{1}{2 \cdot \frac{d}{L} + 1} \cdot C'_{ox}$$

where C'_{ox} is the gate capacitance density.

It is clear that the larger the channel length L, the higher the capacitance density. For example, if $d = 1$ μm, then the capacitance densities of Layout #1, Layout #2, and Layout #3 are: $(1/3) \cdot C'_{ox}$, $(50/51) \cdot C'_{ox}$, and $(5/6) \cdot C'_{ox}$, respectively. Thus, from a capacitance-density point of view, Layout #2 (1/100) has the highest effective capacitance density and Layout #1 (100/1) has the lowest. In addition to drain/source contacts, substrate contacts and interconnect routing also consume silicon area, eroding the effective capacitance density.

This does not mean that Layout #2 (1/100) is the most popular configuration. A practical issue with capacitors is their *equivalent series resistance* (ESR). The series parasitic resistance of capacitors creates unwanted voltage fluctuations, slows down the charging and discharging processes, and therefore reduces the effectiveness of the capacitors.

The main parasitic resistances in a MOSFET decoupling capacitor are the resistance of the inverted channel, as shown in Figure Q47b, and the resistance of the polysilicon gate. The channel resistance is typically the dominant contributor to the total series resistance. Note that both gate resistance and channel resistance can only be accessed and connected from the two ends of the gate and channel, respectively. For example, the gate capacitance associated with the middle of the channel has more channel resistance to the drain/source terminals.

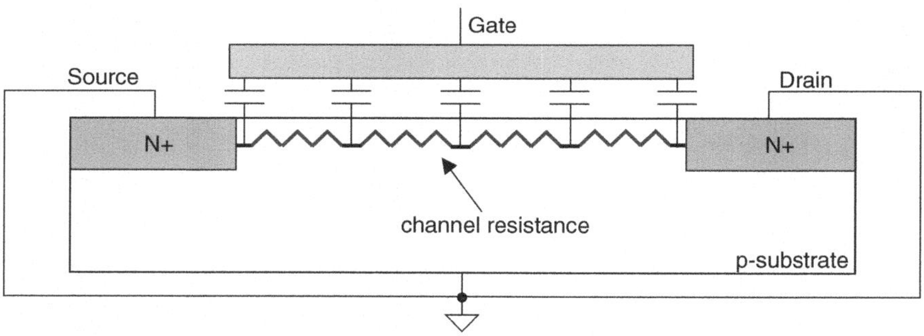

Figure Q47b

The channel resistance can be estimated by:

$$R_{CH} = \frac{L}{W} \cdot R_{\square}$$

where $R_{\square}$ is the sheet resistance of the channel.

Note that in a real implementation the circuit is a distributed RC network. From an ESR point of view, Layout #1 has the highest gate resistance but the lowest channel resistance, and Layout #2 has the highest channel resistance but the lowest gate resistance.

This explains why Layout #3, or similar layouts, are among the most popular arrangements for MOSFET decoupling capacitors – it offers reasonable capacitance density with relatively low series resistance. It is often used as a unit of decoupling capacitors, or a *decoupling cell*, and the square shape makes it easy to fit into various spaces, alone or in arrays.

Short Answer to Question 47

Examine the relative strengths and weaknesses of each approach.

Layout #1 has the highest gate series resistance and the lowest capacitance density.
Layout #2 has the highest channel series resistance and the highest capacitance density.
Layout #3 has a moderate series resistance and a fair capacitance density.

Question 48

A layout consisting of a number of NMOS and PMOS transistors is shown in Figure Q48a. All PMOS transistors are placed in the same N-Well. M1 (NMOS) and M2 (PMOS) are adjacent in the layout. The substrate contacts and N-Well contacts are located far from M1 and M2, respectively. List the potential issues with this layout arrangement and explain in detail why they occur.

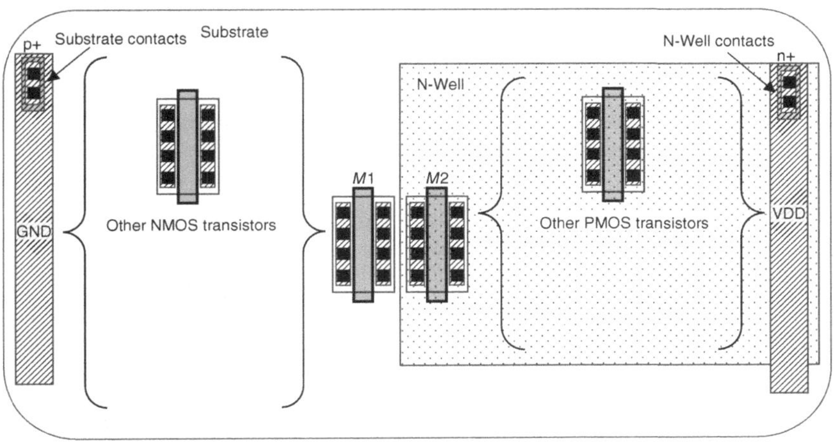

Figure Q48a

Intuitive Analysis: Placing all PMOS transistors in the same N-Well can save layout area. The N-Well contacts, however, should be placed in close proximity to these transistors. The same applies to NMOS transistors; the substrate contacts should be placed near all NMOS devices. To explain why this is important, we need to recall the functions of the substrate and well contacts.

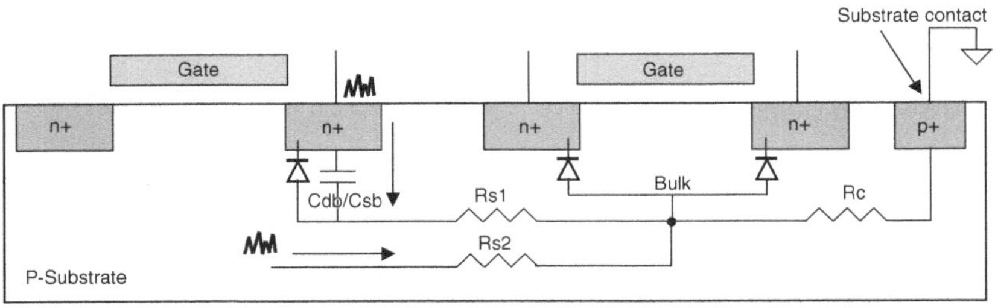

Figure Q48b

Figure Q48b illustrates a cross-section view of two NMOS transistors in a standard CMOS technology. The depletion diodes formed between the n+ of drain/source regions and the p-substrate must be reverse-biased for an NMOS to function properly. These reverse-biased diodes present depletion capacitances between the source/drain and the substrate (C_{sb} and C_{db}). The function of the substrate contacts is to connect the substrate to the lowest potential (GND), preventing the PN junctions at the interface of the source/drain and substrate from being forward biased. However, the substrate is highly resistive, and substrate contacts that are far from the local substrate (bulk) of an NMOS transistor may not be able to maintain the local ground potential. As illustrated in Figure Q48b, noise at the drain/source may couple through the depletion capacitance and contaminate the substrate (remember all NMOS transistors share the same substrate). If the resistance from the substrate contact to the bulk of a transistor (R_c in Figure Q48b) is high, the potential of the bulk is less stable because of the substrate noise. The substrate noise at the bulk can feed through to the drain and source of the transistor and affect the threshold voltage due to bulk-effect (body-effect).

The same also applies to the PMOS transistors inside an N-Well. High resistance between the N-Well contacts and the bulk of a PMOS transistor can make the bulk more susceptible to N-Well noise.

Low-noise analog circuits often require local and low-impedance connectivity between the substrate/N-Well and the supply rails. These contacts should be placed close to the transistors and used in sufficient quantity. Guard rings formed by highly-doped substrate and N-Well contacts can be implemented around sensitive analog circuits to collect substrate/N-Well noise from other parts of the system and suppress noise injection to the surrounding area.

Noise is not the only issue when substrate/well contacts are a distance away from transistors. Another issue is *latch-up*. The physical structure to enable latch-up is the parasitic PNPN junction formed by adjacent NMOS and PMOS transistors. Figure Q48c illustrates how a positive-feedback loop similar to that of a latch is formed by the parasitic BJTs (Bipolar Junction Transistors) of an adjacent NMOS and PMOS transistor pair.

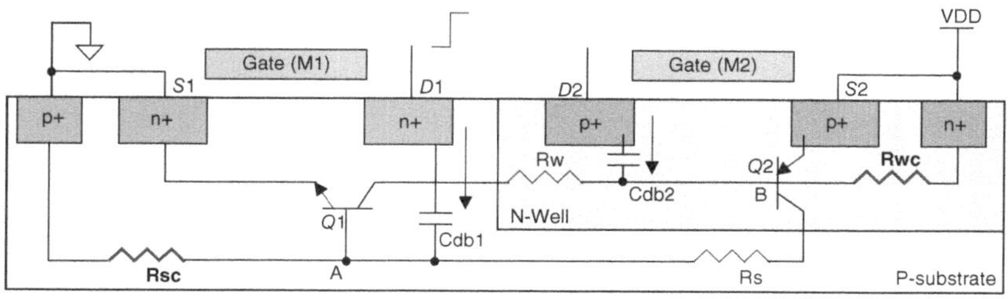

Figure Q48c

Two major parasitic BJTs, Q_1 and Q_2, are formed around M1 and M2, as shown in Figure Q48c. Q_1, a parasitic NPN transistor, is formed by the N-Well, p-substrate, and n+ diffusion (the source of M1). Q_2, a parasitic PNP transistor, is formed by the p+ diffusion (the source of M2), N-Well, and p-substrate. In Figure Q48c, we have assumed that the source of M1 (NMOS) is grounded, and the source of M2 (PMOS) is tied to VDD, which is common in circuits. Note that due to the high resistivity of the substrate and N-Well, there is parasitic resistance between the terminals of the two parasitic BJTs to VDD or GND (R_{wc} and R_{sc}), as well as among the terminals (R_w and R_s). The circuit diagram formed by these parasitic BJTs is shown in Figure Q48d for ease of analysis.

It is clear from Figure Q48d that Q_1, Q_2 and the parasitic resistors form a pair of cross-coupled amplifiers similar to a latch. The base of Q_1 (node A) is the bulk of M1 (NMOS), and the base of Q_2 (node B) is the bulk of M2 (PMOS). Normally the bulk of M1, or node A, is at ground level thanks to the substrate contacts, and Q_1 is turned off. A sharp rising edge at the drain of M1 could feed through to the bulk of M1 (node A) through the drain to bulk capacitance (C_{db1}). If the substrate contacts are far away from the bulk of M1, the parasitic resistance R_{sc} could be too large to hold node A stable. The feedthrough voltage can turn on Q_1, which in turn causes voltage at node B to drop and turns on Q_2. The current through Q_2 maintains the voltage at node A to keeps Q_1 turned on – now the loop is latched up. Note that this is just one of the scenarios that can trigger latch-up. A sharp falling edge at the drain of M2, for example, could do the same trick and bring the loop into a locked condition.

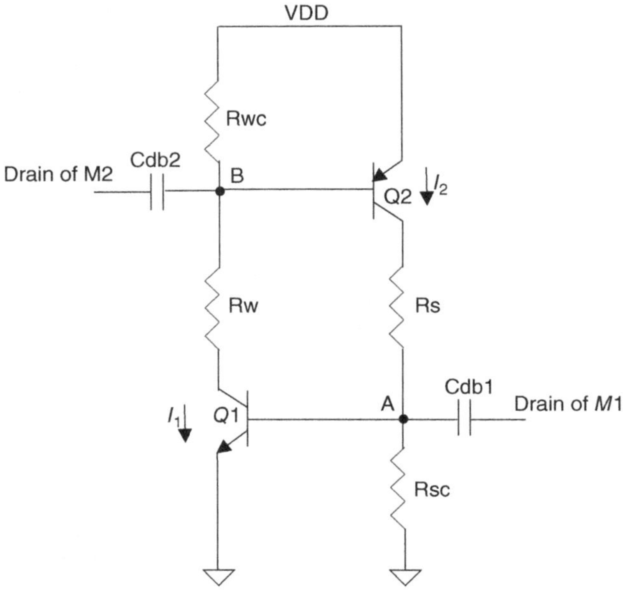

Figure Q48d

Latch-up causes forward-biased PN junctions and currents in the substrate and N-Well. Once the circuit is latched up, the only way to remove the latch-up is to recycle the power supply. It is a harmful condition that results in circuit malfunction. In I/O drivers where devices are generally large, it can cause IC damage due to the large current flow.

The most effective way to lower the chance of latch-up is to reduce the substrate and N-Well contact resistances. As can be seen from Figure Q48d, Q_1 and Q_2 cannot be turned on easily if R_{sc} and R_{wc} are of low resistance. Guard rings around NMOS and PMOS transistors can effectively help to prevent latch-up. Semiconductor foundries usually specify design rules on the maximum distance allowed between transistors and their substrate/well contacts to ensure good latch-up immunity.

Short Answer to Question 48

List the potential issues with this layout arrangement and explain in detail why they occur.

Large substrate or N-Well contact resistance can cause transistor bulk noise issues and latch-up.

Question 49

A metal line carrying a quiet analog signal, A, and another metal line carrying a noisy digital signal, D, are routed in parallel and placed close to each other in the layout implementation, as shown in Figure Q49a (a). The distance and parasitic capacitance between the two metal lines are d and C_C, respectively. **(a)** Describe how to lower the noise coupling from the noisy line D to the quiet line A if no other devices or metal routings can be added. **(b)** What would happen if we place a shielding metal line, S, between them, as shown in Figure Q49a (b)? C_S is the parasitic capacitance between the analog line, A, and the shielding line, S. For the best shielding effect, what should this shielding line be connected to: analog ground, digital ground, or floating? Assume the analog supply (AVDD) and digital supply (DVDD) are separated, each with a dedicated pad. Further assume the analog ground (AGND) and digital ground (DGND) are separated until star-connected at the shared ground pad (GND).

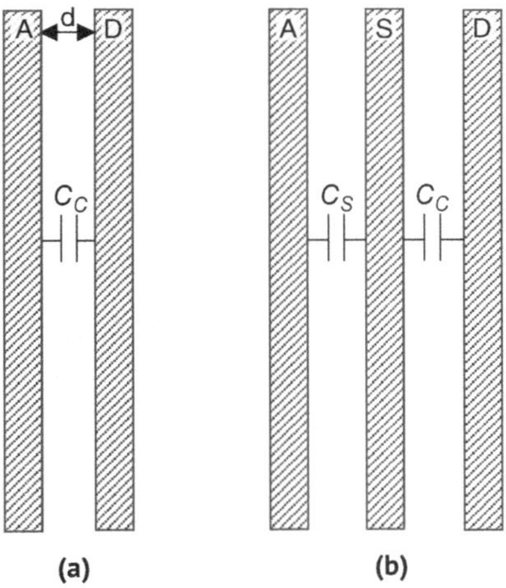

Figure Q49a

Intuitive and Quantitative Analysis: Before analyzing any noise coupling issues at the signal and power rails, it is always a good idea to understand the circuit's power domains and their respective interconnections. Figure Q49b is a simplified circuit diagram that illustrates the coupling between two lines, with reference to the connectivity of the analog and digital power rails. AVDD and AGND are the analog power supply and ground. DVDD and DGND are the digital power supply and ground. To analyze the impact of digital noise on an analog signal, the analog circuit is assumed to be noiseless, with zero routing resistance on the analog power and ground rails. R_{DG} represents the routing resistance from DGND to the shared ground pad, GND, which serves as the common reference ground for both analog and digital circuits. v_n is the noise source on the digital line with an output impedance of r_D. i_{ND} represents the noise current injected into DGND from the rest of the digital circuit, and v_{DGND} is the digital noise voltage on DGND due to i_{ND} and R_{DG}. The analog signal source is quiet (AC = 0) with an output impedance of r_A. And finally, there is no noise current injected from the analog circuits into AGND, so $i_{NA} = 0$.

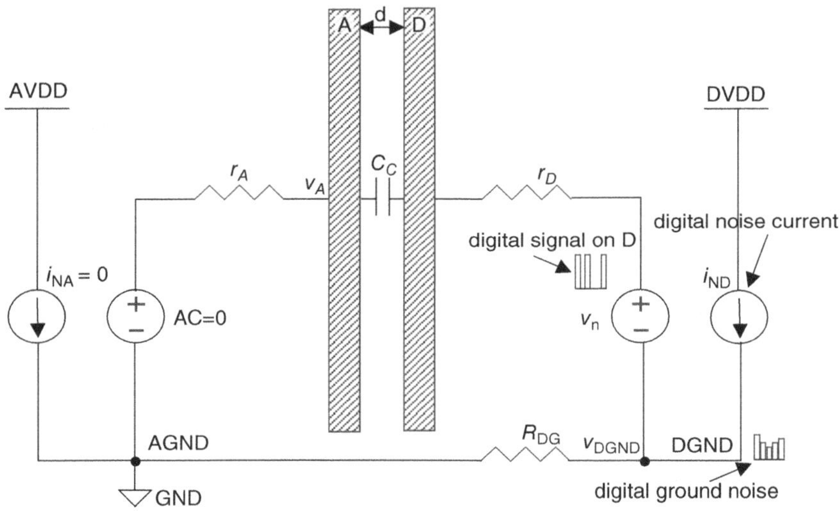

Figure Q49b

Intuitively, the digital noise voltage v_n and v_{DGND} can cause noise on the analog line (v_A) through the coupling capacitance, C_C. To calculate the total noise at v_A, each individual noise source must be evaluated and then aggregated. For the noise source v_n, the circuit forms a (noise) voltage divider and the v_A is given by (in s-domain):

$$v_{A(v_n)} = \frac{r_A}{r_D + R_{DG} + \frac{1}{C_C \cdot s} + r_A} \cdot v_n$$

The noise current from other digital circuits, i_{ND}, mostly flows through the digital ground resistance R_{DG} to ground pad (GND), causing ground bounce (noise) at the digital ground, v_{DGND}. To simplify the analysis, we assume that the entire current i_{ND} flows through R_{DG}. This digital ground noise affects the v_A in the same way as V_n does:

$$v_{A(i_{\text{ND}})} = \frac{r_A}{r_D + R_{\text{DG}} + \dfrac{1}{C_C \cdot s} + r_A} \cdot i_{\text{ND}} \cdot R_{\text{DG}}$$

And the total coupled noise (mean square) at line A is (assuming v_n and i_{ND} are uncorrelated):

$$v_A^2 = \left(\frac{r_A}{r_D + R_{\text{DG}} + \dfrac{1}{C_C \cdot s} + r_A}\right)^2 \cdot \left(v_n^2 + i_{\text{ND}}^2 \cdot R_{\text{DG}}^2\right)$$

Apparently, the smaller the coupling capacitance, C_C, the less noise is coupled onto v_A. The most straightforward way to lower the C_C is to increase the spacing between the two lines. An alternative solution is to route the two lines on different metal layers, as the fringe capacitance between two adjacent metal lines at the same layer could be larger than the plate capacitance when they overlap each other.

With a piece of shielding metal (S) placing between the quiet analog line (A) and the noisy digital line (D), the situation becomes more complicated (Figure Q49c). C_S is the coupling capacitance between the analog line A and shield line S, and R_S is the routing resistance of the shield line when it is connected to a potential (AGND or DGND). The voltage v_S is the noise voltage at the shield line.

The shield line S breaks the direct coupling between the analog line A and digital line D, and helps block the digital noise from the digital line D. The question is: "Which ground should it be connected to?" AGND or DGND?

In Figure Q49c, the shield line is connected to AGND. Intuitively, the lower the routing resistance R_S, the less noise on v_A. If $R_S = 0$, then $v_S = \text{AGND}$ and $v_A = 0$.

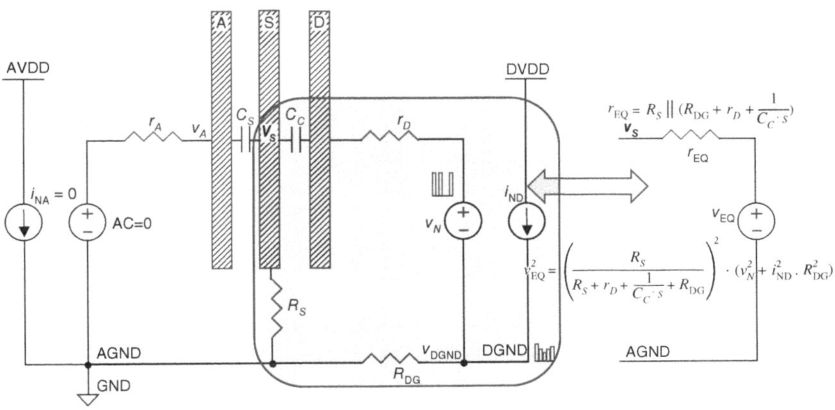

Figure Q49c

Analyzing the v_A quantitatively is more involved than the previous case without shielding. We can apply KCL/KVL to have v_A solved with a bit of effort. A more effective way to analyze a circuit such as this one is to use *Thévenin's theorem* (see *Appendix 2, Section 2, Thévenin's theorem and Norton's theorem*). Thévenin's theorem simplifies a linear circuit with an equivalent two-terminal circuit network consisting of a voltage source, v_{EQ}, and a series impedance, r_{EQ}. v_{EQ} is equal to the voltage across the two terminals when the external load attached to the two terminals are open circuit. And r_{EQ} is equal to the input impedance of the two-terminal network when all internal voltage sources are short $(V = 0)$ and current sources are open $(I = 0)$.

To determine the contribution of the noise sources v_N and i_{ND} to the node voltage v_A, we can use the Thévenin equivalent circuit shown on the right side of Figure Q49c to replace the shaded circuit network consisting of the noise sources v_N and i_{ND}. Between the equivalent noise source v_{EQ} and v_A, the circuit forms a voltage divider circuit similar to that in Figure Q49b. Without further derivation, we realize that when R_s is sufficiently small and negligible, i.e., $R_s = 0$, the noise at node v_S is negligible because $v_{EQ} = 0$, and therefore $v_A = 0$.

This result is important in layout implementations – shield lines with a low-impedance connection to the quiet analog ground provide good protection from adjacent noisy lines, digital or otherwise. But what if we connect the shield line to the noisy digital ground (DGND) instead?

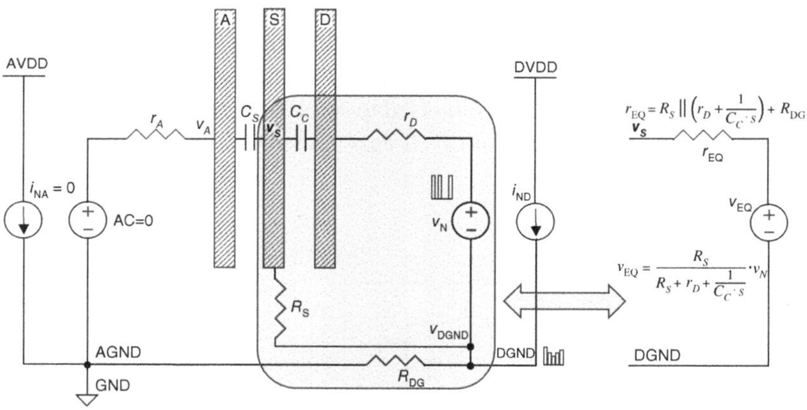

Figure Q49d

The simplified circuit for a shield line tied to digital ground (DGND) is illustrated in Figure Q49d. On the right side of Figure Q49d is the Thévenin's equivalent circuit of the noise source v_N (shaded area). If we can make the resistance between the shield line and DGND negligible $(R_s = 0)$, the equivalent noise voltage, v_{EQ}, is also negligible $(v_{EQ} = 0)$. Intuitively, this is because if $R_s = 0$, the shield line shorts the noise voltage source v_N. This is similar to the situation when the shield line is connected to AGND. However, unlike the previous case, the noise voltage on DGND, v_{DGND}, generated by noise current from other digital circuits (i_{ND}) directly causes noise at v_S through the shield line (R_S). The noise on v_A can be written as:

$$v_A = \frac{r_A \cdot (i_{\text{ND}} \cdot R_{\text{DG}})}{R_S + R_{\text{DG}} + \frac{1}{C_C \cdot s} + r_A}$$

This means that the shield line connected to DGND can bring in digital ground noise through capacitive coupling between the shield line and the "protected" quiet analog line, which is generally a no-go in practice.

In some cases it might be difficult to find a quiet ground for the shield lines to attach to. What happens if we let the shield line float? Figure Q49e illustrates this scenario; it is essentially the same as that of Figure Q49b (direct coupling between noisy digital line and the quiet analog line). The effective coupling capacitance is reduced due to the series connection of two parasitic capacitances, C_S and C_C.

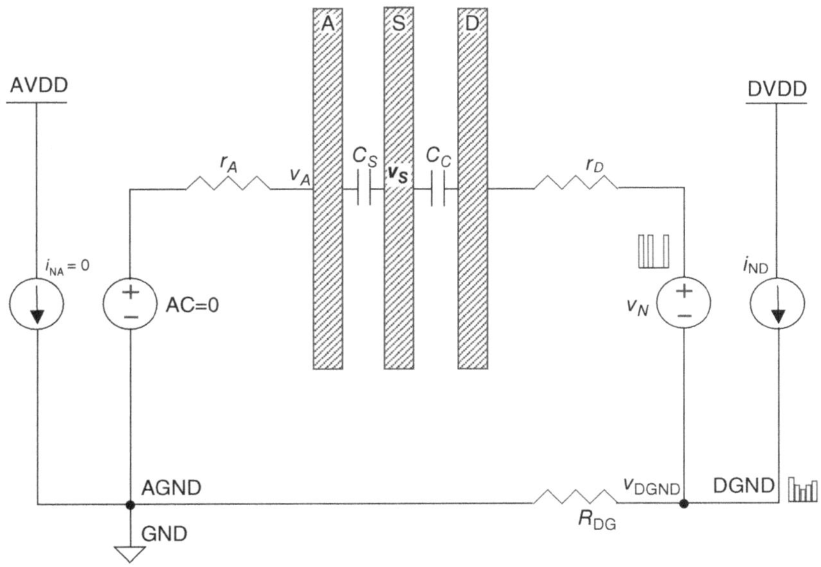

Figure Q49e

It is arguable that as shielding takes additional layout space, we may just use that space to separate the two lines further to reduce noise coupling. Which way is more effective is subject to the relationship of the parasitic capacitance to the physical dimensions of the process technology in use. Also keep in mind that a floating shield line is susceptible to noise coupling. If the floating shield line also has coupling relationships with other noisy circuit parts, it will bring in additional noise to the line that it tries to protect.

In the above analysis, we see that separation of the routings of AGND and DGND helps to prevent digital ground noise from appearing to the analog ground. If the star connection point is at a common ground pad (GND) in a wire-bond pad frame, however, the analog and digital current flow through the same bonding wire to the ground pin at the package. The noise voltage generated by the digital switching current upon the impedance of the bonding wire (inductive) would also affect the AGND, as illustrated in Figure Q49f. In many low-noise applications, digital ground and analog ground are assigned to separate pads, to avoid noise sharing via common bonding wires.

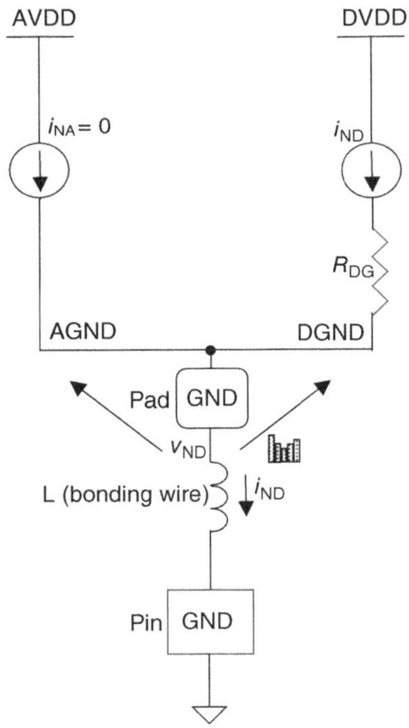

Figure Q49f

Short Answer to Question 49

(a) Describe how to lower the noise coupling from the noisy line D to the quiet line A if no other devices or metal routings can be added. **(b)** What would happen if we place a shielding metal line, S, between them, as shown in Figure Q49a (b)? For the best shielding effect, what should this shielding line be connected to: analog ground, digital ground, or floating?

(a) To lower the noise coupling, we can increase the spacing between the two lines. An alternative solution is to route the two lines on different metal layers.

(b) A shielding line can help isolate the quiet analog line A from the noisy digital line D. The shielding line should be connected to the analog ground (with low impedance) for the best shielding effect.

Question 50

A layout floor plan for a chip-level power rail network is shown in Figure Q50a. It consists of VDD and GND buses with evenly distributed branches so that each block-level circuit has access to nearby VDD and GND rails. The chip allows one VDD pad at the top and one GND pad at the bottom. Two possible VDD pad positions and three possible GND positions at the pad frame are indicated in Figure Q50a. Please explain which one of the following three VDD/GND pad arrangements is preferred and why it is preferred? The three VDD/GND arrangements are: VDD1/GND1, VDD1/GND3, and VDD2/GND2. Assume all circuit blocks consume a similar amount of power.

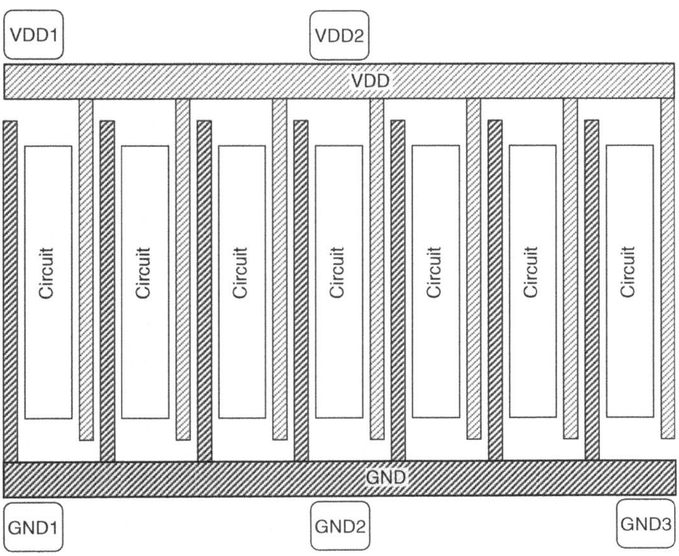

Figure Q50a

Intuitive Analysis: First, we need to understand the major objectives that the chip-level power and ground layout planning aims to achieve. The first target is to lower the overall resistance/impedance on the VDD and GND routings. The routing resistance/impedance results in IR drop and noise on the power and ground rails, affecting noise performance of analog circuits and timing and signal integrity of digital circuits. Keep in mind that at the chip level, the total current (static and dynamic) is usually significantly larger than that of the block-level circuits. Even a seemingly small resistance, such as 1 Ω, could result in a large IR drop of 100 mV, if the total current is 100 mA. Careful planning of chip-level power and ground routing resistance is critical to ensuring the performance of an IC.

At the chip level, we also need to distribute the power and ground buses to all circuits evenly so that there is no weak spot with excessive IR drop on the chip. In this question, we assume that all circuit blocks consume a similar level of power. In practice, this is usually not the case, and priority needs to be given to the circuit blocks that consume large power and/or have low-noise requirements.

Last but not least, for most circuit systems, ground potential differences and ground noise are more detrimental to circuit performance than VDD noise. This is because most circuits are designed with a certain voltage headroom (margin) on VDD, as well as certain power supply rejection to the noise on VDD. The ground, however, is the reference point for most circuit systems and the potential difference and noise on GND are less forgiving.

Based on these considerations, we evaluate the three VDD/GND pad assignments. Figure Q50b is a sketch of pad arrangement #1 (VDD1/GND1) with the longest and shortest current path from VDD to GND identified.

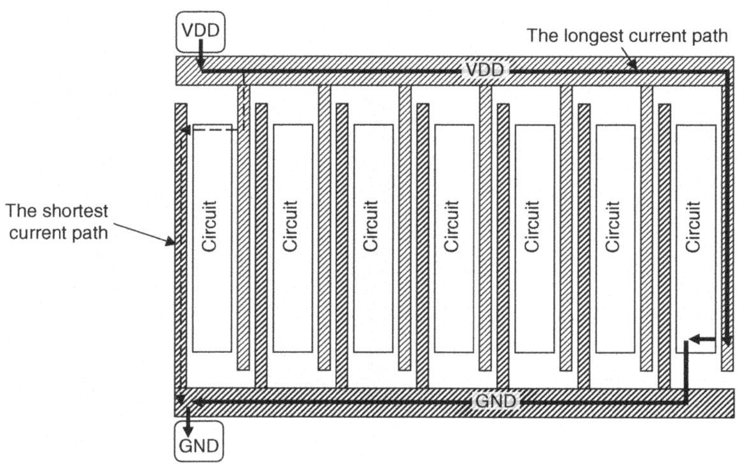

Figure Q50b

The problem with this pad arrangement is evident. If we assume the chip layout is close to a square with the length of each side equal to L, the longest current path from VDD to GND is about $3 \cdot L$, and the shortest path is about $1 \cdot L$. This arrangement would cause a large IR drop on the circuit blocks on the right side of the chip, especially the blocks at the right edge. It also has a large variation in the ground routing length between the lower left and upper right $(\Delta = 2L)$.

Figure Q50c illustrates the pad arrangement #2 (VDD1/GND3). An interesting feature of this arrangement is that the total current path from VDD to GND for all circuit blocks is roughly the same $(2 \cdot L)$ due to the diagonal arrangement of the VDD and GND pads. The benefit of this arrangement is the uniformity of the total parasitic resistance on the VDD and GND routings to all circuit blocks. The major drawback is the asymmetry between the VDD and GND resistances. Ground routing length (resistance) increases from the lower right to the upper left of the chip by a large delta $(\Delta = 2L)$.

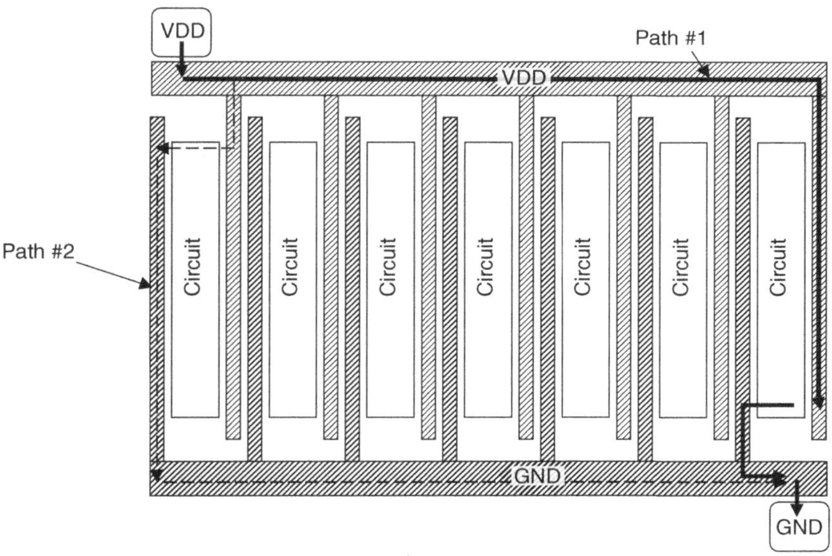

Figure Q50c

For the pad arrangement #3 (VDD2/GND2), the longest path now is approximately $2 \cdot L$, and the shortest path is approximately L (Figure Q50d). Compared to the other two arrangements, the arrangement #3 has the smallest maximum total routing length $(2 \cdot L)$, tied with arrangement #2 and better than arrangement #1 $(3 \cdot L)$. It also has the least overall ground routing length (resistance) and variations ($\Delta = 1.5 \cdot L$, from lower middle to upper left or right), better than both arrangements #1 and #2 ($\Delta = 2 \cdot L$). As mentioned earlier, in many systems, a low and uniform ground resistance (and potential) is a priority. The arrangement #3 is generally preferred over the other two.

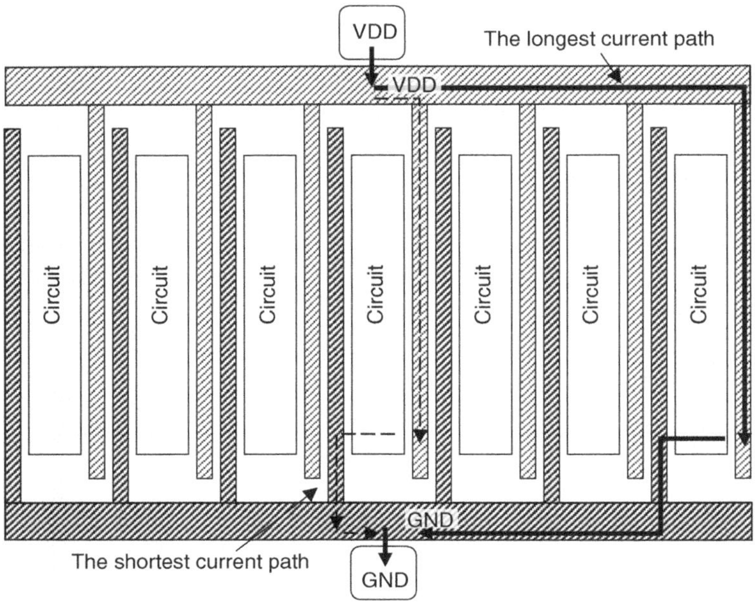

Figure Q50d

In this question there is strictly one ground pad and one power pad allowed. More power and ground pads could help to further lower the power and ground resistance, and improve the IR drop uniformity as well. For example, ground and power pads can be assigned to both sides of the pad frame. Also, at chip level, impedance of the bonding wires and other package interconnectivities on power and ground also cause IR drop and noise, and more power and ground pads would help with the reduction of the inductance of the bonding wires. Other techniques, such as *double/triple bonding* (the use of two or three parallel wire bonds to connect a single die pad to a lead), *exposed pad* (a metal plate on the bottom of an IC package that is soldered directly to the PCB to provide a low-inductance electrical ground connection and a low-resistance thermal path), and *flip chip* packaging (a packaging method where the silicon die is inverted and connected directly to a substrate or PCB using solder bumps, eliminating wire bonds) provide more methods to further lower the inductance/impedance of the interconnectivities at package level. IC packaging is beyond the scope of this book, but interested readers are encouraged to explore more into this important field of knowledge.

Short Answer to Question 50

Please explain which one of the following three VDD/GND pad arrangements is preferred and why it is preferred? The three VDD/GND arrangements are: VDD1/GND1, VDD1/GND3, and VDD2/GND2.

The pad arrangement #3 (VDD2/GND2) has the benefit of the smallest maximum total routing length $(2 \cdot L)$. It also has the smallest overall ground routing length (resistance) and variations $(\Delta = 1.5 \cdot L)$. In many systems, low and uniform ground resistance (and potential) is a priority, so arrangement #3 is generally preferred over the other two.

General References

1 Baker RJ. *CMOS: Circuit Design, Layout, and Simulation*. 3rd ed. Hoboken (NJ): Wiley/IEEE Press; 2010.
2 Hastings A. *The Art of Analog Layout*. 2nd ed. Upper Saddle River (NJ): Pearson/Prentice Hall; 2006.
3 Lienig J and Scheible J. *Fundamentals of Layout Design for Electronic Circuits*. Cham, Switzerland: Springer Nature; 2020.
4 Saint C and Saint J. *IC Mask Design: Layout-driven Performance Tuning*. New York (NY): McGraw-Hill; 2001.

Appendix 1

Device Equations

Section 1. MOSFET (long-channel) equations

Large-signal (DC) equations

- NMOS operating in the saturation region ($V_{GS} > V_{TH}, V_{DS} > V_{GS} - V_{TH}$)

$$I_D = \frac{K_p \cdot W}{2 \cdot L} \cdot (V_{GS} - V_{TH})^2 \cdot \left[1 + \lambda \cdot (V_{DS} - V_{DS,SAT})\right]$$

  K_p: transconductance parameter
  λ: channel length modulation parameter
  V_{TH}: threshold voltage
  $V_{DS,SAT}$: saturation voltage, $V_{DS,SAT} = V_{GS} - V_{TH}$

- NMOS operating in the triode (linear) region ($V_{GS} > V_{TH}, V_{DS} < V_{GS} - V_{TH}$)

$$I_D = \frac{K_p \cdot W}{L} \cdot \left[(V_{GS} - V_{TH}) \cdot V_{DS} - \frac{V_{DS}^2}{2}\right]$$

- PMOS:
  Equations are the same as those of the NMOS, but with V_{GS} replaced by V_{SG} and V_{DS} replaced by V_{SD}.

Small-signal (AC) equations

- NMOS/PMOS in the saturation region (body effect ignored) (Figure A1.1a):

$$g_m = \sqrt{2 \cdot K_p \cdot \frac{W}{L} \cdot I_D}$$

$$r_o = \frac{1}{\lambda \cdot I_{D,SAT}}$$

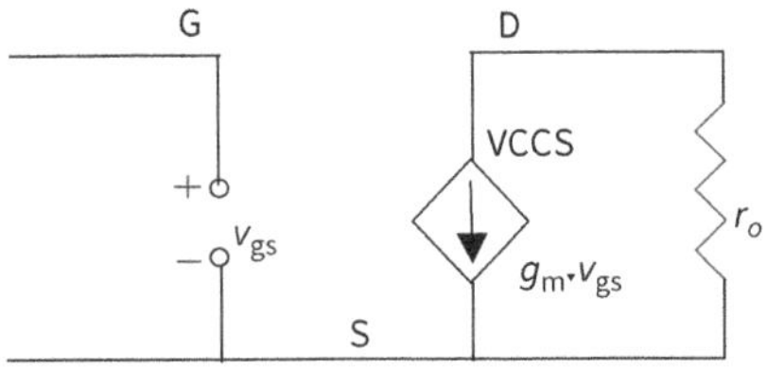

Figure A1.1a

g_m: transconductance
r_o: output resistance
$I_{D,SAT}$: saturation current; $I_{D,SAT} = I_D$ at $V_{DS} = I_{DS,\ SAT}$ (NMOS), or $V_{SD} = V_{SD,SAT}$ (PMOS)

Body effect equations

- NMOS body effect:

$$V_{THN} = V_{THN0} + \gamma_n \cdot \left(\sqrt{2 \cdot |V_{fp}| + V_{SB}} - \sqrt{2 \cdot |V_{fp}|}\right)$$

- PMOS body effect:

$$V_{THP} = V_{THP0} + \gamma_p \cdot \left(\sqrt{2 \cdot |V_{fn}| + V_{BS}} - \sqrt{2 \cdot |V_{fn}|}\right)$$

V_{THN0}/V_{THP0}: zero-bias threshold voltage of NMOS/PMOS.
γ_n/γ_p: body effect coefficient of NMOS/PMOS
V_{fp}/V_{fn}: electrostatic potential of the p-type substrate/N-well
V_{BS}/V_{SB}: bulk-to-source/source-to-bulk voltage

Parasitic capacitances

- NMOS in the saturation region as an example (Figure A1.1b):

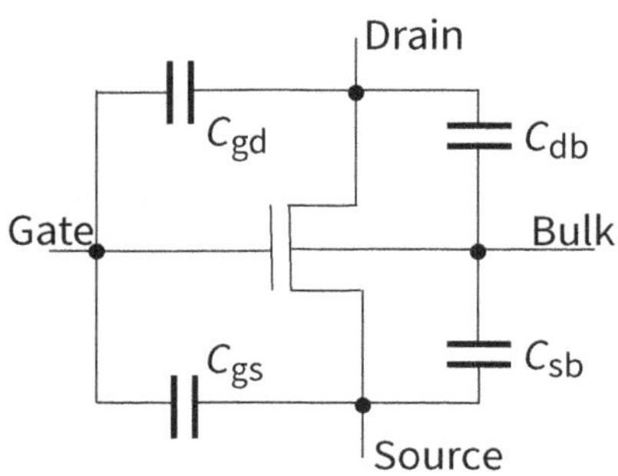

Figure A1.1b

C_{gs}: gate-to-source capacitance

$$C_{gs} = \frac{2}{3} \cdot W \cdot L \cdot C'_{ox}$$

C_{gd}: gate-to-drain capacitance

$$C_{gd} = \mathrm{CGDO} \cdot W$$

C_{sb}: source-to-bulk capacitance (junction capacitance)

$$C_{sb} = \frac{C_{j0} \cdot A_S}{\left[1 + \left(\frac{V_{SB}}{\psi_0}\right)\right]^{mj}} + \frac{C_{j0sw} \cdot P_S}{\left[1 + \left(\frac{V_{SB}}{\psi_{0sw}}\right)\right]^{mj_{sw}}}$$

C_{db}: drain-to-bulk capacitance (junction capacitance)

$$C_{db} = \frac{C_{j0} \cdot A_D}{\left[1 + \left(\frac{V_{DB}}{\psi_0}\right)\right]^{mj}} + \frac{C_{j0sw} \cdot P_D}{\left[1 + \left(\frac{V_{DB}}{\psi_{0sw}}\right)\right]^{mj_{sw}}}$$

C'_{ox}: gate-oxide capacitance per area
W and L: gate width and length, respectively
CGDO: gate-to-drain overlap capacitance per unit width
A_S/A_D: area of source/drain
P_S/P_D: perimeter of source/drain
C_{j0}: junction diode zero-bias capacitance per area for the bottom component
C_{j0sw}: junction diode zero-bias capacitance per length for the sidewall component
ψ_0: junction diode built-in potential for the bottom component
ψ_{0sw}: junction diode built-in potential for the sidewall component
mj: junction diode grading coefficient for the bottom component
mj_{sw}: junction diode grading coefficient for the sidewall component

- NMOS in the triode region:
  Equations for C_{sb} and C_{db} are the same. The gate-to-source and gate-to-drain capacitances are given by:

$$C_{gs} = C_{gd} = \frac{1}{2} \cdot W \cdot L \cdot C'_{ox}$$

Noise equations

- NMOS/PMOS in saturation region (Figure A1.1c):

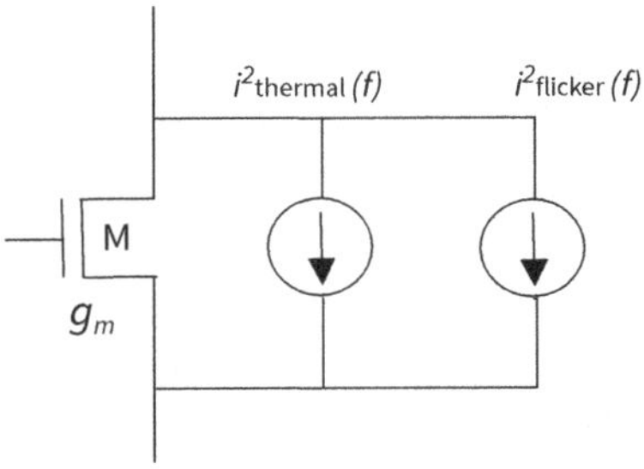

Figure A1.1c

The power spectral density (PSD) of the thermal noise current at the drain is:

$$i_{\text{thermal}}^2(f) = \frac{8 \cdot K \cdot T \cdot g_m}{3}$$

The PSD of the flicker noise current at drain is:

$$i_{\text{flicker}}^2(f) = \frac{\text{KF} \cdot I_D^{\text{AF}}}{(C'_{\text{ox}})^2 \cdot L \cdot W \cdot f}$$

KF: flicker noise coefficient
I_D: drain current
AF: flicker noise exponent (generally close to 1)
C'_{ox}: gate capacitance (C_{ox}) per area
W and L: gate width and length, respectively
K: Boltzmann's constant, $K = 1.38 \cdot 10^{-23}$ (joule/kelvin)
g_m: transconductance
T: absolute temperature (in kelvin)

Section 2. Bipolar transistor (BJT) equations

Large-signal (DC) equations (Figure A1.2a)

- NPN operating in the forward active region:

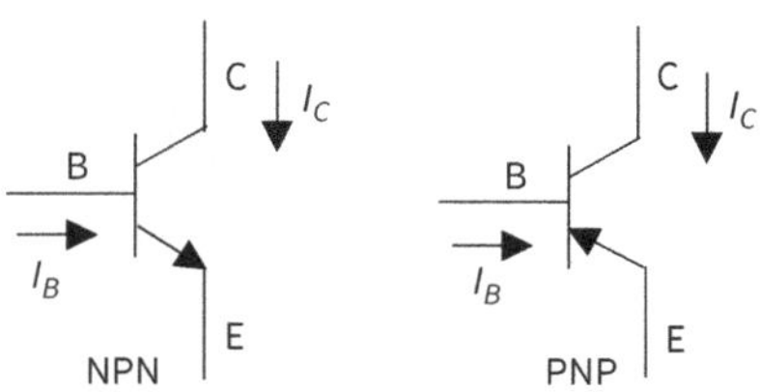

Figure A1.2a

$$I_B = \frac{I_S}{\beta_F} \cdot e^{\frac{V_{BE}}{V_T}}$$

$$I_C = \beta_F \cdot I_B$$

- PNP operating in the forward active region:

$$I_B = -\frac{I_S}{\beta_F} \cdot e^{-\frac{V_{BE}}{V_T}}$$

$$I_C = \beta_F \cdot I_B$$

β_F: forward current gain
I_S: reverse saturation current (scale current)
V_T: thermal voltage, $V_T = K \cdot T/q$

Small-signal (AC) equations

- NPN/PNP operating in the forward active region (Figure A1.2b):

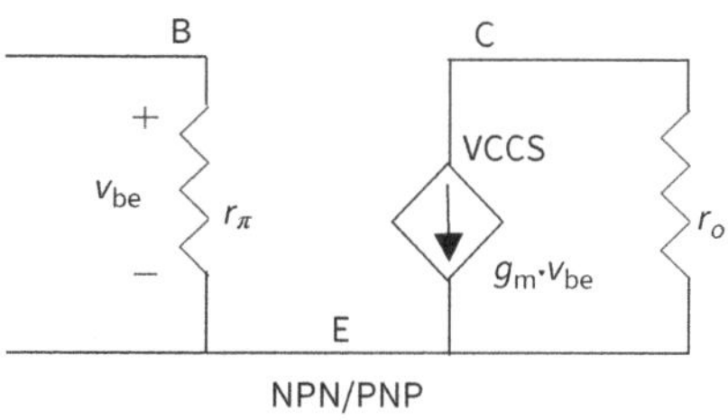

Figure A1.2b

$$g_m = \frac{q \cdot I_C}{K \cdot T} = \frac{I_C}{V_T}$$

$$r_\pi = \frac{\beta}{g_m} = \frac{V_T}{I_B}$$

$$r_o = \frac{V_A}{I_C}$$

q: elementary charge, $q = 1.6 \cdot 10^{-19}$ coulombs
K: Boltzmann's constant, $K = 1.38 \cdot 10^{-23}$ joule/kelvin
T: absolute temperature (in kelvin)

g_m: transconductance
β: small-signal current gain
r_π: input resistance
r_o: output resistance
V_T: thermal voltage, $V_T = K \cdot T/q$
V_A: Early voltage

Parasitic resistances and capacitances (NPN as an example) (Figure A1.2c)

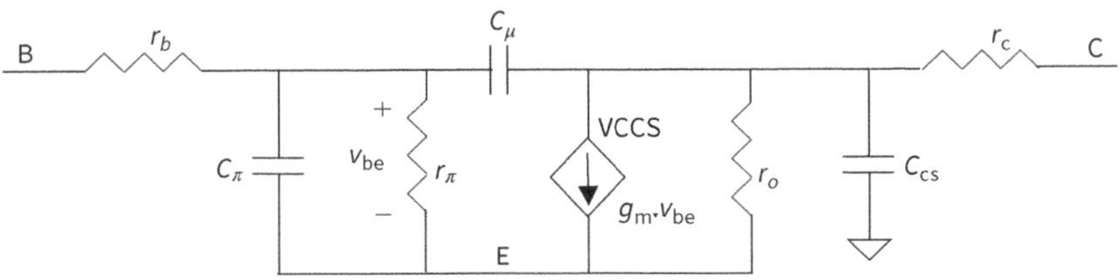

Figure A1.2c

r_b: base parasitic resistance
r_c: collector parasitic resistance
C_π: base-emitter capacitance

$$C_\pi = C_b + C_{je}$$

C_b: base charging capacitance

$$C_b = \tau_F \cdot g_m$$

C_{je}: emitter-base depletion capacitance
C_μ: base-collector capacitance

$$C_\mu = \frac{C_{\mu 0}}{\left[1 + \left(\frac{V_{CB}}{\psi_{0c}}\right)\right]^{m_c}}$$

C_{cs}: collector-substrate capacitance

$$C_{cs} = \frac{C_{cs0}}{\left[1 + \left(\frac{V_{CS}}{\psi_{0s}}\right)\right]^{m_s}}$$

τ_F: base transit time
$C_{\mu 0}$: base-collector junction diode zero-bias capacitance
C_{cs0}: collector-substrate junction diode zero-bias capacitance
ψ_{0c}: base-collector junction diode built-in potential
ψ_{0s}: collector-substrate junction diode built-in potential
m_c: base-collector junction diode grading coefficient
m_s: collector-substrate junction diode grading coefficient

Noise equations

- NPN/PNP in forward active region (Figure A1.2d):

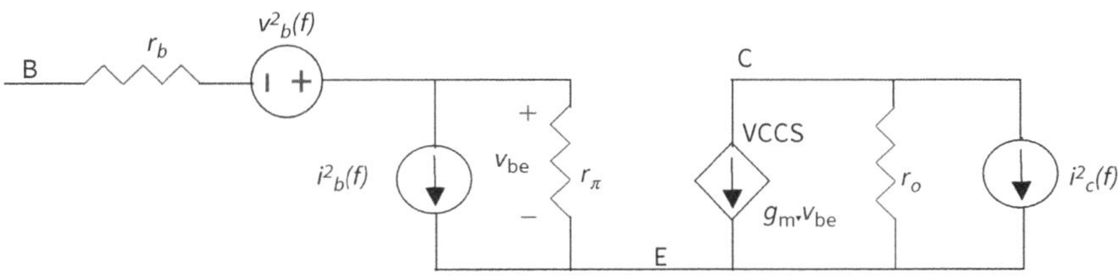

Figure A1.2d

The PSD of the base voltage noise (thermal) is:

$$v_b^2(f) = 4 \cdot K \cdot T \cdot r_b$$

The PSD of the base current noise (shot and flicker) is:

$$i_b^2(f) = 2 \cdot q \cdot I_B + \frac{K_f \cdot I_B^a}{f}$$

The PSD of the collector current noise (shot) is:

$$i_c^2(f) = 2 \cdot q \cdot I_C$$

r_b: base parasitic resistance
K: Boltzmann's constant, $K = 1.38 \cdot 10^{-23}$ joule/kelvin
T: absolute temperature (in kelvin)
q: elementary charge, $q = 1.6 \cdot 10^{-19}$ coulombs
K_f: flicker noise coefficient
a: flicker noise exponent (in the range 0.5–2)
I_B: base current (DC)
I_C: collector current (DC)

Section 3. Diode (PN junction) equations

I–V relationship

$$I_D = I_S \cdot \left(e^{\frac{V_D}{n \cdot V_T}} - 1\right)$$

$$I_D \approx I_S \cdot e^{\frac{V_D}{n \cdot V_T}}; \text{ forward biasing}$$

$$I_D \approx -I_S; \text{ reverse biasing}$$

V_T: thermal voltage, $V_T = K \cdot T/q$
I_S: reverse saturation current (scale current)
n: ideality factor (also referred to as the emission coefficient in SPICE models, typically 1 to 2)

Noise equations (Figure A1.3a)

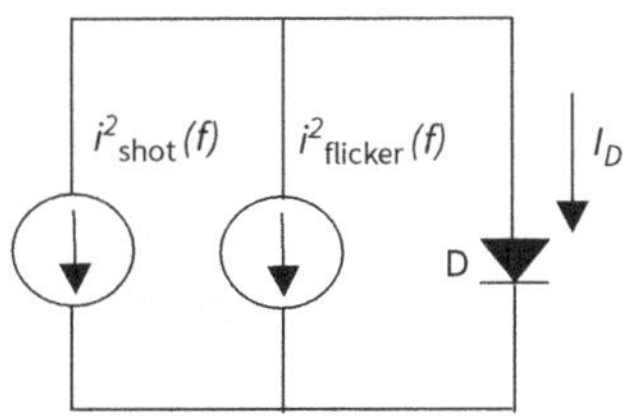

Figure A1.3a

The PSD of the shot noise current is:

$$i^2_{\text{shot}}(f) = 2 \cdot q \cdot I_D$$

The PSD of the flicker noise current is:

$$i^2_{\text{flicker}}(f) = \frac{K_f \cdot I_D^a}{f}$$

q: elementary charge, $q = 1.6 \cdot 10^{-19}$ coulombs
K_f: flicker noise coefficient
I_D: diode current (DC)
a: flicker noise exponent (in the range 0.5–2)

Section 4. Resistor, capacitor, and inductor equations

Resistor

- *I–V* relationship:

$$V = I \cdot R$$

- Thermal noise PSD (Figure A1.4a):

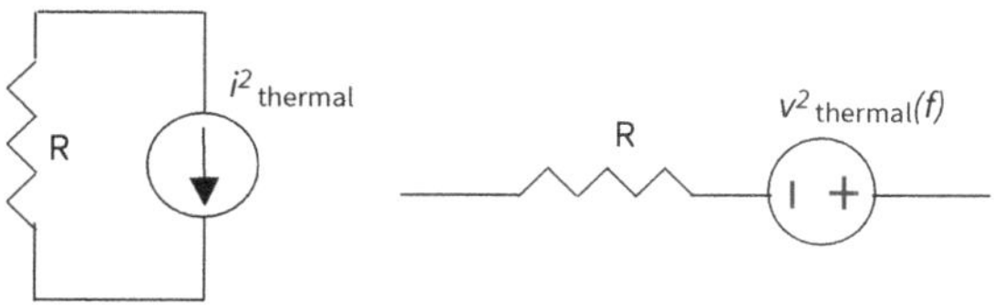

Figure A1.4a

$$i^2_{\text{thermal}}(f) = \frac{4 \cdot K \cdot T}{R}; \text{or } v^2_{\text{thermal}}(f) = 4 \cdot K \cdot T \cdot R$$

K: Boltzmann's constant, $K = 1.38 \cdot 10^{-23}$joule/kelvin
T: absolute temperature (in kelvin)

Capacitor

- *I–V* relationship:

$$I = C \cdot \frac{dV}{dt}$$

The voltage across a capacitor cannot change instantaneously.

Inductor

- *I–V* relationship:

$$V = L \cdot \frac{dI}{dt}$$

The current through an inductor cannot change instantaneously.

Appendix 2

Circuit Theorems

Section 1. Kirchhoff's circuit laws

Kirchhoff's current law (KCL)

The sum of currents flowing into any circuit node equals the sum of the currents flowing out of that node.

For example, for the circuit node A in Figure A2.1a, we have the following equation:

$$I_1 + I_3 = I_2 + I_4$$

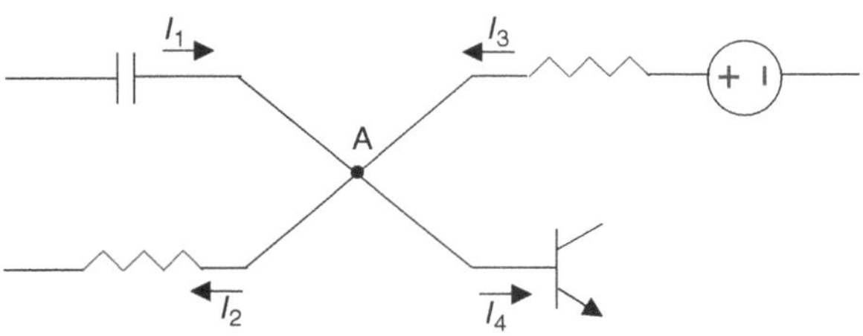

Figure A2.1a

Note that KCL applies not only to a single circuit node, but also to a circuit network. In the circuit in Figure A2.1b, the following equation holds:

$$I_1 + I_3 = I_2 + I_4$$

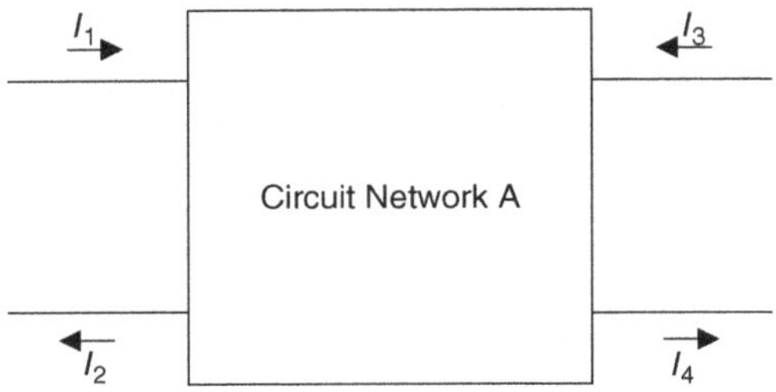

Figure A2.1b

An example of this is the terminal current relationship of a bipolar transistor, as shown in Figure A2.1c.

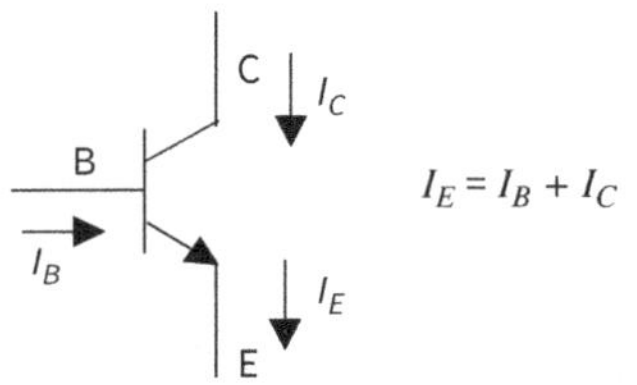

Figure A2.1c

Kirchhoff's voltage law (KVL)

The directed algebraic sum of the voltages around any closed circuit loop is zero.

For example, for the closed loop ABCD in Figure A2.1d, we have the following equation:

$$-V_1 + V_2 + V_3 + V_4 = 0$$

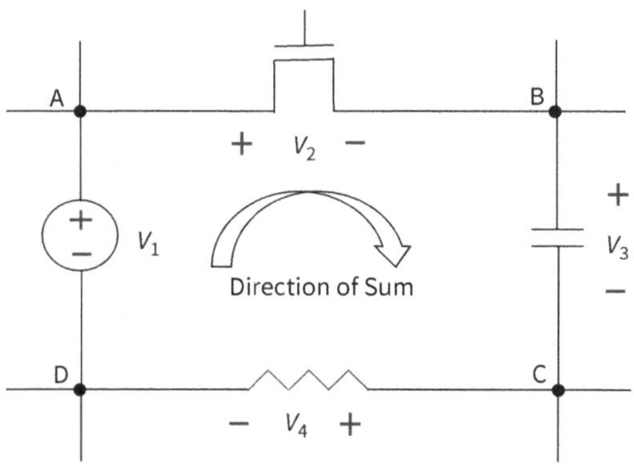

Figure A2.1d

When applying KVL, pay attention to the signs of the voltages: assign a positive sign to voltages in the same direction as the directed sum, and a negative sign to voltages opposite to the directed sum.

Section 2. Thévenin's theorem and Norton's theorem

Thévenin's theorem

Any linear electrical one-port network can be replaced at its terminals by an equivalent combination of a voltage source V_{th} in series with a resistance R_{th}.

Besides linear resistors, Thévenin's theorem also applies to linear AC circuits consisting of reactive (inductive and capacitive) impedances. Essentially, the theorem applies to networks made of linear time-invariant (LTI) components: linear impedances, independent voltage and current sources, dependent voltage and current sources, etc. (Figure A2.2a).

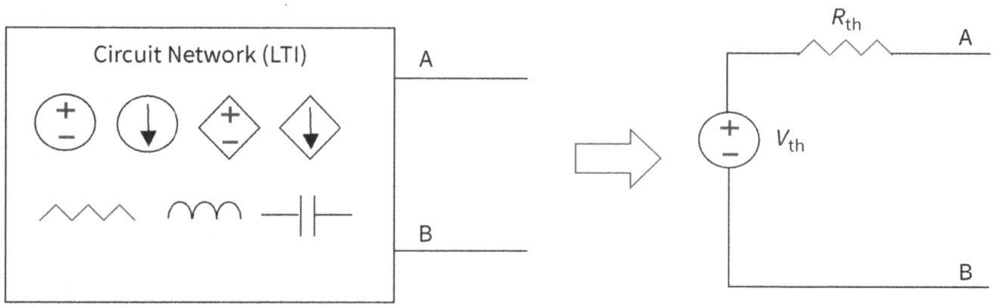

Figure A2.2a

Method to determine V_{th} and R_{th}:

- V_{th}: the voltage at terminals (A/B) of the network when the terminals (A/B) are open circuited.
- R_{th}: the resistance (or impedance) seen between terminals (A/B) when all ideal voltage sources in the circuit are replaced by short circuits and all ideal current sources are replaced by open circuits. All dependent voltage and current sources remain unchanged.

To determine the equivalent resistance R_{th} when there are dependent sources in the circuit network, a test voltage source (V_t) can be applied at terminals (A/B), and the resulting current flowing into the network (I_t) is calculated. The ratio V_t/I_t is R_{th}. Likewise, we can inject a test current (I_t) and calculate the resulting terminal voltage (V_t). The quantitative analysis of Question 1, *Input and output resistance of a transimpedance amplifier (TIA)*, is an example of calculating an equivalent resistance using a test source.

Norton's theorem

Norton's theorem is the dual of Thévenin's theorem. Instead of using an equivalent voltage source in series with a resistance to represent the circuit network, Norton's theorem employs an equivalent current source I_{no} in parallel with a resistance R_{no} (Figure A2.2b).

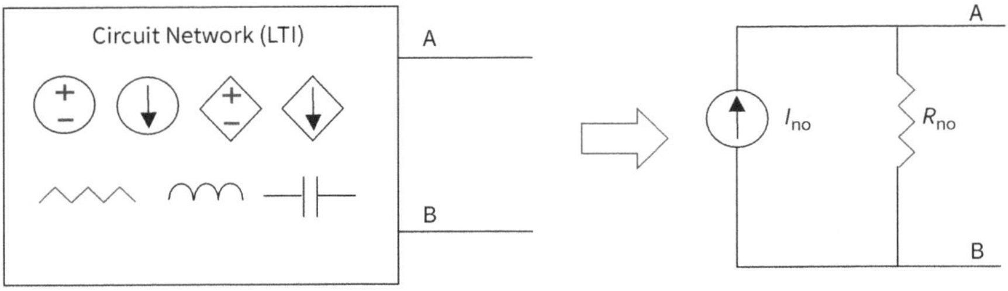

Figure A2.2b

To find the Norton equivalent I_{no} and R_{no}:

- I_{no}: the current flowing at terminals (A/B) of the original circuit when the terminals are short-circuited (i.e., zero resistance or impedance between the terminals).
- R_{no}: found using the same method as R_{th} in Thévenin's equivalent.

A Norton equivalent circuit is related to the Thévenin equivalent by the following equations (Figure A2.2c):

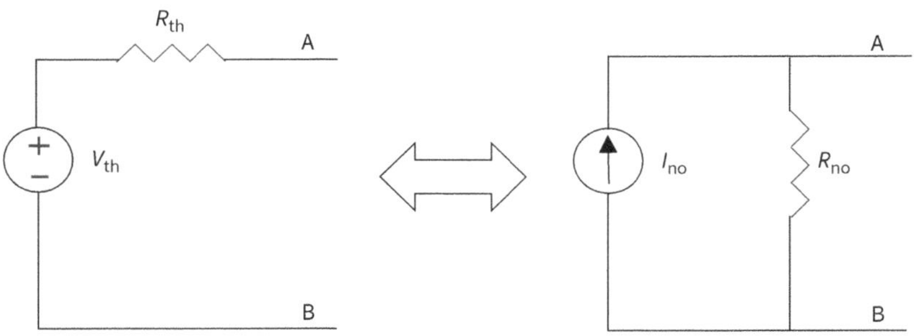

$$V_{th} = I_{no} \cdot R_{no}$$
$$R_{th} = R_{no}$$

Figure A2.2c

Section 3. The superposition theorem

The superposition theorem states that *for a linear circuit with multiple independent voltage or current sources, the total response (voltage or current) in a circuit element is the sum of the responses from each individual source acting alone.*

Steps for using the superposition theorem (Figure A2.3a):

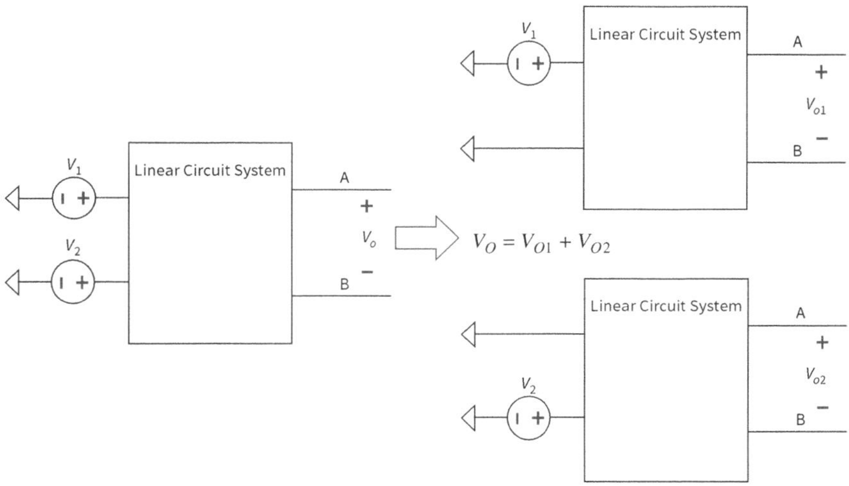

Figure A2.3a

Step 1. Keep only one independent voltage or current source at a time. Replace all other independent voltage sources with short circuits and all other independent current sources with open circuits. Dependent sources (e.g., voltage-controlled voltage source, voltage-controlled current source, etc.) remain unchanged.

Step 2. Calculate the voltages and currents due to the independent source chosen in Step 1.

Step 3. Repeat Steps 1 and 2 for each independent voltage and/or current source in the circuit.

Step 4. Superimpose the individual voltages and currents to obtain the total response.

Note that for DC and AC currents and voltages, an algebraic sum is used for superposition. For random sources, such as uncorrelated noise sources, the powers (or equivalently the variances) are summed instead. Question 10, *Device noise contribution in a folded cascode amplifier* is an example of using superposition for uncorrelated noise sources in circuits.

Section 4. Miller's theorem

The Miller's theorem states that *in a linear circuit, a branch with impedance Z connecting two nodes with node voltages V_1 and V_2 can be replaced by two branches connecting the corresponding nodes to ground. The impedances of the two branches are:*

$$Z_1 = \frac{Z}{1 - A}$$

and

$$Z_2 = \frac{Z}{1 - \frac{1}{A}}$$

where Z_1 and Z_2 are the impedance connecting the node with voltage V_1 and the node with voltage V_2 to ground, respectively. A is the voltage gain from V_1 to V_2 ($A = V_2/V_1$).

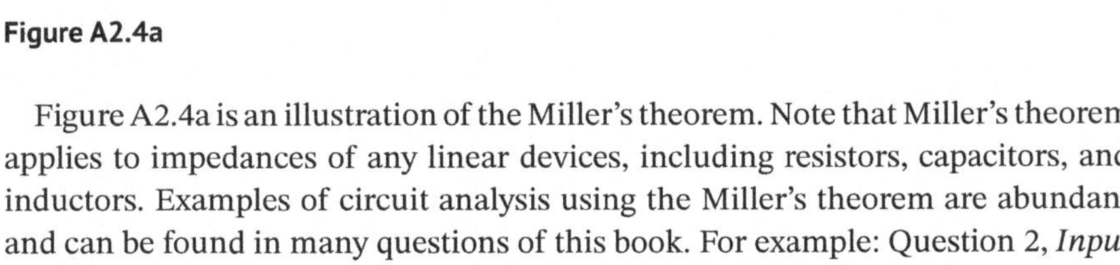

Figure A2.4a

Figure A2.4a is an illustration of the Miller's theorem. Note that Miller's theorem applies to impedances of any linear devices, including resistors, capacitors, and inductors. Examples of circuit analysis using the Miller's theorem are abundant and can be found in many questions of this book. For example: Question 2, *Input capacitance of a source follower*, Question 3, *Input capacitance of a differential pair*, Question 5, *Folded cascode amplifier and input Miller effect reduction*, and Question 8, *Compensation of a two-stage op-amp, pole splitting*.

The Miller's theorem is used to find the effective impedances to ground seen at the input and output terminals of a linear network. In many amplifier circuits, an impedance branch serves a dual role as both a feedback and a feedforward path. When the theorem is applied using a simplified gain approximation, the crucial high-frequency zero introduced by the feedforward path is often ignored and must be calculated separately. A good example can be found in Question 8, *Compensation of a two-stage op-amp, pole splitting*.

Appendix 3

Circuit Analysis in the Complex Frequency Domain (*s*-domain)

Section 1. Laplace transform and transfer functions

Laplace transform and complex frequency domain (*s*-domain)

The Laplace transform converts a time-domain function into an *s*-domain function:

$$F(s) = \mathcal{L}\left[f(t)\right] = \int_{0}^{\infty} f(t) \cdot e^{-s \cdot t} dt$$

The parameter *s* belongs to the complex frequency domain:

$$s = \sigma + j \cdot \omega$$

The real part, σ, is the *damping factor* and is related to the time constant, while the imaginary part, ω, represents the angular frequency. The former is associated with the transient response, while the latter is associated with the steady-state (frequency) response.

One of the most important applications of the Laplace transform is solving time-domain differential equations. Laplace transform can turn time-domain differential equations into *s*-domain algebraic equations due to its derivative property:

$$\mathcal{L}\left[\frac{df(t)}{dt}\right] = s \cdot F(s) - f(0)$$

$$\mathcal{L}\left[\frac{d^2f(t)}{dt^2}\right] = s^2 \cdot F(s) - s \cdot f(0) - \left.\frac{df(t)}{dt}\right|_{t=0}$$

$$\mathcal{L}\left[\frac{d^n f(t)}{dt^n}\right] = s^n \cdot F(s) - \sum_{k=1}^{n} s^{n-k} \cdot \frac{df^{k-1}(t)}{dt^{k-1}}\Bigg|_{t=0}$$

The $f(0)$, $df(t)/dt|t = 0$, etc., are initial values when $t = 0$. Applying the Laplace transform converts an nth-order linear ordinary differential equation (ODE) in the time domain (t) into an nth-degree algebraic equation in the s-domain. This transformation simplifies differential equation analysis and supports the use of root locus and other analytical methods. These tools predict time-domain performance by mapping the poles and zeros of the transfer function onto the complex s-plane. s-domain analysis has proven to be an effective tool for circuit analysis. These studies are described in *control systems theory* and beyond the scope of this book. In this appendix, however, we will introduce some basic yet useful concepts, equations, and results from s-domain analysis for circuit designers without control systems background.

S-domain impedance of resistors, capacitors, and inductors

Resistor

- I–V relationship:

$$V(t) = I(t) \cdot R$$

- s-domain impedance:

$$Z(s) = \frac{V(s)}{I(s)} = R$$

Capacitor

- I/V relationship:

$$I(t) = C \cdot \frac{dV(t)}{dt}$$

- S-domain impedance:

$$Z(s) = \frac{V(s)}{I(s)} = \frac{1}{C \cdot s}$$

Inductor

- I/V relationship:

$$V(t) = L \cdot \frac{dI(t)}{dt}$$

- s-domain impedance:

$$Z(s) = \frac{V(s)}{I(s)} = L \cdot s$$

Note that when calculating s-domain impedances (and transfer functions which will be discussed later), the initial value at $t = 0$ is assumed to be zero.

When using s-domain impedances to calculate node voltages and branch currents, we treat them as if they were resistance. The following example illustrates how.

Figure A3.1a shows a passive RC filter. Let us use s-domain impedances to determine the relationship between the output voltage $V_{out}(s)$ and the input voltage $V_{in}(s)$.

First, all circuit components are replaced by their s-domain impedances, denoted by Z. The resulting circuit is shown in Figure A3.1b. Next, we treat all

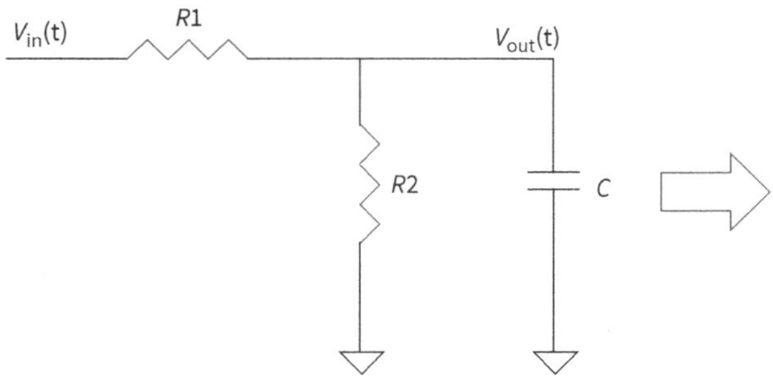

Figure A3.1a

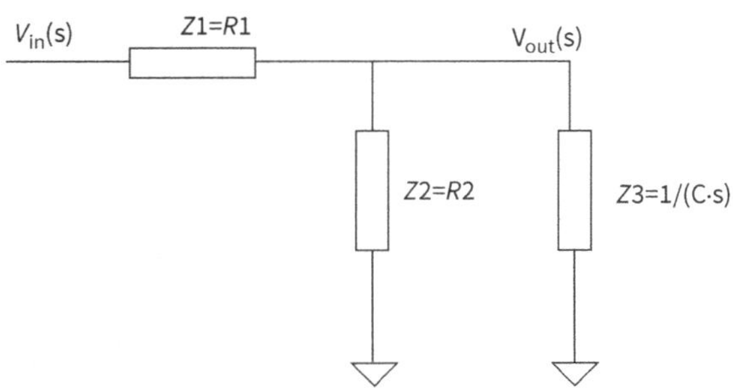

Figure A3.1b

s-domain impedances as if they were resistances, and calculate the output voltage $V_{out}(s)$.

It is clear that Z_1, Z_2, and Z_3 form a voltage divider, and the output voltage $V_{out}(s)$ can be calculated by:

$$V_{out}(s) = \frac{Z_2 \parallel Z_3}{Z_1 + Z_2 \parallel Z_3} \cdot V_{in}(s) = \frac{R_2}{R_1 \cdot R_2 \cdot C \cdot s + R_1 + R_2} \cdot V_{in}(s)$$

Transfer functions

The (s-domain) transfer function of a system is defined as its output divided by its input, i.e.,

$$G(s) = \frac{V_{out}(s)}{V_{in}(s)}$$

Mathematically, the general form of an nth-order transfer function is:

$$G(s) = \frac{b_m \cdot s^m + b_{m-1} \cdot s^{m-1} + \ldots + b_1 \cdot s + b_0}{a_n \cdot s^n + a_{n-1} \cdot s^{n-1} + \ldots + a_1 \cdot s + a_0}; m < n$$

If we set the denominator polynomial equal to zero, we obtain the *characteristic equation*:

$$a_n \cdot s^n + a_{n-1} \cdot s^{n-1} + \ldots + a_1 \cdot s + a_0 = 0$$

The solutions (roots) of the characteristic equation are called poles. The solutions of the numerator polynomial set to zero are called zeros. Poles and zeros of a system determine the time response of the system, just like the coefficients of an ordinary differential equation determine its solutions.

Note that poles and zeros are complex numbers that can be plotted in the complex plane, which consists of a real axis and an imaginary axis. A pole or zero located at the left side of the imaginary axis is called a left-hand-plane pole or zero (LHP pole or zero). The real part of a left-hand-plane pole or zero is a negative number. Similarly, a pole or zero located at the right side of the imaginary axis is called a right-hand-plane pole or zero (RHP pole or zero).

Given the transfer function and the input, the output of a system can now be calculated as:

$$V_{out}(s) = G(s) \cdot V_{in}(s)$$

To translate an s-domain signal back to the time domain, we use the inverse Laplace transform. In engineering practice, the Laplace transforms of common functions are tabulated to facilitate analytical efficiency and streamline the solution of complex systems. Table A3.1a lists Laplace transform results of some common types of electrical signals. Note that this table serves a dual purpose – from left to right, Laplace transform; from right to left, inverse Laplace transform.

Table A3.1a

Signal Function Name	*f(t)*	*F(s)*	Note
Impulse	$\delta(t)$	1	
Unit step (Heaviside)	$u(t)$	$1/s$	time: $[0, \infty)$
Ramp	$t \cdot u(t)$	$1/s^2$	
Power	$t^n \cdot u(t)$	$n!/s^{n+1}$	
Exponential	$e^{-a \cdot t} \cdot u(t)$	$1/(s+a)$	
Sinusoidal	$\sin(\omega_0 \cdot t) \cdot u(t)$	$\omega_0/\left(s^2+\omega_0^2\right)$	
Cosinusoidal	$\cos(\omega_0 \cdot t) \cdot u(t)$	$s/\left(s^2+\omega_0^2\right)$	
Damped sinusoidal	$e^{-a \cdot t} \cdot \sin(\omega_0 \cdot t) \cdot u(t)$	$\omega_0/\left[(s+a)^2+\omega_0^2\right]$	
Damped cosinusoidal	$e^{-a \cdot t} \cdot \cos(\omega_0 \cdot t) \cdot u(t)$	$(s+a)/\left[(s+a)^2+\omega_0^2\right]$	
Damped power	$t^n \cdot e^{-a \cdot t} \cdot u(t)$	$n!/(s+a)^{n+1}$	

Section 2. First-order systems

The standard form of the transfer function of a stable first-order system is:

$$G(s) = \frac{K}{\tau \cdot s + 1}$$

where K is the DC gain and τ is the time constant of the system.

The step-response settling time T_s (for a 5% error band) is approximately:

$$T_s \approx 3 \cdot \tau$$

The time constant τ of a first-order system is related to its -3 dB bandwidth (or cutoff frequency), $f_{-3\text{dB}}$, by:

$$\tau = \frac{1}{2 \cdot \pi \cdot f_{-3\text{dB}}} = \frac{1}{\omega_{-3\text{dB}}}$$

where f_{-3dB} is the −3 dB linear frequency, and ω_{-3dB} is the −3 dB angular frequency.

The relationship between the time constant τ and the pole p of a first-order system (in angular frequency) is given by:

$$\tau = -\frac{1}{p}$$

Circuit example: the transfer function of the circuit in Figure A3.1a is given by:

$$G(s) = \frac{V_{out}(s)}{V_{in}(s)} = \frac{R_2}{R_1 \cdot R_2 \cdot C \cdot s + R_1 + R_2}$$

Rearranging it into the standard form gives:

$$G(s) = \frac{\frac{R_2}{R_1 + R_2}}{\frac{R_1 \cdot R_2}{R_1 + R_2} \cdot C \cdot s + 1}$$

The DC gain, time constant, and −3 dB cutoff frequency of the circuit are:

$$\text{DC gain} = \frac{R_2}{R_1 + R_2}$$

$$\tau = \frac{R_1 \cdot R_2}{R_1 + R_2} \cdot C = (R_1 \parallel R_2) \cdot C$$

$$f_{-3dB} = \frac{1}{2 \cdot \pi \cdot \tau} = \frac{1}{2 \cdot \pi \cdot \frac{R_1 \cdot R_2}{R_1 + R_2} \cdot C} = \frac{1}{2 \cdot \pi \cdot (R_1 \parallel R_2) \cdot C}$$

This is a first-order low-pass filter with the DC gain less than 1 and cutoff frequency determined by R_1, R_2, and C.

In contrast to its stable counterpart, the transfer function of an *unstable first-order system* is defined by a positive real pole, taking the form:

$$G(s) = \frac{K}{\tau \cdot s - 1}$$

An example of such a system is the regenerative latch (see *Question 33, Comparator – regenerative latch*).

Section 3. Transfer function of feedback systems

The closed-loop transfer function of a feedback system can be derived from that of its open-loop.

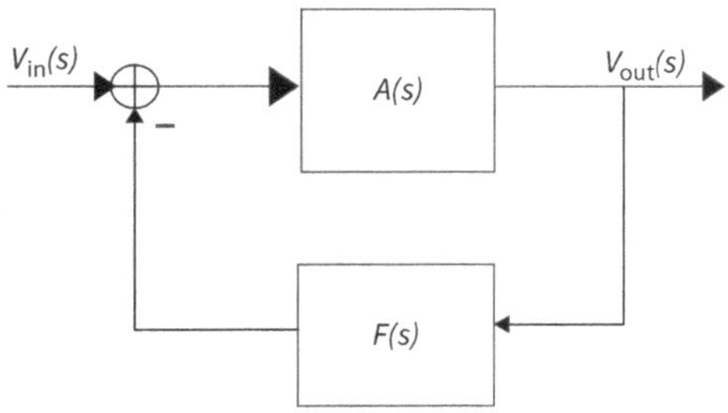

Figure A3.3a

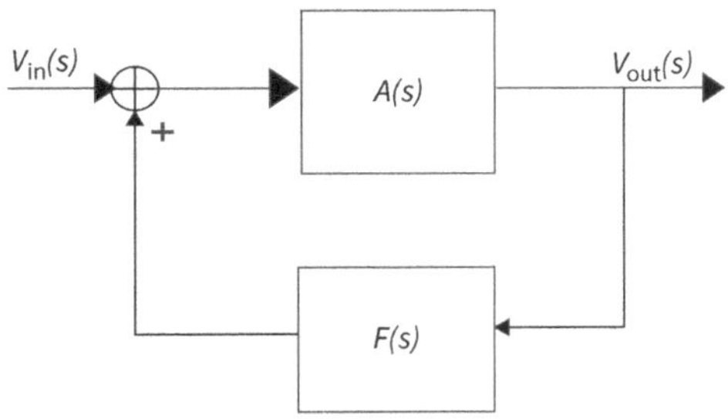

Figure A3.3b

For a negative-feedback system as shown in Figure A3.3a, the open-loop transfer function is:

$$G(s) = A(s) \cdot F(s)$$

$A(s)$: the transfer function of the forward path.
$F(s)$: the transfer function of the feedback path.

The closed-loop transfer function is given by:

$$\varnothing(s) = \frac{A(s)}{1 + A(s) \cdot F(s)}$$

For a unity-gain feedback system with $F(s) = 1$, the open-loop transfer function is the same as that of the forward path:

$$G(s) = A(s)$$

And the closed-loop transfer function becomes:

$$\varnothing(s) = \frac{A(s)}{1 + A(s)}$$

For a positive-feedback system, as shown in Figure A3.3b, the closed-loop transfer function is:

$$\varnothing(s) = \frac{A(s)}{1 - A(s) \cdot F(s)}$$

Circuit example: Assume a one-stage amplifier has a left-hand-plane pole (LHP pole) with absolute value of p_1 (in angular frequency, or ω) and DC gain of A. Derive the closed-loop DC gain and -3 dB bandwidth if the amplifier is used in a unity-gain feedback loop.

The open-loop transfer function of an amplifier with unity-gain feedback is the transfer function of the amplifier itself, which can be written as:

$$G(s) = A(s) = \frac{A}{\frac{1}{p_1} \cdot s + 1}$$

The closed-loop transfer function with unity-gain feedback is:

$$\varnothing(s) = \frac{A(s)}{1 + A(s)} = \frac{A}{\frac{1}{p_1} \cdot s + A + 1} = \frac{\frac{A}{A+1}}{\frac{1}{p_1 \cdot [A+1]} \cdot s + 1} \approx \frac{1}{\frac{1}{p_1 \cdot A} \cdot s + 1}$$

With unity-gain feedback, the closed-loop DC gain drops from A to approximately 1, assuming $A \gg 1$. The time constant of the closed-loop system is reduced by a factor of approximately A (the open-loop gain of the amplifier), and the -3 dB bandwidth of the closed-loop system increases by a factor of approximately A.

Section 4. Second-order systems

The standard format of the transfer function of a stable second-order system is:

$$G(s) = \frac{K \cdot \omega_n^2}{s^2 + 2 \cdot \zeta \cdot \omega_n \cdot s + \omega_n^2}$$

K: DC gain

ζ: damping ratio. For a stable second-order system, $\zeta \geq 0$, because if $\zeta < 0$ the system has right-hand-plane poles and becomes unstable.

ω_n: natural frequency, $\omega_n > 0$.

The damping ratio ζ (Greek letter "zeta") plays an important role in the dynamics of a second-order system. In fact, the discriminant, D(s), of the characteristic equation of the standard second-order system is:

$$D(s) = (2 \cdot \zeta \cdot \omega_n)^2 - 4 \cdot \omega_n^2 = 4 \cdot \omega_n^2 \cdot (\zeta^2 - 1)$$

This means that as long as $\zeta \geq 1$, the poles of the system are real numbers. The poles become complex numbers and have imaginary parts when $\zeta < 1$. Depending on the damping ratio ζ, a second-order system can fall into one of the four categories:

$\zeta > 1$: the system is *overdamped*. No overshoot in step response and no peaking in magnitude frequency response. The larger the ζ, the slower the time-domain settling. The system has two real poles.

$\zeta = 1$: the system is *critically damped*. No overshoot in step response and no peaking in magnitude frequency response. The system has double real poles.

$0 < \zeta < 1$: the system is *underdamped*. Overshoot and ringing occur in step response, and peaking occurs in the magnitude frequency response for $\zeta < 0.6$. The smaller the ζ, the larger the overshoot/ringing and peaking. The time-domain ringing is caused by the imaginary part of the poles when the damping ratio is less than 1.

$\zeta = 0$: the system is *undamped and marginally stable* with undamped oscillation of constant amplitude. The frequency of the oscillation is the natural frequency ω_n. Both poles of the system are pure imaginary and a complex conjugate pair.

A rule-of-thumb estimation of the step-response settling time T_s (for a 5% error band) of a second-order system when $\zeta < 0.8$ is given by:

$$T_s = \frac{3.5}{\zeta \cdot \omega_n}; \text{ for } \zeta < 0.8$$

Note that the above equation can only be used when the damping ratio is less than 0.8. The settling time of an overdamped second-order system is typically longer than that of a slightly underdamped system. A damping ratio of approximately 0.707 is often considered optimal because it provides the best trade-off between settling time and overshoot/ringing for a second-order step response.

The −3 dB bandwidth of an underdamped second-order system is given by:

$$f_{-3\text{dB}} = \frac{\omega_n}{2 \cdot \pi} \cdot \sqrt{(1 - 2 \cdot \zeta^2) + \sqrt{(1 - 2 \cdot \zeta^2)^2 + 1}}$$

When $\zeta = 0.707$ (optimal damping ratio), the −3 dB bandwidth is $\omega_n/(2 \cdot \pi)$. For a given damping ratio, the larger the natural frequency, the wider the −3 dB bandwidth.

Circuit example: Assume a two-stage op-amp has two real left-hand-plane poles (LHP poles) with absolute values of p_1 and p_2 (in angular frequency) associated with the first and second stage, and the DC gain of the two stages are A_1 and A_2, respectively. Assume $p1$ is the dominant pole ($p_1 \ll p_2$) and the op-amp approximates a first-order system up to unity-gain frequency. Derive the closed-loop DC gain and −3 dB bandwidth if the amplifier is used in a unity-gain feedback loop.

For a unity-gain amplifier, the open-loop transfer function reduces to that of the op-amp, given by:

$$G(s) = A(s) = \frac{A_1}{\frac{1}{p_1} \cdot (s + p_1)} \cdot \frac{A_2}{\frac{1}{p_2} \cdot (s + p_2)}$$

The closed-loop transfer function with unity-gain feedback is:

$$\varnothing(s) = \frac{A(s)}{1 + A(s)} = \frac{A_1 \cdot A_2 \cdot p_1 \cdot p_2}{s^2 + (p_1 + p_2) \cdot s + (1 + A_1 \cdot A_2) \cdot p_1 \cdot p_2}$$

$$\approx \frac{A_1 \cdot A_2 \cdot p_1 \cdot p_2}{s^2 + (p_1 + p_2) \cdot s + A_1 \cdot A_2 \cdot p_1 \cdot p_2}$$

The damping ratio and natural frequency are:

$$\zeta = \frac{p_1 + p_2}{2 \cdot \sqrt{A_1 \cdot A_2 \cdot p_1 \cdot p_2}}$$

$$\omega_n = \sqrt{A_1 \cdot A_2 \cdot p_1 \cdot p_2}$$

If the nondominant pole p_2 is located near the open-loop unity-gain frequency (angular frequency), i.e.,

$$p_2 = A_1 \cdot A_2 \cdot p_1$$

the system parameters are approximated as:

$$\zeta = \frac{p_1 \cdot (1 + A_1 \cdot A_2)}{2 \cdot A_1 \cdot A_2 \cdot p_1} \approx 0.5; \omega_n = A_1 \cdot A_2 \cdot p_1$$

The closed-loop system is underdamped with a damping ratio of 0.5, and significant overshoot and ringing can be expected at the closed-loop step response.

If the nondominant pole p_2 is two times the open-loop unity-gain frequency (angular frequency), i.e.,

$$p_2 = 2 \cdot A_1 \cdot A_2 \cdot p_1$$

the system parameters are approximated as:

$$\zeta = \frac{p_1 \cdot (1 + 2 \cdot A_1 \cdot A_2)}{2 \cdot \sqrt{2} \cdot A_1 \cdot A_2 \cdot p_1} \approx 0.707; \omega_n \approx 1.414 \cdot A_1 \cdot A_2 \cdot p_1$$

The closed-loop system is underdamped with an optimal damping ratio of 0.707, and has an overall optimized settling time and overshoot.

If we further push the nondominant pole p_2 to four times the open-loop unity-gain frequency (angular frequency), i.e.,

$$p_2 = 4 \cdot A_1 \cdot A_2 \cdot p_1$$

the system parameters are approximated as:

$$\zeta = \frac{p_1 \cdot (1 + 4 \cdot A_1 \cdot A_2)}{2 \cdot 2 \cdot A_1 \cdot A_2 \cdot p_1} \approx 1; \omega_n = 2 \cdot A_1 \cdot A_2 \cdot p_1$$

Now the closed-loop system is critically damped with a damping ratio of 1. No overshoot and ringing in step response.

We notice that when we push the nondominant pole p_2 to higher frequencies, the damping ratio of the closed-loop system increases, and time-domain overshoot and ringing are reduced and eventually eliminated. This observation aligns with the principle that increasing the separation between the dominant and non-dominant poles enhances the phase margin, thereby resulting in a more stable closed-loop system.

Another interesting observation is that in a two-stage amplifier, even though the two open-loop poles generated by the amplifier *RC* circuits are real numbers, the closed-loop poles can be complex numbers depending on the values of the gains and poles of the two stages.

It should be noted that the practical two-stage op-amps often have more than two poles, and one or more zeros (e.g., zero caused by compensation capacitor). These additional poles and zeros also impact the time-domain response of the closed-loop amplifier. Nevertheless, higher-order systems can often be approximated by first- or second-order models, provided that one or two dominant poles significantly influence the transient response.

Section 5. *S*-domain and frequency domain

The relationship between Laplace transform and Fourier transform

In the Laplace transform of a function, if we replace s with $j \cdot \omega$ (i.e., let $\sigma = 0$ and $s = j \cdot \omega$), the result is the Fourier transform of the same function. Mathematically, the Fourier transform is a special case of the Laplace transform. When we discard the real part σ (damping factor) and keep the imaginary part ω (angular frequency) in the Laplace transform, the domain changes from the two-dimensional s-plane to the single $j \cdot \omega$-axis. The result is the steady-state frequency characteristics of the function.

In other words, the frequency response of a system can be derived from its transfer function by replacing s with $j \cdot \omega$:

$$G(j \cdot \omega) = G(s \mid s = j \cdot \omega) = a(\omega) + b(\omega) \cdot j$$

The magnitude frequency response is:

$$|G(j \cdot \omega)| = \sqrt{a(\omega)^2 + b(\omega)^2}$$

The phase frequency response is:

$$\angle G(j \cdot \omega) = \arctan\left[\frac{b(\omega)}{a(\omega)}\right]$$

The frequency response (Bode plot) of some typical elements

- Integrator element (Figure A3.5a):

$$G(s) = \frac{1}{s}$$

Bode Plot:

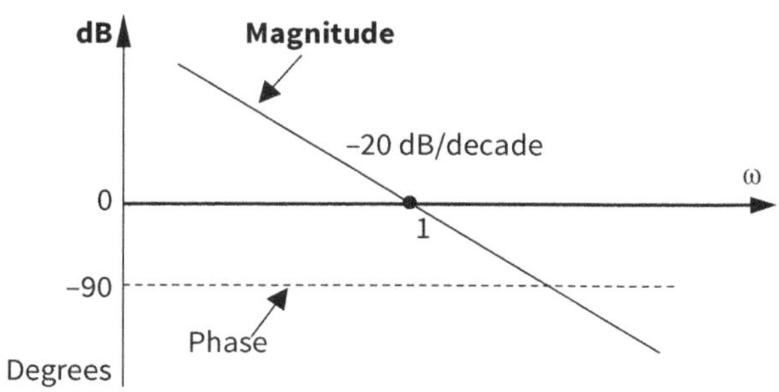

Figure A3.5a

- Differentiator element (Figure A3.5b):

$$G(s) = s$$

Bode Plot:

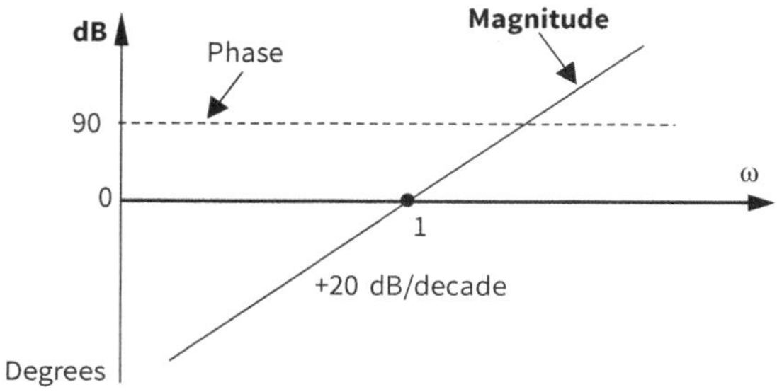

Figure A3.5b

- First-order (inertial) element (Figure A3.5c):

$$G(s) = \frac{1}{\tau \cdot s + 1}; \tau > 0$$

Bode Plot:

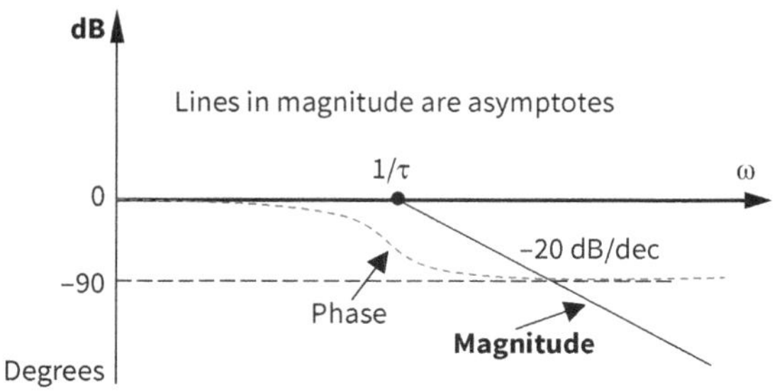

Figure A3.5c

- First-order differentiator element (an example of left-hand-plane zero) (Figure A3.5d):

$$G(s) = 1 + \tau \cdot s; \tau > 0$$

Bode Plot:

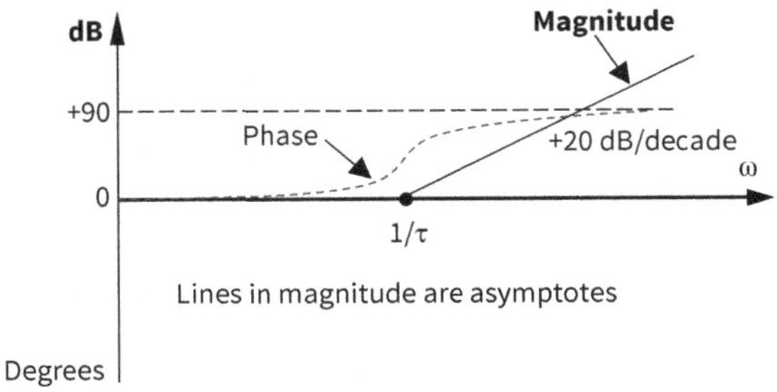

Figure A3.5d

- Unstable first-order differentiator element (An example of right-hand-plane zero) (Figure A3.5e):

$$G(s) = 1 - \tau \cdot s; \tau > 0$$

Bode Plot:

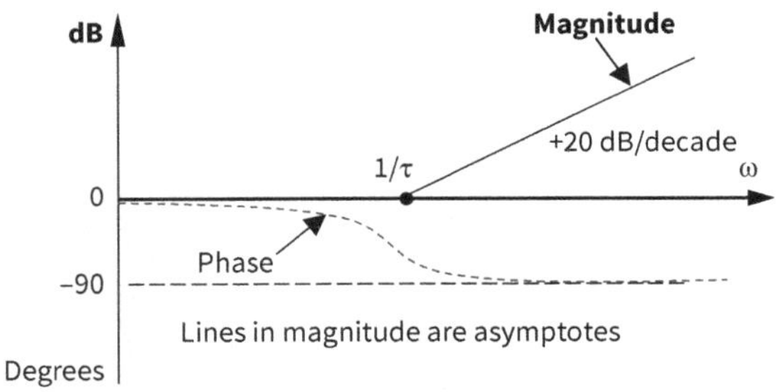

Figure A3.5e

- Second-order (oscillation) element (Figure A3.5f):

$$G(s) = \frac{\omega_n^2}{s^2 + 2 \cdot \zeta \cdot \omega_n \cdot s + \omega_n^2}$$

$$\zeta \geq 0, \omega_n > 0$$

Bode Plot:

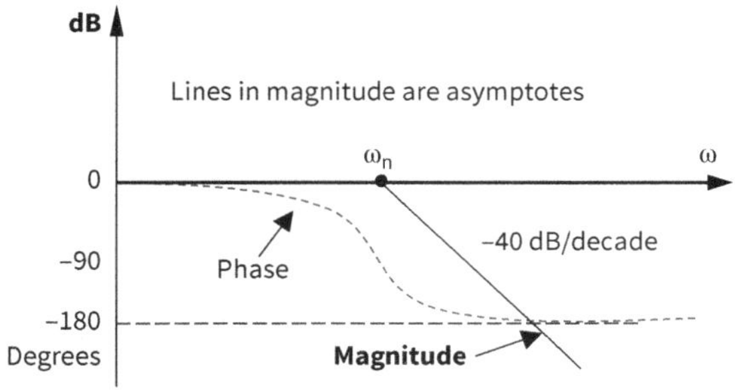

Figure A3.5f

Appendix 4

Basic Circuit Analysis Examples

Section 1. Voltage and current dividers

Example 1: Resistor voltage divider

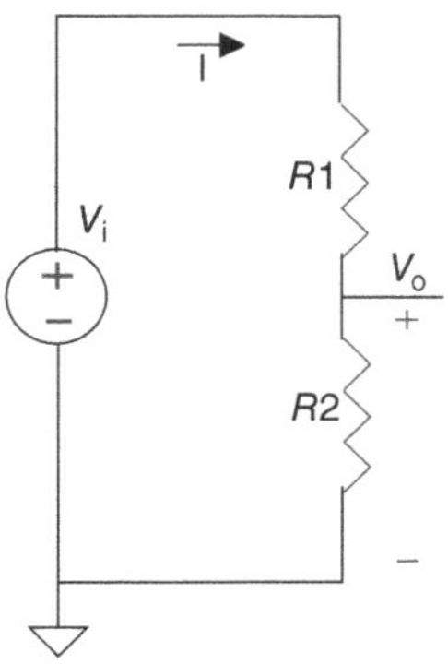

Figure A4.1a

Voltage dividers are one of the most basic circuit topologies. For a resistor divider as shown in Figure A4.1a, we can apply KVL (see *Appendix 2, Section 1, Kirchhoff's circuit laws*) and Ohm's Law:

$$\begin{cases} I \cdot R_1 + I \cdot R_2 = V_i \\ V_o = I \cdot R_2 \end{cases}$$

The output voltage, V_o, can be written as:

$$V_o = \frac{R_2}{R_1 + R_2} \cdot V_i$$

This means a larger resistance results in a larger voltage drop. The result can be applied to other types of impedance as well, e.g., capacitor dividers.

Example 2: Capacitor voltage divider

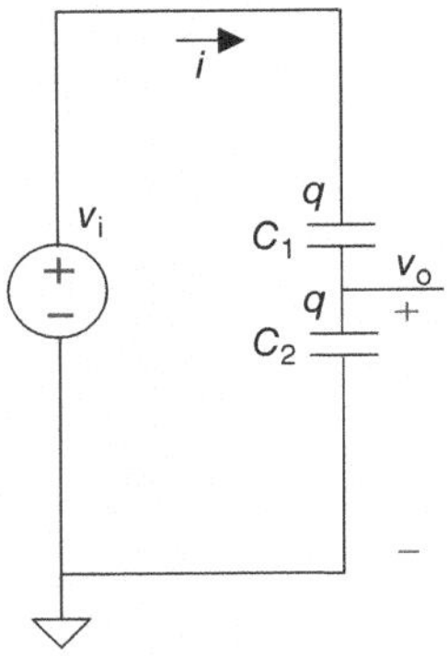

Figure A4.1b

We can again apply KVL and the basic capacitor equations to obtain the output voltage v_o.

$$\begin{cases} \dfrac{q}{C_1} + \dfrac{q}{C_2} = v_i \\ v_o = \dfrac{q}{C_2} \end{cases}$$

Note that capacitors in series have equal charge (charge conservation – one way to think about this is to look at the isolated node at v_o in Figure A4.1b). The output voltage is:

$$v_o = \frac{C_1}{C_1 + C_2} \cdot v_i$$

The result is similar to that of the resistor divider – larger impedance (smaller capacitance) has larger voltage drop. In fact, we can treat capacitors as resistors by using their s-domain impedance (see *Appendix 3, Section 1, Laplace transform and transfer functions*), and obtain the same result:

$$v_o(s) = \frac{Z_2}{Z_1 + Z_2} \cdot v_i(s) = \frac{\dfrac{1}{C_2 \cdot s}}{\dfrac{1}{C_1 \cdot s} + \dfrac{1}{C_2 \cdot s}} \cdot v_i(s) = \frac{C_1}{C_1 + C_2} \cdot v_i(s)$$

An example of a capacitor divider can be found in Question 32, Switched-capacitor common-mode feedback control (CMFB) circuits.

In general, the voltage divider equation can be applied to any type (or combination of types) of impedance (Figure A4.1c):

$$v_o(s) = \frac{Z_2}{Z_1 + Z_2} \cdot v_i(s)$$

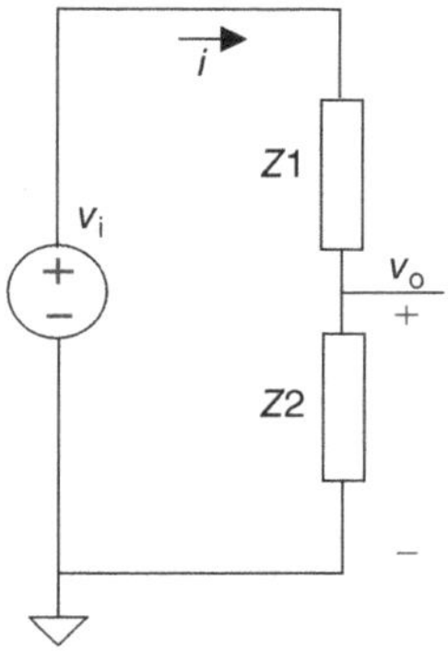

Figure A4.1c

Example 3: Resistor current divider

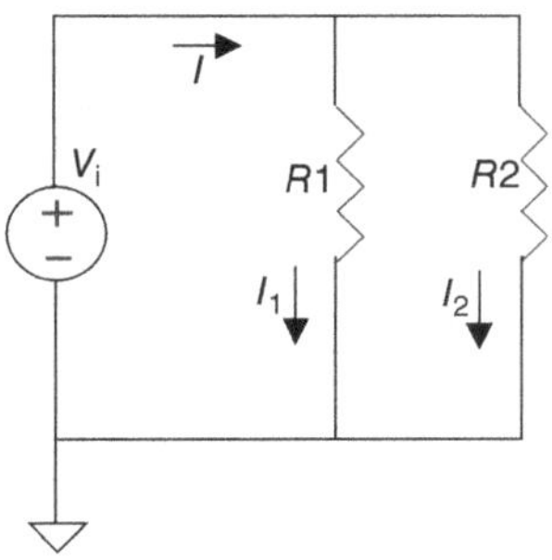

Figure A4.1d

A current divider diverts current into two or more branches as shown in Figure A4.1d. We can apply KCL and Ohm's Law, noting that resistors in parallel share the same terminal voltage:

$$\begin{cases} I_1 \cdot R_1 = I_2 \cdot R_2 = V_i \\ I = I_1 + I_2 \end{cases}$$

The system of equations yields the following relationship:

$$\frac{I_1}{I_2} = \frac{R_2}{R_1}; \text{ and } I_1 + I_2 = I$$

This result shows that more current flows through the branch with lower resistance. Similar to voltage dividers, this result can be applied to other types of impedance (e.g., capacitance, inductance). In general, we have the following equations for current dividers (Figure A4.1e):

$$\frac{i_1}{i_2} = \frac{Z_2}{Z_1}; \text{ and } i_1 + i_2 = i$$

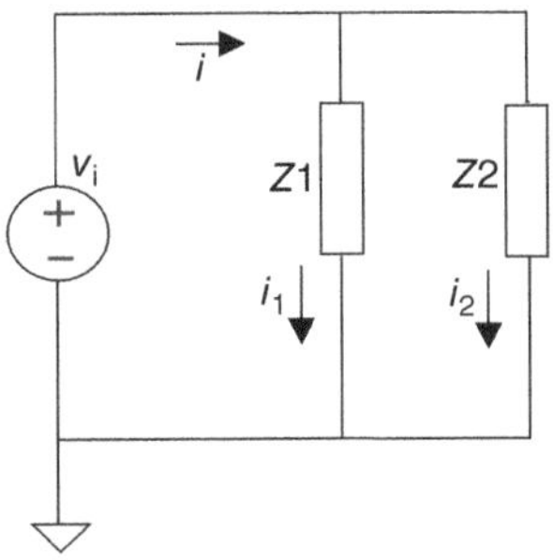

Figure A4.1e

In the example of an application of current dividers as shown in Figure A4.1f, the current (I) from the master bias circuit is divided between two identical local bias generators, each receiving half the original current.

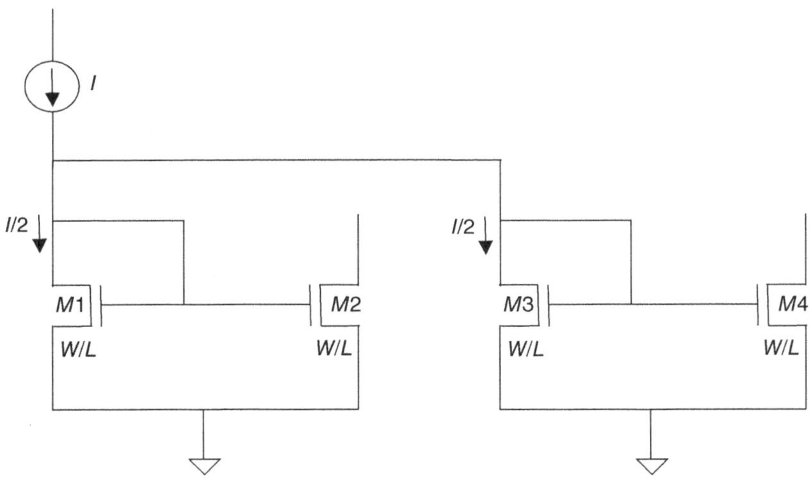

Figure A4.1f

Section 2. *RC* circuits – transient and frequency response

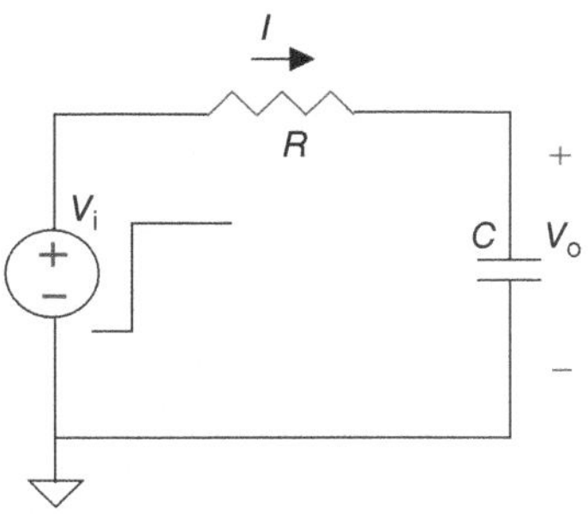

Figure A4.2a

Example 1: First-order low-pass *RC* circuits – step response

A step input is defined as:

$$\begin{cases} V_i(t) = 0 \text{ at } t < 0 \\ V_i(t) = V \text{ at } t \geq 0 \end{cases}$$

where V is the magnitude of the step input.

To derive the step response of the first-order *RC* circuit as shown in Figure A4.2a, we can use KVL/KCL and the time-domain equations of the resistor and capacitor. This leads to solving a differential equation. A more convenient method is to use s-domain impedances, as shown in Example 2 of the previous section (Section 1), and treat the *RC* circuit as a voltage divider between the R and C:

$$V_o(s) = \frac{\frac{1}{C \cdot s}}{R + \frac{1}{C \cdot s}} \cdot V_i(s) = \frac{1}{R \cdot C \cdot s + 1} \cdot V_i(s)$$

The *transfer function* of the *RC* circuit is:

$$H(s) = \frac{V_o(s)}{V_i(s)} = \frac{1}{R \cdot C \cdot s + 1}$$

This is a standard first-order system with time constant $\tau = R \cdot C$ (see *Appendix 3, Section 2, First-order systems*). The step response of the output voltage is:

$$V_o(s) = \frac{1}{(R \cdot C \cdot s + 1)} \cdot V_i(s) = \frac{1}{(R \cdot C \cdot s + 1)} \cdot \frac{V}{s}$$

Notice that the Laplace transform of the step input is:

$$V_i(s) = \frac{V}{s}$$

The time-domain transient response can be found by performing inverse Laplace transform (see *Appendix 3, Section 1, Laplace transform and transfer functions*):

$$V_o(t) = V \cdot \left(1 - e^{-\frac{t}{\tau}}\right) = V \cdot \left(1 - e^{-\frac{t}{R \cdot C}}\right)$$

Note that we have assumed the initial condition of $V_o(0) = 0$ (i.e., the initial voltage of the capacitor is zero).

The step response of the first-order *RC* circuit is shown in Figure A4.2b.

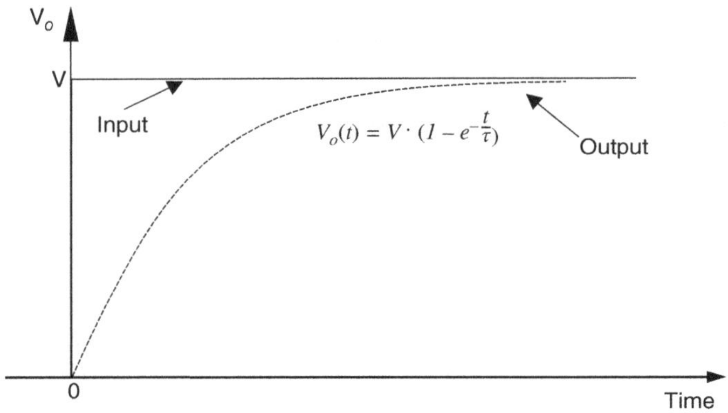

Figure A4.2b

The time constant ($\tau = R \cdot C$) determines the settling behavior of the step response. The smaller the time constant, the faster the output settles. Table A4.2a lists the settling percentage versus time constant (up to $5 \cdot \tau$):

Table A4.2a

Time Constant (Unit = τ)	**Settling (Input = 100%)**	**Error (Input = 100%)**
1	63.2%	36.8%
2	86.5%	13.5%
3	95.0%	5.0%
4	98.2%	1.8%
5	99.3%	0.7%

Understanding an *RC* circuit's time constant and step response is important in circuit design and analysis. For example, Question 29 studies the time constant of a charge-transfer track-and-hold (T/H) circuit to meet the precision requirement. Another example is Question 39, in which the time constant of digital buffers and physical routings is studied for calculating the propagation delays and rise/fall times.

Example 2: First-order low-pass *RC* circuits – frequency response

If we apply a sinusoidal signal to a *linear* circuit such as the first-order *RC* circuit shown in Example 1, the output, after reaching the steady state, will also be a sinusoidal signal with the same frequency as that of the input (Figure A4.2c). The relationship between the steady-state magnitude/phase of the output signal and the input frequency is called the frequency response.

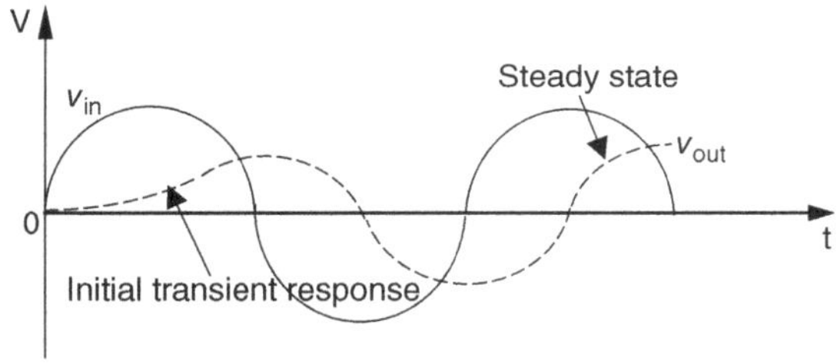

Figure A4.2c

The frequency response of a system can be obtained from its transfer function by replacing s with $j \cdot \omega$ (see *Appendix 3, Section 5, S-domain and frequency domain*):

$$G(j \cdot \omega) = G(s \mid s = j \cdot \omega) = a(\omega) + b(\omega) \cdot j$$

The magnitude frequency response is:

$$|G(j \cdot \omega)| = \sqrt{a(\omega)^2 + b(\omega)^2}$$

The phase frequency response is:

$$\angle G(j \cdot \omega) = \arctan\left[\frac{b(\omega)}{a(\omega)}\right]$$

This is because by making the real part of the complex frequency (s) zero, we ignore the transient component and the remaining part is the steady-state frequency response. Note that ω is the angular frequency, with units of rad/s. The relationship between the angular frequency and linear frequency (in Hz) is:

$$\omega = 2 \cdot \pi \cdot f$$

The transfer function of the first-order *RC* circuit is:

$$H(s) = \frac{V_o(s)}{V_i(s)} = \frac{1}{R \cdot C \cdot s + 1}$$

By replacing s with $j \cdot \omega$, we obtain the frequency response of the *RC* circuit:

$$H(j \cdot \omega) = \frac{1}{R \cdot C \cdot j \cdot \omega + 1} = \frac{1}{R^2 \cdot C^2 \cdot \omega^2 + 1} - \frac{R \cdot C \cdot \omega}{R^2 \cdot C^2 \cdot \omega^2 + 1} \cdot j$$

The magnitude frequency response is:

$$|H(j \cdot \omega)| = \sqrt{\frac{1}{R^2 \cdot C^2 \cdot \omega^2 + 1}} = \sqrt{\frac{1}{(2 \cdot \pi)^2 \cdot R^2 \cdot C^2 \cdot f^2 + 1}}$$

The phase frequency response is:

$$\angle H(j \cdot \omega) = \text{arctan}\left[-R \cdot C \cdot \omega\right] = \text{arctan}\left[-2 \cdot \pi \cdot R \cdot C \cdot f\right]$$

The magnitude and phase frequency response of the first-order *RC* circuit are shown in Figure A4.2d:

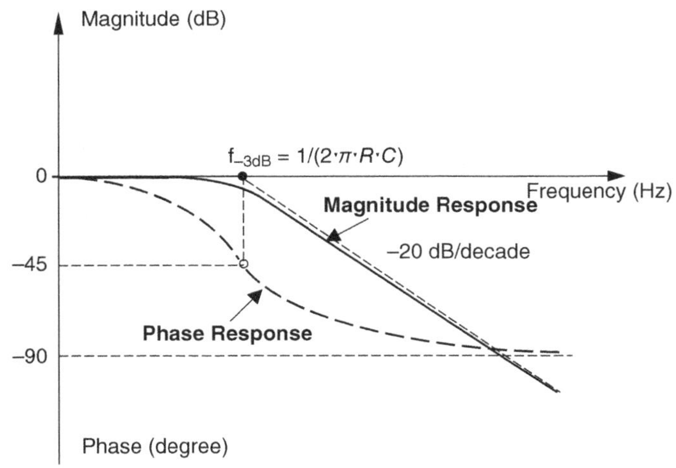

Figure A4.2d

In the frequency-response plot, both the x-axis (frequency) and the magnitude on the *y*-axis are on a logarithm (base-10) scale. In particular, the magnitude frequency response is measured in dB (decibel):

$$|H(j \cdot \omega)|(\text{dB}) = 20 \cdot \log_{10}|H(j \cdot \omega)|$$

One of the benefits of using dB for magnitude is that the magnitude response of multiple cascaded circuits can be derived by simply adding each magnitude response together (product rule). Note that the total phase shift of a cascaded circuit system is the arithmetic sum of the phase shift of each stage; consequently, phase is additive rather than logarithmic and is expressed in degrees or radians.

The frequency-response plot shows that the first-order *RC* circuit in this example is a *low-pass* filter. At DC, the gain magnitude is 0 dB (1 V/V), and phase shift is 0°. The gain (magnitude) drops and phase lag increases with increasing frequency. The degradation of magnitude is slow at the beginning. A critical turning point happens at cutoff frequency (also known as −3 dB frequency, $f_{-3\text{dB}}$, or −3 dB bandwidth), at which the magnitude is −3 dB (≈70%) and phase shift is −45°. Higher than this frequency, the magnitude response drops at a rate approximating −20 dB/decade (meaning a magnitude drop of 20 dB for every 10-fold increase in frequency). The phase shift approaches −90° at high frequency.

For a quick plotting of the frequency response, the magnitude response can be approximated by two lines instead: from DC to $f_{-3\text{dB}}$, a horizontal line of 0 dB; after $f_{-3\text{dB}}$, a line with the slope of −20 dB/decade (i.e., the oblique asymptote of the actual curve). This type of plot is called a *Bode plot.*

The −3 dB frequency is given by:

$$f_{-3\text{dB}} = \frac{1}{2 \cdot \pi \cdot R \cdot C} = \frac{1}{2 \cdot \pi \cdot \tau}$$

The −3 dB frequency is directly related to the time constant. The shorter the time constant, the faster the circuit reacts, and the higher the −3 dB frequency (bandwidth). The −3 dB frequency is also referred to as the *pole* of the first-order *RC* circuit, because it is the root of the characteristic equation (i.e., letting the denominator polynomial equal zero, see *Appendix 3, Section 1, Laplace transform and transfer functions*).

The frequency response of *RC* circuits is essential knowledge for circuit analysis. For instance, it is widely used in amplifier bandwidth and stability analysis. Question 7 and Question 8 are good examples of amplifier closed-loop stability analysis using information of its open-loop poles.

Section 3. Basic amplifiers

Example 1: Simple common-source amplifier

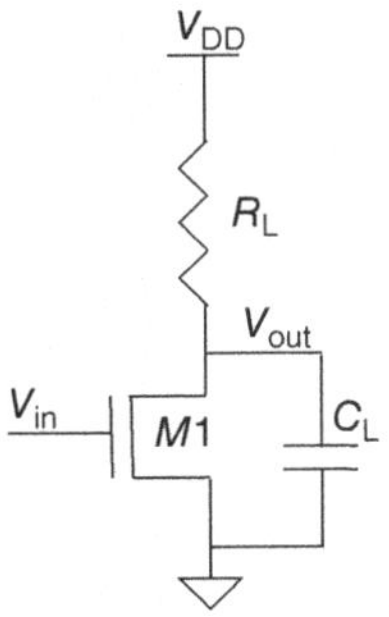

Figure A4.3a

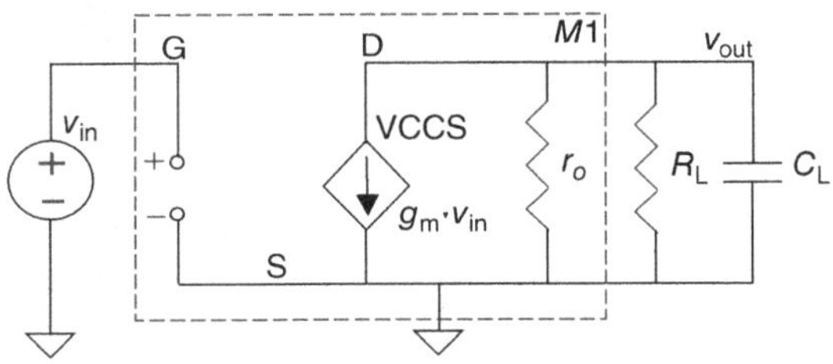

Figure A4.3b

A simple common-source amplifier is shown in Figure A4.3a. R_L is the load resistor and C_L is the load capacitor. In the following analysis, we will ignore the parasitic capacitance of M1 and assume $R_L \ll r_o$, where r_o is the output resistance of M1. We will analyze the following of the amplifier: DC gain, −3 dB bandwidth, input resistance, and output resistance.

Small-signal analysis of the amplifier is based on its small-signal equivalent circuit, as shown in Figure A4.3b. Transistor M1 is modeled as a voltage-controlled current source with transconductance of g_m, and an output resistance of r_o (see *Appendix 1, Section 1, MOSFET (long-channel) equations*). In small-signal analysis, VDD and ground are both considered AC grounds because they are DC potentials. Notice that we use lowercase letters for small (AC) signals (e.g., v_{in}, and v_{out}) in Figure A4.3b – a convention that we use throughout the book.

Applying KVL and KCL, we have the following equations:

$$v_{\text{out}} = -g_m \cdot v_{\text{in}} \cdot \left(r_o \parallel R_L \parallel \frac{1}{C_L \cdot s} \right) \approx \frac{-g_m \cdot R_L}{R_L \cdot C_L \cdot s + 1} \cdot v_{\text{in}}$$

The gain of the amplifier is:

$$A(s) = \frac{v_{\text{out}}(s)}{v_{\text{in}}(s)} \approx \frac{-g_m \cdot R_L}{R_L \cdot C_L \cdot s + 1}$$

This is a first-order transfer function (see *Appendix 3, Section 2, First-order systems*). The DC gain is:

$$A(0) = A(s = 0) \approx -g_m \cdot R_L$$

The negative sign means the common-source amplifier is an inverting amplifier. The time constant of the circuit is:

$$\tau \approx R_L \cdot C_L$$

And the −3 dB frequency (bandwidth) is:

$$f_{-3\text{dB}} = \frac{1}{2 \cdot \pi \cdot \tau} \approx \frac{1}{2 \cdot \pi \cdot R_L \cdot C_L}$$

To determine the input (output) resistance of the amplifier, we can put a test voltage source at the input (output) and calculate the current flowing into the input (output) (see *Appendix 2, Section 2, Thévenin's theorem and Norton's theorem* or the Quantitative Analysis section of Question 1, *Input and output resistance of a transimpedance amplifier (TIA)*). Going through this process, it is not difficult to conclude that the input and output resistances of the simple common-source amplifier are:

$$r_{\text{in}} = \infty$$

$$r_{\text{out}} = r_o \parallel R_L \approx R_L$$

Common-source amplifiers are the most popular voltage amplifier topology and used in many applications. Most questions in Chapter 2 involve common-source amplifiers except Question 20, *Frequency response of a source follower*. In particular, Question 12 studies distortions in a simple common-source amplifier and Question 13 deals with the emitter degeneration of a common-emitter amplifier (the bipolar equivalent of the common-source amplifier).

Example 2: Simple common-drain amplifier (source follower)

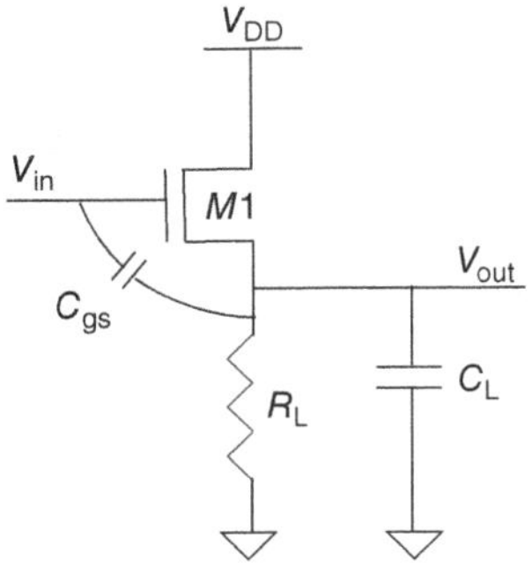

Figure A4.3c

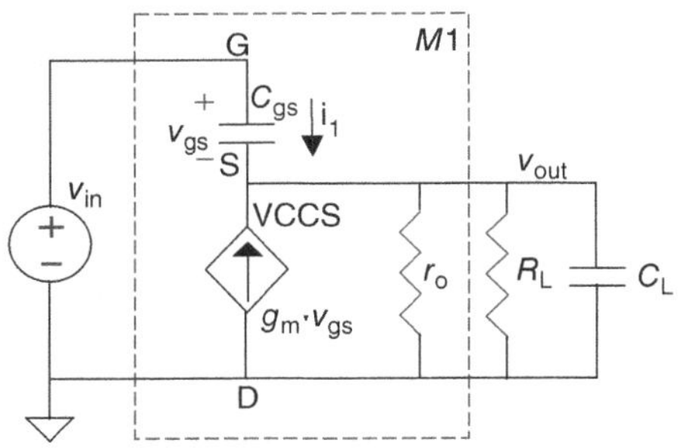

Figure A4.3d

A simple common-drain amplifier is shown in Figure A4.3c. R_L is the load resistor and C_L is the load capacitor. C_{gs} is the gate-to-source capacitance of the transistor M1, which plays an important role in the frequency response of the common-drain amplifier. Other parasitic capacitances of M1 are ignored, such as the gate-to-drain capacitance C_{gd}. Assume $R_L \ll r_o$, where r_o is the output resistance of M1. We will analyze the following of the amplifier: DC gain, −3 dB bandwidth, input resistance, and output resistance.

The small-signal model of the common-drain amplifier is shown in Figure A4.3d. Let us first derive the transfer function (the gain) of the common-drain amplifier. The current flowing through C_{gs}, i_1, can be written as:

$$i_1 = -g_m \cdot (v_{in} - v_{out}) + \frac{v_{out}}{r_o \parallel R_L \parallel \frac{1}{C_L \cdot s}} \approx -g_m \cdot (v_{in} - v_{out}) + \frac{v_{out}}{R_L \parallel \frac{1}{C_L \cdot s}}$$

The output voltage, v_{out}, can be written as:

$$v_{out} = v_{in} - v_{gs} = v_{in} - i_1 \cdot \frac{1}{C_{gs} \cdot s}$$

The transfer function is:

$$A(s) = \frac{v_{out}(s)}{v_{in}(s)} \approx \frac{C_{gs} \cdot R_L \cdot s + g_m \cdot R_L}{\left(C_{gs} + C_L\right) \cdot R_L \cdot s + g_m \cdot R_L + 1} \approx \frac{C_{gs} \cdot s + g_m}{\left(C_{gs} + C_L\right) \cdot s + g_m};$$
$$\text{if } g_m \cdot R_L \gg 1$$

The DC gain $(s = 0)$ is:

$$A(0) = \frac{v_{out}(0)}{v_{in}(0)} = \frac{g_m \cdot R_L}{g_m \cdot R_L + 1} \approx 1; \text{if } g_m \cdot R_L \gg 1$$

The frequencies of the pole and zero (left-hand-plane zero, or LHP zero) are (see *Appendix 3, Section 1, Laplace transform and transfer functions*):

$$\omega_p = \frac{g_m}{C_{gs} + C_L}; f_p = \frac{g_m}{2 \cdot \pi \cdot \left(C_{gs} + C_L\right)}$$

$$\omega_z = \frac{g_m}{C_{gs}}; f_z = \frac{g_m}{2 \cdot \pi \cdot C_{gs}}$$

The magnitude frequency response is shown in Figure A4.3e:

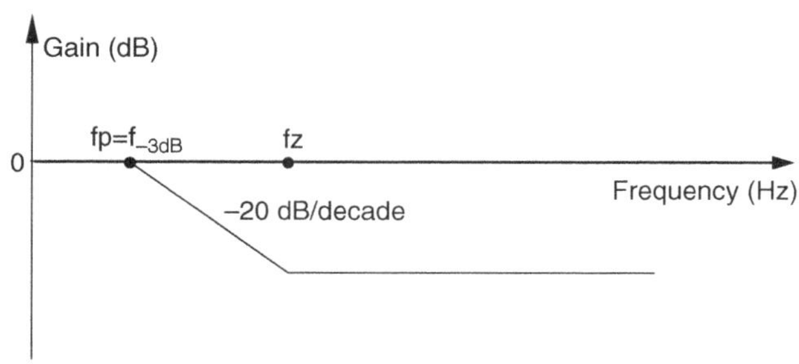

Figure A4.3e

The −3 dB bandwidth is equal to the pole frequency:

$$f_{-3\text{dB}} = f_p = \frac{g_m}{2 \cdot \pi \cdot \left(C_{gs} + C_L\right)}$$

The input resistance of the common-drain amplifier is considered infinite at DC because the gate of MOSFET prevents a direct conduction path:

$$r_{in} = \infty$$

To analyze the output resistance, we insert a test source v_{Test} at the output of the amplifier, and calculate the current flowing into the circuit (i_{Test}), as shown in Figure A4.3f. Note that when calculating the equivalent impedance, the independent voltage source (v_{in}) should be treated as short circuit.

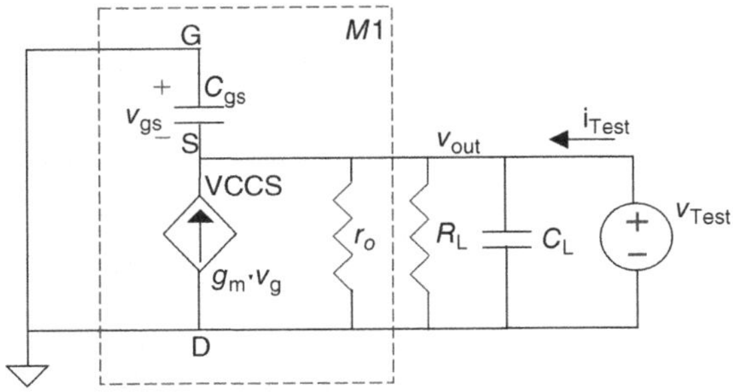

Figure A4.3f

Applying KCL, we have:

$$i_{\text{Test}} = \frac{v_{\text{Test}}}{r_o \parallel R_L \parallel \dfrac{1}{C_L \cdot s} \parallel \dfrac{1}{C_{gs} \cdot s}} + g_m \cdot v_{\text{Test}}$$

The output impedance is:

$$Z_o = \frac{v_{\text{Test}}}{i_{\text{Test}}} = \frac{1}{\dfrac{1}{\left(r_o \parallel R_L \parallel \dfrac{1}{(C_L + C_{gs}) \cdot s}\right)} + g_m} \approx \frac{1}{\dfrac{1}{R_L \parallel \dfrac{1}{(C_L + C_{gs}) \cdot s}} + g_m}$$

Let $s = 0$ to ignore the capacitive reactance (C_L and C_{gs}), and the output resistance is:

$$r_o \approx \frac{1}{\dfrac{1}{R_L} + g_m} = \frac{R_L}{g_m \cdot R_L + 1} \approx \frac{1}{g_m}; \text{if } g_m \cdot R_L \gg 1$$

Common-drain amplifiers have high −3 dB bandwidth, high input resistance, and low output resistance, with a DC gain close to 1 and hence the name "source follower." They are often used as unity-gain voltage buffers and drivers. Note that real implementations have additional poles in their transfer functions due to the series resistance at the gate of M1 and other parasitic capacitances, which cause gain roll-off at high frequency.

Question 2 studies the input capacitance of source followers, and Question 20 explores source followers with a current source type of load.

Example 3: Simple common-gate amplifier

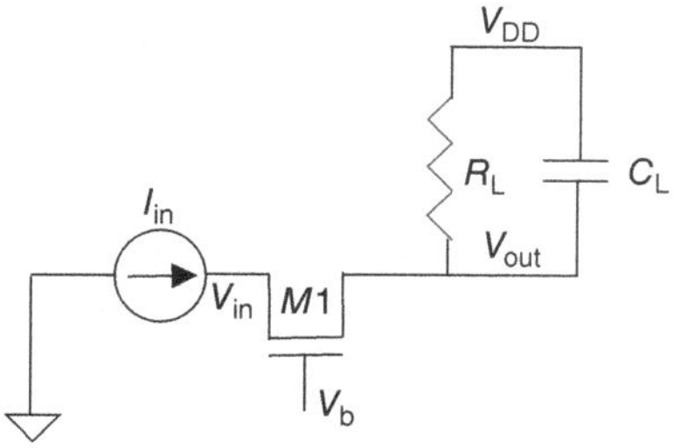

Figure A4.3g

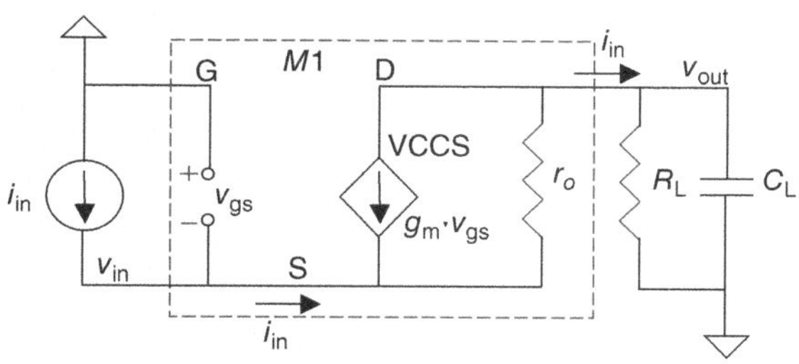

Figure A4.3h

A simple common-gate amplifier is shown in Figure A4.3g. R_L is the load resistor and C_L is the load capacitor. The input is a current, i_{in}, from a current source. The voltage at the input (i.e., the source of M1) is v_{in}. Assume $R_L \ll r_o$, where r_o is the output resistance of M1, and ignore the parasitic capacitances of M1. We will evaluate the following characteristics of the amplifier: DC gain, −3 dB bandwidth, input resistance, and output resistance.

The small-signal model of the common-drain amplifier is shown in Figure A4.3h. Note that the gate voltage of M1 (V_b) is a DC bias voltage and therefore an AC ground. The input current, i_{in}, flows through transistor M1 and the load resistor R_L and capacitor C_L. The output voltage v_{out} can be written as:

$$v_{out} = i_{in} \cdot \left(R_L \parallel \frac{1}{C_L \cdot s} \right) = \frac{i_{in} \cdot R_L}{R_L \cdot C_L \cdot s + 1}$$

For an amplifier with a current input and a voltage output, the gain is a *transimpedance:*

$$A(s) = \frac{v_{out}}{i_{in}} = \frac{R_L}{R_L \cdot C_L \cdot s + 1}$$

This is a first-order transfer function. The transimpedance (DC) is:

$$A(0) = R_L$$

The −3 dB bandwidth is:

$$f_{-3\text{dB}} = \frac{1}{2 \cdot \pi \cdot R_L \cdot C_L}$$

To find the input resistance, we can use i_{in} as the test input current and calculate the resulting voltage at the input (v_{in}):

$$v_{\text{in}} = (i_{\text{in}} - v_{\text{in}} \cdot g_m) \cdot r_o + i_{\text{in}} \cdot \left(R_L \parallel \frac{1}{C_L \cdot s}\right)$$

This yields:

$$Z_{\text{in}}(s) = \frac{v_{\text{in}}}{i_{\text{in}}} = \frac{r_o + \dfrac{R_L}{R_L \cdot C_L \cdot s + 1}}{1 + g_m \cdot r_o}$$

The input resistance is:

$$r_{\text{in}} = Z_{\text{in}}(0) = \frac{r_o + R_L}{1 + g_m \cdot r_o} \approx \frac{1}{g_m}; \text{if } g_m \cdot r_o \gg 1 \text{ and } r_o \gg R_L$$

To find the output resistance, we insert a test voltage source v_{Test} at the output of the amplifier, as shown in Figure A4.3i:

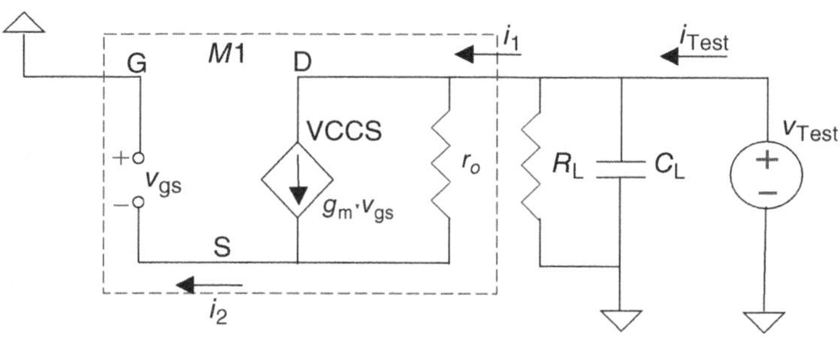

Figure A4.3i

When calculating the equivalent impedance, independent current source (i_{in}) should be treated as open circuit. Based on the small-signal model, it follows that:

$$i_2 = i_1 = 0$$

The output impedance is:

$$Z_{\text{out}} = \frac{v_{\text{Test}}}{i_{\text{Test}}} = R_L \parallel \frac{1}{C_L \cdot s}$$

And the output resistance is:

$$r_{\text{out}} = R_L$$

This analysis assumes that the amplifier is driven by an ideal current source with infinite output impedance. Practical current sources have finite output impedance. Assume the current source has an output resistance of r_{o1} (Figure A4.3j) and this forms a *cascode current source* structure. The output resistance of the amplifier changes to:

$$r_{\text{out}} = R_L \parallel (r_{o1} + r_o + r_{o1} \cdot r_o \cdot g_m) \approx R_L \parallel r_{o1} \cdot r_o \cdot g_m \approx R_L, \text{if}\, R_L \ll r_{o1} \cdot r_o \cdot g_m$$

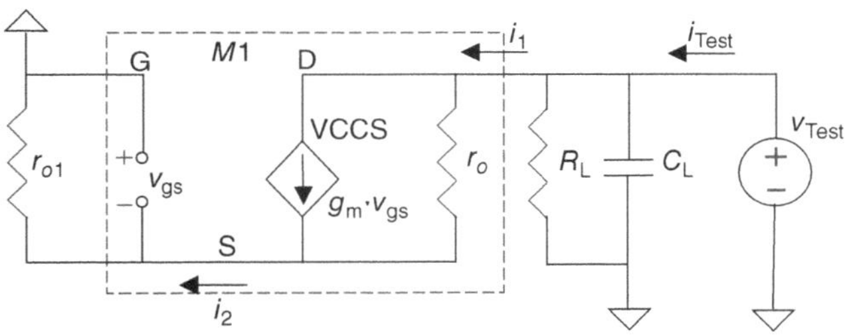

Figure A4.3j

The derivation is left as an exercise for interested readers (see Quantitative Analysis of Question 4, *Output resistance of a tied-gate current source* for a demonstration of the derivation of the cascode current source output resistance).

Characterized by low input resistance and high output resistance, common-gate amplifiers are ideal for current-input applications such as current buffers. Additionally, the common-gate stage is an essential building block of the cascode amplifier.

Figure A4.3k is the circuit diagram of a fully differential telescopic amplifier with cascode NMOS and PMOS loads. Transistors M3 and M4 function as common-gate amplifiers; they provide relatively low input impedance at the drains of M1 and M2 to accommodate the signal currents, while maintaining high output impedance at V_{op} and V_{on} to achieve large voltage gain.

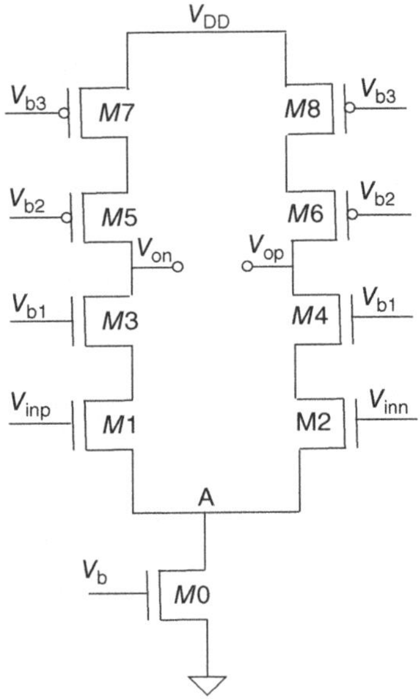

Figure A4.3k

More cascode amplifier examples can be found in Question 5, Folded cascode amplifier and input Miller effect reduction, Question 10, Device noise contribution in a folded cascode amplifier and Question 15, Common-mode feedback control (CMFB) in fully differential amplifiers. The latter two questions offer different perspectives on common-gate amplifiers in cascode topologies: Question 10 evaluates the impact of source degeneration on the amplifier's input-referred noise performance, while Question 15 examines the nondominant and dominant poles at the input and output of the common-gate amplifiers.

Section 4. Basic op-amp based closed-loop circuits

An operational amplifier (also known as op-amp or op amp) is an amplifier with a differential input, a single-ended or differential output (also known as fully differential op-amp), and a substantially high DC gain. It is most often used in closed-loop circuits and provides stable gain and other electrical characteristics, thanks to negative feedback.

Internally, op-amps often have two or more stages of amplification due to the high DC gain requirement. The design of op-amp internal circuits is discussed throughout Chapter 2 (*Amplifiers – Open-loop and Closed-loop*), with analysis of questions on topics such as gain and gain-bandwidth product, stability, slew rate, noise, common-mode feedback control (for fully differential op-amps), etc. In this section, we focus on the analysis of some basic op-amp based closed-loop circuits.

Example 1: Inverting amplifiers

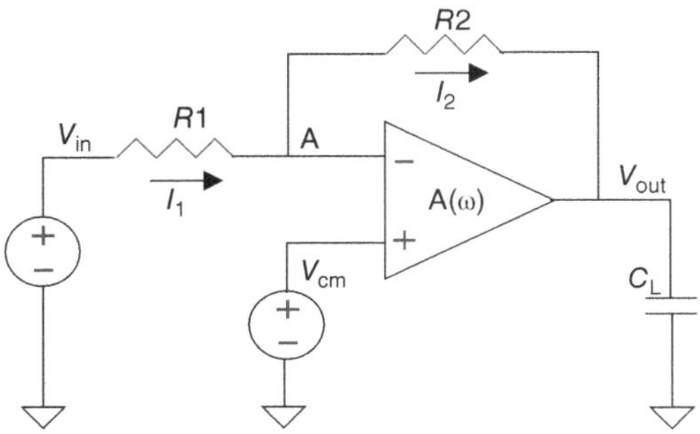

Figure A4.4a

A CMOS op-amp based inverting amplifier is shown in Figure A4.4a. The op-amp has a gain $A(\omega)$: its DC gain is $A(0)$ and the gain decreases at high frequency according to its frequency-response characteristics $A(\omega)$ (see *Appendix 3, Section 5, S-domain and frequency domain*). C_L is the load capacitor, and V_{cm} is a DC bias voltage setting the midpoint of the input range (e.g., VDD/2). In the following analysis, we ignore the input parasitic capacitances of the op-amp and assume the output resistance of the op-amp is zero. We will now evaluate the closed-loop gain, input resistance, and output resistance of the inverting amplifier.

Let us start with a quick intuitive analysis of the circuit in Figure A4.4a. An important concept of op-amp based closed-loop circuits is "virtual ground." As the gain of the op-amp is substantially large (e.g., 80 dB or 10^4 V/V) and the output voltage of the op-amp is finite (limited by supply rails, e.g., 5 V), the voltage between the inverting and non-inverting input of the op-amp is very small (e.g., 0.5 mV). For a rough estimation, we can consider the potentials at the two op-amp input terminals equal in a closed-loop circuit.

This means the voltage of node A in Figure A4.4a, V_A, equals V_{cm}. As V_{cm} is a DC bias voltage and essentially an AC ground, node A is often referred to as a "virtual ground."

As the dynamic (AC) characteristic of the amplifier is of interest here, the circuit diagram in Figure A4.4a is redrawn as its small-signal model in Figure A4.4b. the DC voltage V_{cm} is considered an AC ground, and all signals are AC quantities.

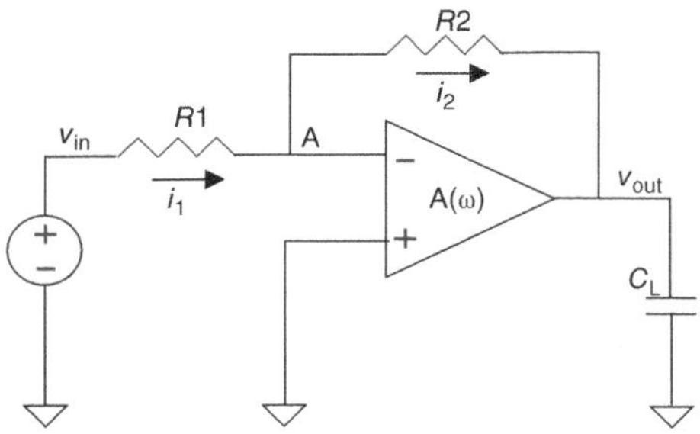

Figure A4.4b

The current in R_1 is:

$$i_1 = \frac{v_{in} - v_A}{R_1} \approx \frac{v_{in}}{R_1}$$

Because the input resistance of the CMOS op-amp is very large (assumed infinite here) and there is no input capacitance at node A, the current in R_1 equals that in $R_2 (i_1 = i_2)$. This yields:

$$v_{out} = v_A - i_2 \cdot R_2 = 0 - i_1 \cdot R_2 = -\frac{R_2}{R_1} \cdot v_{in}$$

The closed-loop gain is:

$$A_{CL} = \frac{v_{out}}{v_{in}} = -\frac{R_2}{R_1}$$

The output is an inverting version of the input with a gain of R_2/R_1, hence the name inverting amplifier.

The input resistance is R_1 since node A is at virtual ground. The output resistance is zero because the op-amp has zero output resistance (so v_{out} is essentially connected to a voltage source).

The quick gain estimation result of the inverting amplifier, however, does not reveal how the characteristic of the op-amp would impact the closed-loop gain, because we have assumed that $A(\omega) = \infty$ in the previous analysis. When considering the finite gain of the op-amp, we have the following system of equations:

$$\begin{cases} \dfrac{v_{in} - v_A}{R_1} = \dfrac{v_A - v_{out}}{R_2} \\ v_{out} = -A(\omega) \cdot v_A \end{cases}$$

which yields:

$$v_{out} = -\frac{R_2}{R_1 + \dfrac{R_1 + R_2}{A(\omega)}} \cdot v_{in}$$

The closed-loop gain is:

$$A_{CL} = \frac{v_{\text{out}}}{v_{\text{in}}} = -\frac{R_2}{R_1 + \dfrac{R_1 + R_2}{A(\omega)}}$$

The magnitude of the inverting amplifier DC gain is slightly less than R_2/R_1 due to the finite op-amp DC gain $A(0)$. The closed-loop gain is also affected by the frequency response of the op-amp and would roll down at high frequency. Question 17, *Feedback factors of inverting amplifiers and unity-gain buffers*, further explores the relationship between the closed-loop gain and −3 dB bandwidth relative to the open-loop characteristics of the op-amp in inverting amplifiers.

Example 2: Non-inverting amplifiers

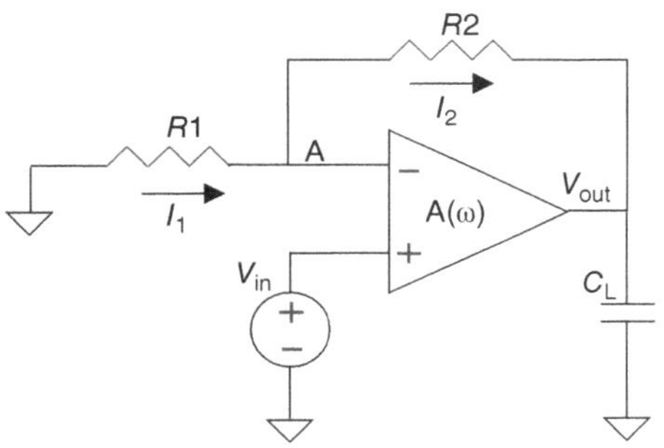

Figure A4.4c

Figure A4.4c shows a CMOS op-amp based non-inverting amplifier. Assume the op-amp is the same as that in Example 1 (Figure A4.4a). We will again evaluate the closed-loop gain, input resistance, and output resistance of the amplifier.

The major difference between the topology in Figure A4.4c and that of the inverting amplifier in Figure A4.4a is that the input signal is connected to the non-inverting (i.e., positive) terminal of the op-amp. If we assume the op-amp is an ideal one with an infinite gain, the voltage of node A would be equal to that of the input signal:

$$v_A = v_{\text{in}}$$

The current in R_1 is:

$$i_1 = \frac{0 - v_A}{R_1} \approx \frac{-v_{\text{in}}}{R_1}$$

As $i_1 = i_2$, this yields:

$$v_{\text{out}} = v_A - i_2 \cdot R_2 = v_{\text{in}} - i_1 \cdot R_2 = \left(1 + \frac{R_2}{R_1}\right) \cdot v_{\text{in}}$$

The closed-loop gain is:

$$A_{CL} = \frac{v_{\text{out}}}{v_{\text{in}}} = 1 + \frac{R_2}{R_1}$$

The output is a non-inverting version of the input with a gain of $1 + R_2/R_1$, hence the name non-inverting amplifier.

The input resistance is effectively infinite, as it looks directly into one of the CMOS op-amp input terminals. The output resistance is zero due to the zero output resistance of the op-amp.

When considering the finite gain of the op-amp, we have the following system of equations:

$$\begin{cases} \dfrac{-v_A}{R_1} = \dfrac{v_A - v_{\text{out}}}{R_2} \\ v_{\text{out}} = A(\omega) \cdot (v_{\text{in}} - v_A) \end{cases}$$

which yields:

$$v_{\text{out}} = \frac{R_1 + R_2}{R_1 + \dfrac{R_1 + R_2}{A(\omega)}} \cdot v_{\text{in}}$$

The closed-loop gain is:

$$A_{CL} = \frac{v_{\text{out}}}{v_{\text{in}}} = \frac{R_1 + R_2}{R_1 + \dfrac{R_1 + R_2}{A(\omega)}}$$

Same as that of the inverting amplifier, the magnitude of the non-inverting amplifier's DC gain is less than its ideal value due to the finite op-amp DC gain $A(0)$, and the closed-loop gain is also affected by the frequency response of the op-amp and would roll down at high frequency. Question 17, *Feedback factors of inverting amplifiers and unity-gain buffers*, further explores the relationship between the closed-loop gain and −3 dB bandwidth relative to the open-loop characteristics of the op-amp in non-inverting amplifiers.

Op-amp based inverting and non-inverting amplifiers are widely used in analog amplification and filtering circuit systems, in both continuous-time and discrete-time applications. More circuit examples and analyses can be found in Question 16, *Frequency response of two op-amp based low-pass filters* and Question 17, *Feedback factors of inverting amplifiers and unity-gain buffers*, as well as Question 29, *Requirement of the amplifier in a charge-transfer track/hold circuit* and Question 30, *Impact of the amplifier input capacitance in switched-capacitor circuits*.

Section 5. Types of negative feedback in amplifier circuits

There are four types of negative feedback in amplifier circuits. From the output side, feedback can be categorized as either voltage-sampling or current-sampling. At the input side, feedback can be either shunt-mixing or series-mixing. The definition of these terms is as follows:

Voltage-sampling: the output voltage is sampled and fed back to the input of the amplifier.
Current-sampling: the output current is sampled and fed back to the input of the amplifier.
Shunt-mixing: the feedback signal is in parallel with the input signal, mixing via current.
Series-mixing: the feedback signal is in series with the input signal, mixing via voltage.

We will illustrate how to identify amplifier feedback types through some circuit examples.

Example 1: Voltage-sampling, shunt-mixing

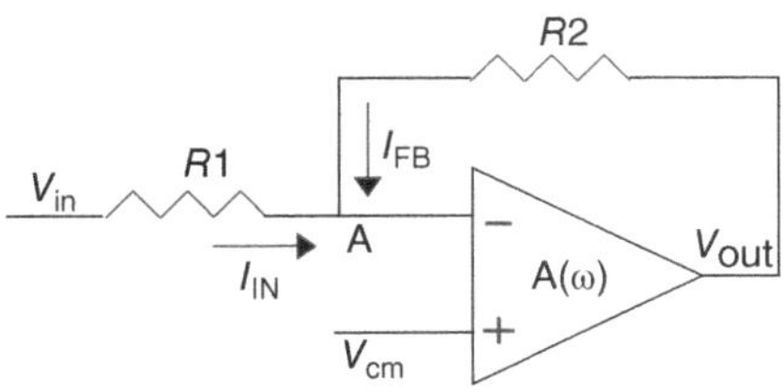

Figure A4.5a

An inverting amplifier is shown in Figure A4.5a. Both input and output are voltages. From the output side, the output voltage V_{out} is sampled and fed back. At the input side, the sampled output voltage (V_{out}) is applied in parallel with the input voltage (V_{in}) by resistors R_2 and R_1. Note that the only way for two voltages to be applied in parallel is through current mixing, which is exactly what happens in inverting amplifiers. As illustrated in Figure A4.5a, the input and output voltages are summed at the inverting node through resistors R_1 and R_2, where the resulting currents I_{IN} and I_{FB} combine. The feedback type is therefore voltage-sampling, shunt-mixing (or voltage-shunt in short).

Another example of an amplifier with voltage-sampling, shunt-mixing feedback is a transimpedance amplifier (TIA) as shown in Figure A4.5b. TIA converts an input current, i_{in}, to an output voltage, v_{out}. It is not difficult to conclude that the DC transfer function of the amplifier is:

$$v_{out} = -R \cdot i_{in}$$

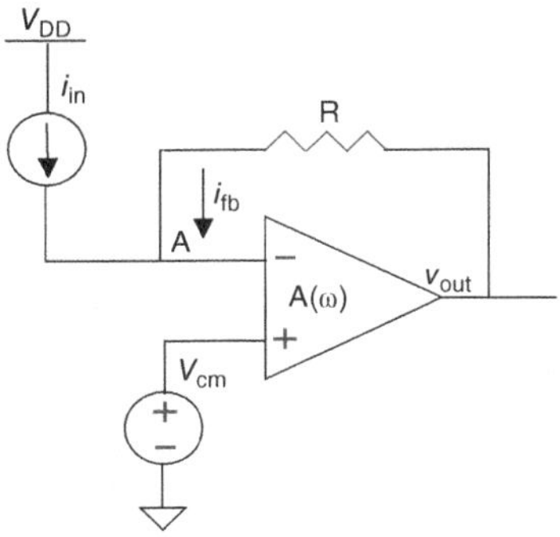

Figure A4.5b

The feedback scheme of the TIA is the same as that of the inverting amplifier in Figure A4.5a. Here the output voltage v_{out} is sampled and applied to the input node A through the feedback current i_{fb} (generated by resistor R), which is in parallel with the input current i_{in}.

Example 2: Voltage-sampling, series-mixing

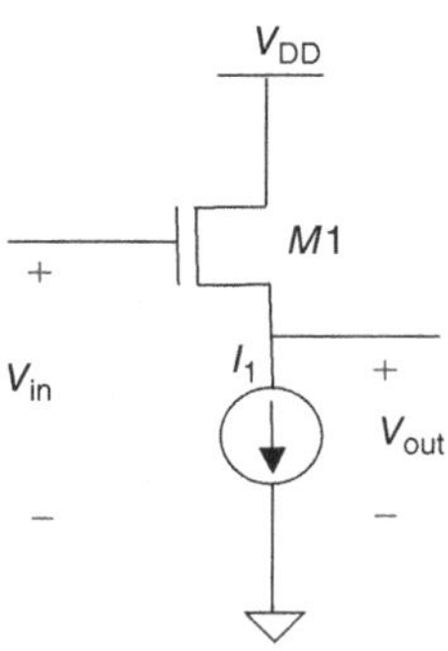

Figure A4.5c

A voltage source follower circuit is shown in Figure A4.5c. Both input and output are voltages. From the output side, the output voltage V_{out} is sampled and applied in series with the input voltage V_{in}. The feedback type is voltage-sampling, series-mixing (or voltage-series).

Other circuit examples that employ voltage-sampling, series-mixing feedback include unity-gain buffers and non-inverting amplifiers, as shown in Figure A4.5d and Figure A4.5e, where the output voltage V_{out}, or a portion of it, is fed back to the inverting input of the op-amp and applied in series with the input voltage V_{in} at the non-inverting input of the op-amp.

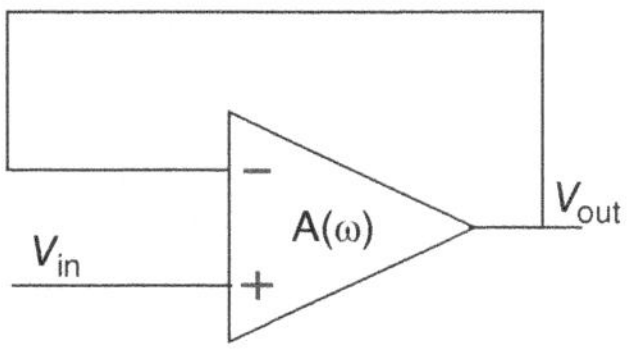

Figure A4.5d

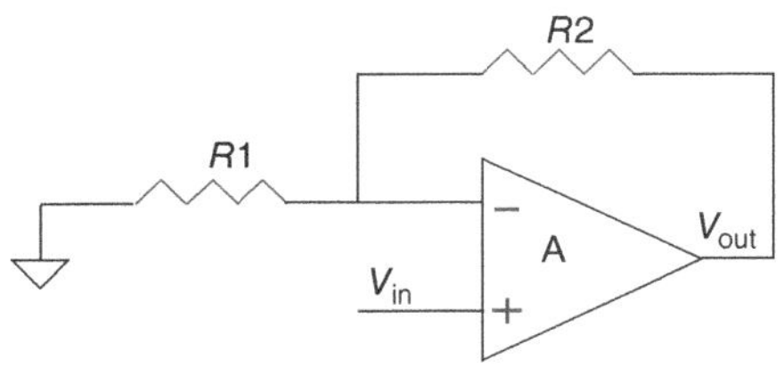

Figure A4.5e

Example 3: Current-sampling, series-mixing

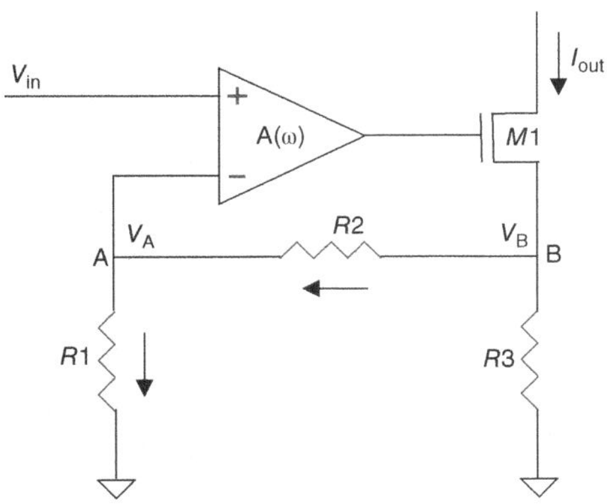

Figure A4.5f

A *transconductance amplifier* is shown in Figure A4.5f. The circuit takes a voltage input (V_{in}), and the output is a current (I_{out}). Resistor R_3 samples the output current and converts it into a voltage (V_B), which is fed back through the resistors R_1 and R_2 to the inverting input of the op-amp (node A). This feedback voltage and the input voltage V_{in} are in series at the input terminals of the op-amp. The feedback type is therefore current-sampling, series-mixing (or current-series).

Using the analytical methods discussed in *Section 4. Basic op-amp based closed-loop circuits*, we can conclude that the *transconductance* of the amplifier is:

$$G_m = \frac{I_{\text{out}}}{V_{\text{in}}} \approx \frac{1 + \frac{R_2}{R_1}}{R_3}; \text{ assuming } A(\omega) \gg 1$$

Transconductance amplifiers are used in applications where a high-impedance current output is needed (instead of a low-impedance voltage output) with a voltage input.

Example 4: Current-sampling, shunt-mixing

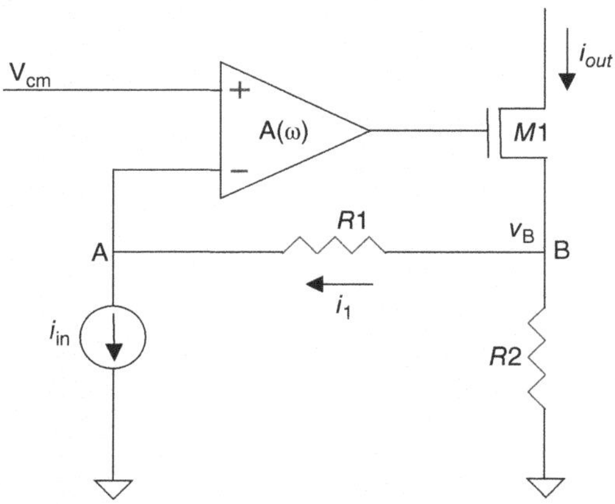

Figure A4.5g

Figure A4.5g shows an op-amp based *current amplifier*. V_{cm} is a DC bias voltage to maintain the proper common-mode voltage for the inputs of the op-amp. The input current (i_{in}) is amplified by the circuit and the output current is the drain current of M1 (i_{out}). The current gain of the amplifier is:

$$A_c = \frac{i_{\text{out}}}{i_{\text{in}}} \approx 1 + \frac{R_1}{R_2}; \text{ assuming } A(\omega) = \infty$$

At the output side, similar to Example 3, the output current i_{out} is sampled by the resistor R_2 and converted into a voltage v_B. This voltage (v_B) is applied to the input node A and mixed with the input current (i_{in}) in parallel by resistor R_1 (i.e., through i_1). Another way to view this is that the resistors R_1 and R_2 form a current divider, and a part of the output current flows back to the input node A through resistor $R_1(i_1)$. Either way suggests a current-sampling, shunt-mixing feedback (or current-shunt).

Negative feedback impacts many aspects of the electrical characteristics of amplifiers. For example, negative feedback results in a reduction of an amplifier's closed-loop gain but an increase of the amplifier's closed-loop −3 dB bandwidth. Another impact of negative feedback on amplifiers is the change in their input and output resistance.

The change of the output resistance of an amplifier depends on the output sampling type of its negative feedback network. Voltage-sampling lowers the output resistance, while current-sampling increases the output resistance. Intuitively, when the output voltage is sampled and involved in negative feedback, the feedback loop works to make the output voltage nearly constant regardless of the output current, or in other words, closer to that of a voltage source with low output resistance. Similarly, with current-sampling the feedback loop works to make output current constant in regardless of the output voltage, making the circuit output closer to that of a current source with high output resistance.

The type of input mixing affects the input resistance. Shunt-mixing reduces the input resistance, and series-mixing increases the input resistance. This is because shunt-mixing adds parallel paths for the input current, while series-mixing effectively adds more resistance into the input current path.

Table A4.5a summarizes the four types of negative feedback in amplifier circuits and their impact on the input or output resistance of amplifiers.

Table A4.5a

Type of Feedback	Input Mixing	Output Sampling	Impact on r_{in}/r_{out}
Voltage-sampling		Voltage	$r_{out}\downarrow$
Current-sampling		Current	$r_{out}\uparrow$
Shunt-mixing	Current, in parallel		$r_{in}\downarrow$
Series-mixing	Voltage, in series		$r_{in}\uparrow$

Examples of different feedback types in amplifier circuits can be found in many questions in the book: Question 1, *Input and output resistance of a transimpedance amplifier (TIA)*, Question 8, *Miller compensation in a two-stage op-amp*, Question 17, *Feedback factors of inverting and non-inverting amplifiers*, Question 18, *Power supply rejection ratio (PSRR) of a voltage regulator*, and Question 29, *Op-amp requirements for a charge-transfer track-and-hold (T/H) circuit*.

Section 6. Circuit noise fundamentals

Before we start noise analysis of a few basic circuit examples, let us first review some important terminology about electrical noise. We may not need to remember all the mathematical equations behind these terminologies, but a study of these concepts can help us better understand the behavior of noise in circuits.

Noise in time domain

Electrical noise is often considered a random signal, meaning it cannot be predicted by a deterministic mathematical function. Figure A4.6a illustrates a noise voltage changing over time.

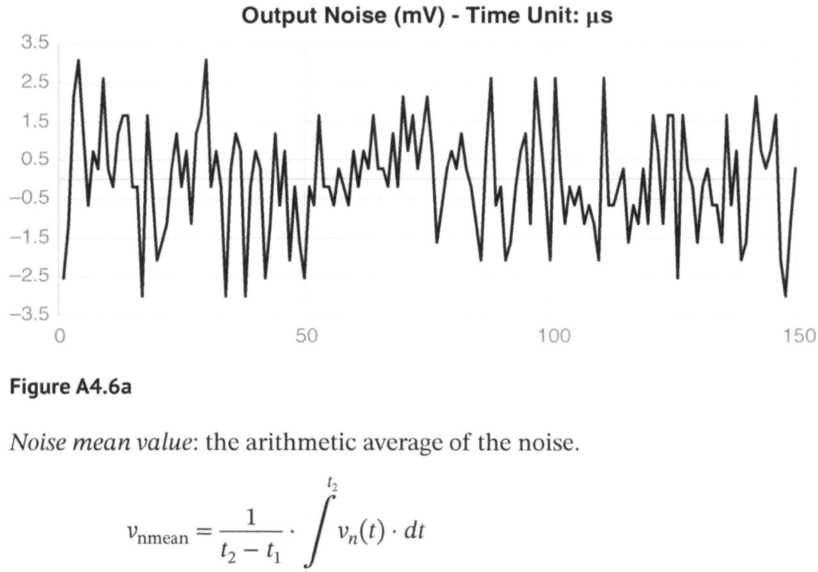

Figure A4.6a

Noise mean value: the arithmetic average of the noise.

$$v_{\text{nmean}} = \frac{1}{t_2 - t_1} \cdot \int_{t_1}^{t_2} v_n(t) \cdot dt$$

where $[t_1, t_2]$ denotes the duration of time during which the noise is observed, and $t_1 < t_2$.

In Figure A4.6a, the noise mean is zero volt. Most of time we are only concerned with the fluctuating part of the noise. If the noise mean is not zero, we often remove the DC part from the noise and make its mean value zero. Throughout the remainder of this section, we assume the noise process is zero mean.

Noise peak-to-peak: the algebraic difference between the maximum and minimum instantaneous noise values within a specified observation interval. In Figure A4.6a, the peak-to-peak noise is about 6 mV.

Noise root mean square (RMS): the square root of the arithmetic mean of the squared instantaneous noise values. Physically, this represents the effective noise power delivered to a load:

$$v_{\mathrm{nRMS}} = \sqrt{\frac{1}{t_2 - t_1} \cdot \int_{t_1}^{t_2} v_n^2(t) \cdot dt}$$

Noise mean square (MS): a measure of the total average power of noise:

$$v_{\mathrm{nMS}} = \frac{1}{t_2 - t_1} \cdot \int_{t_1}^{t_2} v_n^2(t) \cdot dt$$

Noise from a statistic perspective

For any given time window, we can calculate the noise peak-to-peak value. The problem is, however, that we cannot rule out the possibility, or probability, of occurrences with larger peak values (outliers) if we continue to make observation. The random and unpredictable nature of noise calls for a statistics approach.

Assuming a noise process is *stationary*, i.e., its statistical properties do not change over time, we can take a long time observation of the noise to estimate its distribution function. The *probability density function* (PDF) is based on histogram and used to describe the probability of noise falling within a specific range of values. Figure A4.6b shows a noise PDF plot.

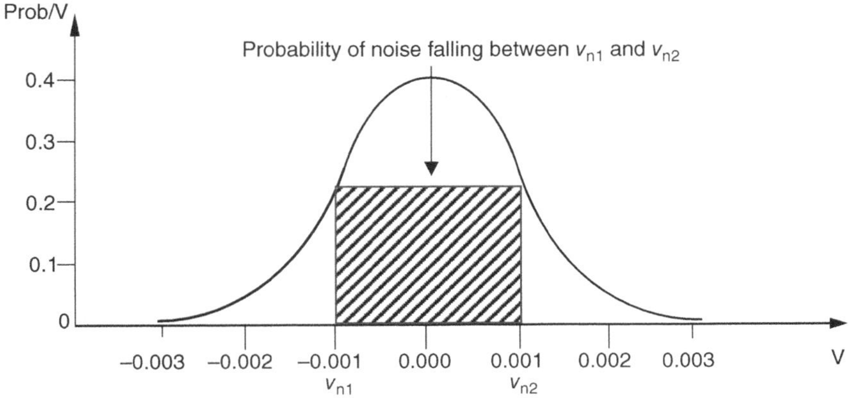

Figure A4.6b

Typically we would take a long time window (or a large number of samples if discrete-time) for PDF plot. The *x*-axis of the PDF plot represents noise value (voltage in this example). The *y*-axis is probability density, and its unit is probability per *x*-axis unit (e.g., probability/Volt in this example). The probability for noise voltage falling within a range between v_{n1} and v_{n2} ($v_{n1} < v_{n2}$) can be calculated by integrating the PDF over the said range:

$$P(v_{n1}, v_{n2}) = \int_{v_{n1}}^{v_{n2}} \text{PDF}(v) \cdot dv$$

where PDF(v) is the probability density function of the noise voltage v.

In other words, the probability of noise falling within a specific range of values is the area under the PDF curve over the range. The total area under the entire PDF curve is normalized to 1.

The PDF of many electrical noises, such as thermal noise, can be approximated to a standard mathematical distribution type called *Gaussian (or normal) distribution*. The probability function of a Gaussian distribution is as follows:

$$\text{PDF}(v) = \frac{1}{\sqrt{2 \cdot \pi \cdot \sigma^2}} \cdot e^{\left[-\frac{(v-\mu)^2}{2 \cdot \sigma^2}\right]}$$

where σ is *standard deviation*, and μ is *arithmetic mean* (in our example $\mu = 0$), and they both share the same units as the noise (i.e., Volts in this example).

Figure A4.6c illustrates the PDF of a noise with Gaussian distribution and a mean of zero.

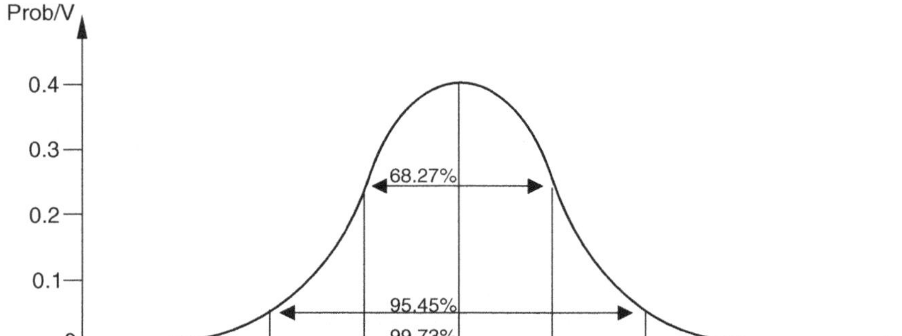

Figure A4.6c

The standard deviation is a key parameter in Gaussian distribution. The probability for the noise voltage falling within $\pm 1\sigma$ is 68.27%. It increases to 95.45% for $\pm 2\sigma$, and 99.73% for $\pm 3\sigma$. This is known as the *68-95-99.7 rule (3-sigma rule).* Importantly, for a noise with Gaussian distribution and zero mean value ($\boldsymbol{\mu} = \mathbf{0\,V}$), the standard deviation is the RMS value of the noise:

$$\sigma = \text{RMS}; \text{if}\,\mu = 0$$

With the help of PDF, we can define the peak-to-peak value of a noise more precisely. For example, we can use a value equal to six times a noise's standard deviation ($6 \cdot \sigma$, or $+/-3 \cdot \sigma$) as its peak-to-peak value, providing a statistical confidence level of 99.7% that the value remains within these bounds. If we use $6.6 \cdot \sigma$, the probability increases to 99.9%.

Noise in the frequency domain

One of the limitations of the PDF is that it does not reveal the noise distribution in the frequency domain. Sometimes, for example, when designing a noise filter, it is important to know the frequency content of a noise. The power spectral density (PSD) describes how the power of a signal or noise is distributed over frequency.

Figure A4.6d illustrates the PSD of a noise voltage generated by a circuit.

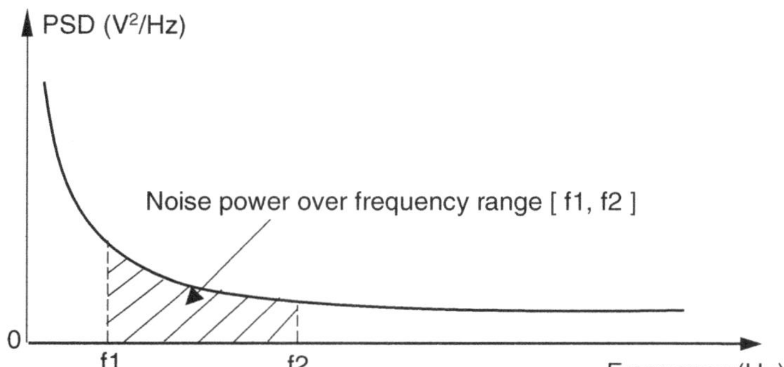

Figure A4.6d

The x-axis of the PSD is frequency, and the y-axis is the power density, with units of V^2/Hz. The noise power over a certain frequency range can be calculated by:

$$P(f_1, f_2) = \int_{f_1}^{f_2} \text{PSD}(f) \cdot df$$

where PSD(f) is the power spectral density function, and $f_1 < f_2$. Note that we assume the power spectral density function is a *one-sided* PSD (commonly used in engineering).

The total spectral noise power is given by:

$$P_{\text{total}} = \int_0^{\infty} \text{PSD}(f) \cdot df$$

The square root of the total spectral noise power in frequency domain is equal to the noise RMS in time domain.

$$v_{\text{nRMS}} = \sqrt{\int_0^{\infty} \text{PSD}(f) \cdot df}$$

where PSD(f) is the power spectral density function of the noise.

This is true because of *energy conservation*, and is mathematically expressed by *Parseval's Theorem*, which is beyond the scope of this book. Interested readers can explore further about the relationship between the signal and/or noise power calculations in the time domain and the frequency domain.

PSD is widely used to model the noise characteristics of electronic devices. The noise PSD functions of most commonly used devices, such as CMOS and bipolar transistors, diodes and resistors, are listed in *Appendix 1, Device Equations*. These noise models form the basis for circuit noise estimation and analysis.

Example 1: KT/C noise

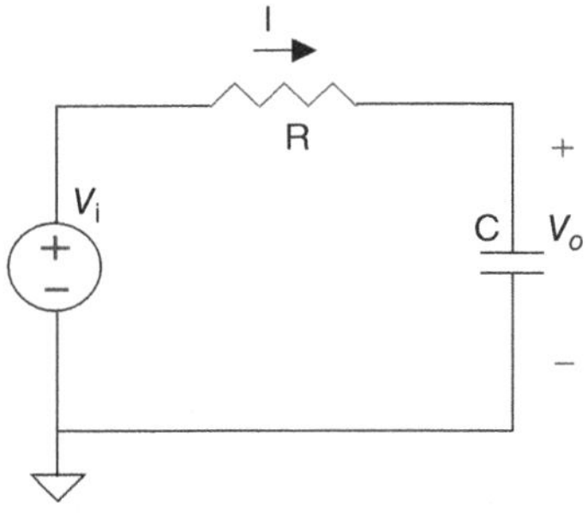

Figure A4.6e

In the first example, we will calculate the output voltage noise (RMS) of a simple *RC* circuit as shown in Figure A4.6e. The signal source V_i is assumed noise-free.

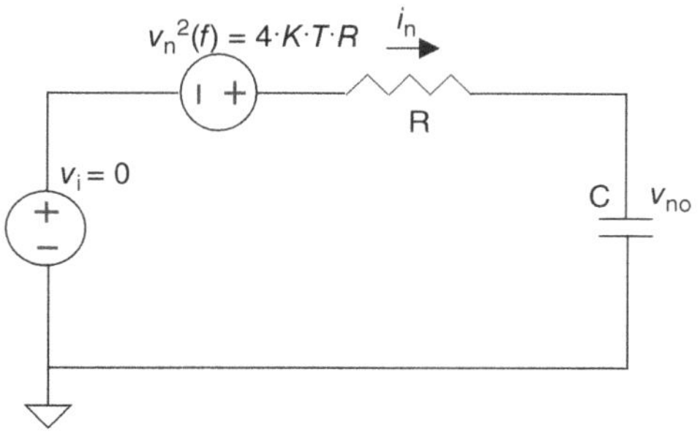

Figure A4.6f

The first step in circuit noise analysis is to identify the noise sources. In this example, the resistor R is the noise source, generating thermal noise which affects the output voltage.

Next, we insert the noise models (PSDs) of all noise sources into the circuit. The thermal noise of a resistor can be modeled as either a voltage noise in series with the resistor or a current noise in parallel with the resistor (see *Appendix 1, Device Equations*). The two models are essentially equivalent. For example, the current noise model can be equivalently converted to the voltage noise model by Thévenin's theorem (see *Appendix 2, Section 2, Thévenin's theorem and Norton's theorem*). We use the voltage noise model for this analysis (Figure A4.6f). Note that the signal voltage source (v_i) is considered a short circuit in the noise analysis. The PSD of the resistor thermal noise (voltage) is:

$$v_n^2(f) = 4 \cdot K \cdot T \cdot R$$

As the noise PSD is a function of frequency, it is convenient to use the (magnitude) frequency response of the *RC* circuit to calculate the output noise power. The magnitude frequency response of the first-order *RC* circuit is given by (see *this appendix, Section 2, RC circuits – transient and frequency response*):

$$|H(j \cdot \omega)| = \sqrt{\frac{1}{R^2 \cdot C^2 \cdot \omega^2 + 1}} = \sqrt{\frac{1}{(2 \cdot \pi)^2 \cdot R^2 \cdot C^2 \cdot f^2 + 1}}$$

The output noise power is:

$$v_{\text{no}}^2 = \int_0^\infty v_n^2(f) \cdot |H(j \cdot \omega)|^2 \cdot df = \int_0^\infty \frac{4 \cdot K \cdot T \cdot R}{(2 \cdot \pi)^2 \cdot R^2 \cdot C^2 \cdot f^2 + 1} \cdot df = \frac{K \cdot T}{C}$$

Hint: The following integration formula is useful in the above derivation:

$$\int \frac{1}{1 + x^2} \cdot dx = \arctan(x) + C$$

The RMS of the output noise is given by:

$$v_{no}(\text{RMS}) = \sqrt{\frac{K \cdot T}{C}}$$

This is the well-known *KT/C noise* – the noise RMS only depends on the capacitance value (and temperature). Intuitively, this is because an increase of resistance raises the thermal noise PSD but at the same time lowers the −3 dB bandwidth of the *RC* filter.

KT/C noise is an important type of circuit noise in many circuit systems, such as switched-capacitor sampling circuits. Question 31 is an example of *KT/C* noise in a switched-capacitor track-and-hold (T/H) circuit.

Example 2: Noise of a simple common-source amplifier

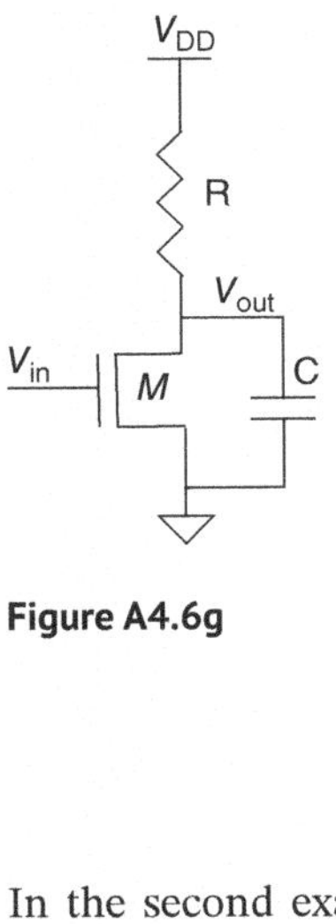

Figure A4.6g

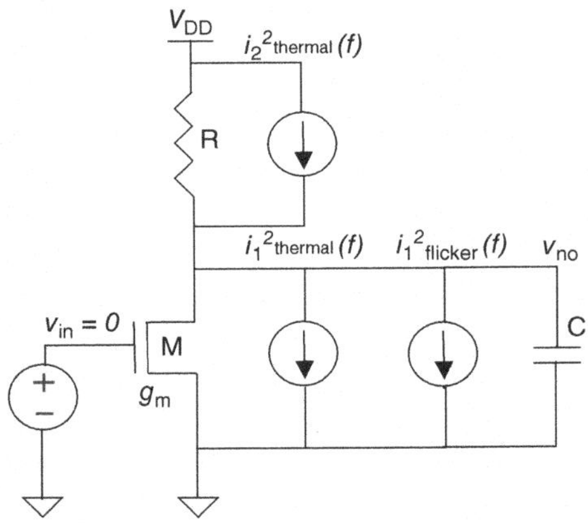

Figure A4.6h

In the second example, we will calculate the output voltage noise of a simple common-source amplifier as shown in Figure A4.6g. The load resistor and capacitance of the amplifier are R and C, respectively. Assume $R \ll r_o$, where r_o is the output resistance of transistor M.

Three noise sources, $i^2_{2_{\text{thermal}}}(f)$, $i^2_{1_{\text{thermal}}}(f)$, and $i^2_{1_{\text{flicker}}}(f)$, are added to the circuit (Figure A4.6h), representing thermal noise of resistor R, thermal noise of transistor M, and flicker noise of transistor M, respectively (see *Appendix 1, Device Equations*). The PSD of these noise sources are:

$$i^2_{2_{\text{thermal}}}(f) = \frac{4 \cdot K \cdot T}{R}$$

$$i^2_{1_{\text{thermal}}}(f) = \frac{8 \cdot K \cdot T \cdot g_m}{3}$$

$$i^2_{1_{\text{flicker}}}(f) = \frac{\text{KF} \cdot I_D^{\text{AF}}}{(C'_{\text{ox}})^2 \cdot L \cdot W \cdot f}$$

To determine the total output noise voltage, we can evaluate the contribution of each independent noise source and combine them using the *superposition theorem* (see *Appendix 2, Section 3, The superposition theorem*). A more convenient way is to apply *Thévenin's theorem* (see *Appendix 2, Section 2, Thévenin's theorem and Norton's theorem*) and transform the circuit into a noise voltage source driving a *RC* filter.

Figure A4.6i is the small-signal circuit diagram for noise analysis purpose. Notice that the input signal is assumed to be zero in the noise analysis and therefore there is no signal current at the drain of transistor M. The output resistance (r_o) of transistor M is neglected, as the load resistance (R) dominates the overall output resistance of the amplifier. The Thévenin's equivalent circuit is shown in Figure A4.6j. Here we treat capacitor C as an external load, and the internal circuit can be modeled as an equivalent noise voltage source (v_n^2) and an output resistance (r_{out}).

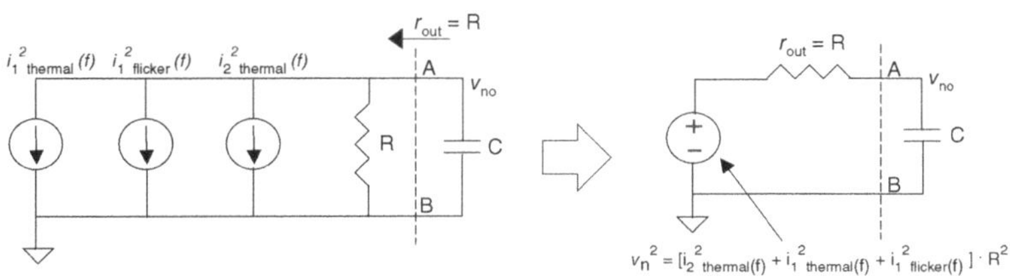

Figure A4.6i-j

The equivalent noise voltage can be derived by calculating the output voltage when the two output terminals (node A and node B) are open circuited (i.e., without any load device):

$$v_n^2(f) = \left[i^2_{2_{\text{thermal}}}(f) + i^2_{1_{\text{thermal}}}(f) + i^2_{1_{\text{flicker}}}(f) \right] \cdot R^2$$

It is important to note that uncorrelated noise sources (voltage or current) must be summed by their mean-square values—such as power (V^2) or PSD (V^2/Hz)—rather than their instantaneous magnitudes, owing to their random, independent nature.

And the output resistance is the resistance looking into the two terminals of the circuit in Figure A4.6i. Independent current sources are treated as open circuit, and independent voltage source treated as short circuit. This yields:

$$r_{out} = R$$

The equivalent circuit in Figure A4.6j is essentially a noise voltage source driving a first-order *RC* filter, and we can use the same method as shown in Example 1 to calculate the output voltage noise:

$$v_{no}^2 = \int_0^{\infty} v_n^2(f) \cdot |H(j \cdot \omega)|^2 \cdot df = \int_0^{\infty} \frac{\left[i_{2_{thermal}}^2(f) + i_{1_{thermal}}^2(f) + i_{1_{flicker}}^2(f) \right] \cdot R^2}{(2 \cdot \pi)^2 \cdot R^2 \cdot C^2 \cdot f^2 + 1} \cdot df$$

This yields:

$$v_{no}^2 = \frac{K \cdot T}{C} + \int_0^{\infty} \frac{\left[\frac{8 \cdot K \cdot T \cdot g_m}{3} + \frac{KF \cdot I_D^{AF}}{(C'_{ox})^2 \cdot L \cdot W \cdot f} \right] \cdot R^2}{(2 \cdot \pi)^2 \cdot R^2 \cdot C^2 \cdot f^2 + 1} \cdot df$$

The RMS value of the output noise is:

$$v_{no}(RMS) = \sqrt{\frac{K \cdot T}{C} + \int_0^{\infty} \frac{\left[\frac{8 \cdot K \cdot T \cdot g_m}{3} + \frac{KF \cdot I_D^{AF}}{(C'_{ox})^2 \cdot L \cdot W \cdot f} \right] \cdot R^2}{(2 \cdot \pi)^2 \cdot R^2 \cdot C^2 \cdot f^2 + 1} \cdot df}$$

The total output noise contains a KT/C component (generated by load resistor *R*), but also depends on the load resistance value as well as the parameters of transistor M1 (e.g., g_m, W, L, C'_{ox}, KF, and so on).

Noise analysis is important in circuit systems such as data converters, low-noise amplifiers, and precision references. Question 10 studies input-referred noise in a folded cascode op-amp. Question 23 investigates the impact of resistor thermal noise in current mirrors. Question 31 explores thermal noise in switched-capacitor circuits.

Section 7. Basic digital circuits

CMOS inverters are the most basic building blocks of CMOS digital logic systems. More complex logic elements, such as NAND gates, NOR gates, RS latches, *D* flip-flops are mostly based on inverters. A good understanding of inverters is important for studying digital logic circuits. In this section, we provide three examples examining some of the basic characteristics of CMOS inverters.

Example 1: CMOS inverter rise/fall times and propagation delays

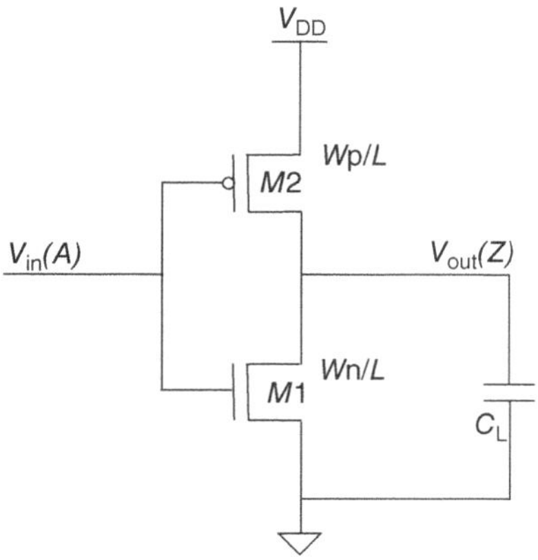

Figure A4.7a

A CMOS inverter as shown in Figure A4.7a consists of an NMOS transistor (M1) and a PMOS transistor (M2). C_L represents the total load capacitance. W_p and W_n are the widths of the PMOS and NMOS transistors, respectively. Both transistors have the same length, usually the minimum length allowed in the process technology. The logic relationship between the input A and output Z is:

$$Z = \overline{A}$$

This is an *inverting* function, also known as *complementary* or *negated* function. Table A4.7a is the truth table of the inverter.

Table A4.7a

Input (A)	**Output (Z)**
0 (Logic Low – 0V)	1 (Logic High – VDD)
1 (Logic High – VDD)	0 (Logic Low – 0V)

The most important electrical characteristics for an inverter are rise and fall times, propagation delays, and switching point. In this example, we analyze the rise/fall times and propagation delays of CMOS inverters. The switching point will be examined in the next example.

Figure A4.7b and Figure A4.7c illustrate the circuit model for estimating the rise/fall times and propagation delays of an inverter. Transistors M1 and M2 are modeled as switches, S_n and S_p, with on-resistance, R_n and R_p, respectively.

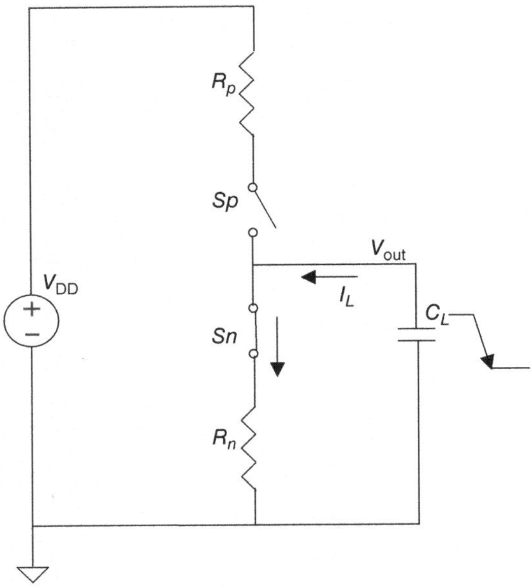

Figure A4.7b

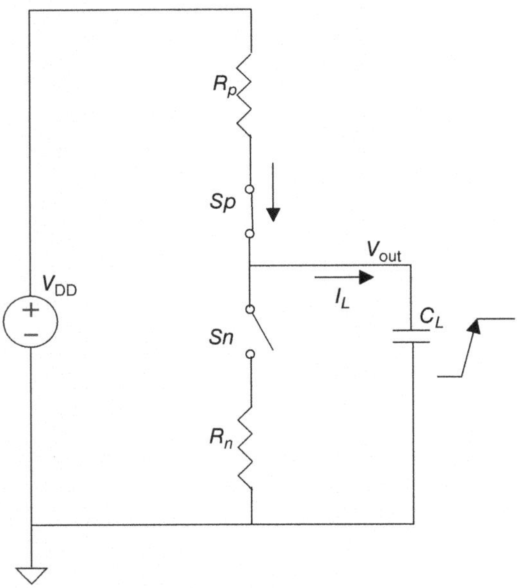

Figure A4.7c

Assume initially the output of the inverter is at VDD and the input is at 0 V (ground). When the input voltage transitions from 0 to VDD, transistor M1 is turned on, and transistor M2 is turned off (Figure A4.7b). The load capacitance C_L is discharged from VDD to ground through the on-resistance of M1. The fall time is determined by the time constant of the *RC* circuit formed by R_n and C_L. Similarly, the rise time is determined by the time constant of the *RC* circuit formed by R_p and C_L (Figure A4.7c).

We need to determine the values of R_n and R_p to calculate the time constant associated with the rise and fall transitions of the inverter. This evaluation, however, is more complex than it initially appears.

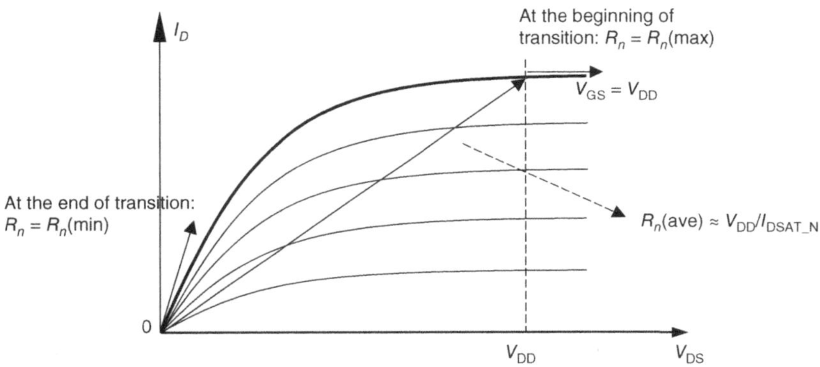

Figure A4.7d

To simplify the analysis, let us assume the input transitions from 0V to VDD instantaneously. Transistor M1, at the beginning of the output transient response, is in saturation region (i.e., $V_{GS} = V_{DS} = \text{VDD}$), and M2 shuts off. M1 behaves like a current source and the output resistance is high. With the output voltage (the V_{DS} of M1) drops from VDD towards ground, M1 transitions from saturation to triode region, and the output resistance of M1 reduces towards that of a linear resistor. This is illustrated in Figure A4.7d.

So the R_n changes during the fall time of the inverter and it is difficult to model the effective on-resistance as a fixed value. One estimation method is to use the DC resistance of M1 when both its V_{GS} and V_{DS} are equal to VDD to represent the *average* resistance during the transition (Figure A4.7d):

$$R_n = \frac{\text{VDD}}{\frac{K_{pn} \cdot W_n}{2 \cdot L} \cdot (\text{VDD} - V_{\text{THN}})^2}$$

Similarly, the R_p can be estimated by:

$$R_p = \frac{\text{VDD}}{\frac{K_{pp} \cdot W_p}{2 \cdot L} \cdot (\text{VDD} - V_{\text{THP}})^2}$$

where K_{pn} and K_{pp} are the transconductance parameters of NMOS and PMOS transistors, respectively. V_{THN} and V_{THP} are the threshold voltage of NMOS and PMOS transistors, respectively.

With the on-resistances of M1 and M2 modeled, we now determine the transition time characteristics. Figure A4.7e illustrates the definition of rise/fall times (T_r and T_f) and propagation delays (T_{dHL} and T_{dLH}) of an inverter. Note that propagation delays are measured at the switching point of the input and output of the inverter, usually around 50% of the signal swing (VDD/2), while the rise and fall times are measured from 10% to 90% of the swing.

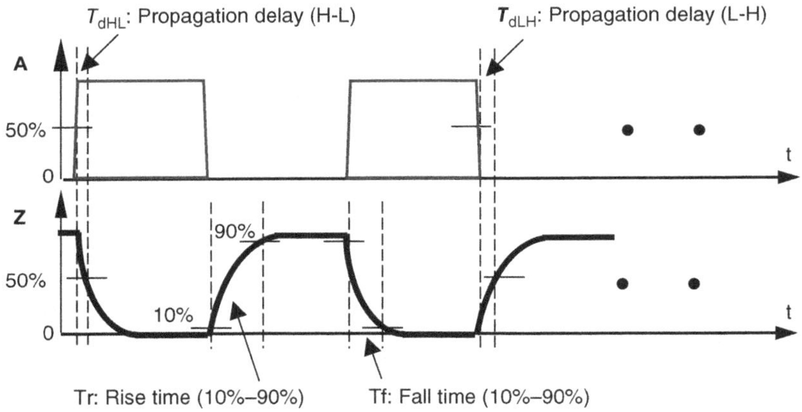

Figure A4.7e

Knowing the time constants of the inverter for output rising and falling transitions, we can use the equations of first-order *RC* circuits transient response (see *Appendix 4, Section 2, RC circuits – transient and frequency response*) to calculate the rise/fall times and propagation delays.

The propagation delay when an inverter output goes from low to high is:

$$T_{\text{dLH}} \approx 0.7 \cdot R_p \cdot C_L$$

The propagation delay when an inverter output goes from high to low is:

$$T_{\text{dHL}} \approx 0.7 \cdot R_n \cdot C_L$$

As a sanity check, it takes one time constant (i.e., $R_n \cdot C_L$, or $R_p \cdot C_L$) to reach 63.2% of its final steady-state value.

The rise time (10%–90%) of the inverter is:

$$T_r \approx 2.2 \cdot R_p \cdot C_L$$

The fall time (10%–90%) of the inverter is:

$$T_f \approx 2.2 \cdot R_n \cdot C_L$$

Example 2: CMOS inverter switching point

In the analysis of Example 1, we have assumed that the input to the inverter can change state instantaneously. In reality, the input signal has certain rise/fall times. During the input transition from 0V to VDD, for example, M1 goes through three operating regions sequentially: cutoff, saturation, and triode (linear). At the same time, M2 in turn goes through triode (linear), saturation, and cutoff. The change in the operating regions of M1 and M2 as a function of the input voltage is shown in Figure A4.7f.

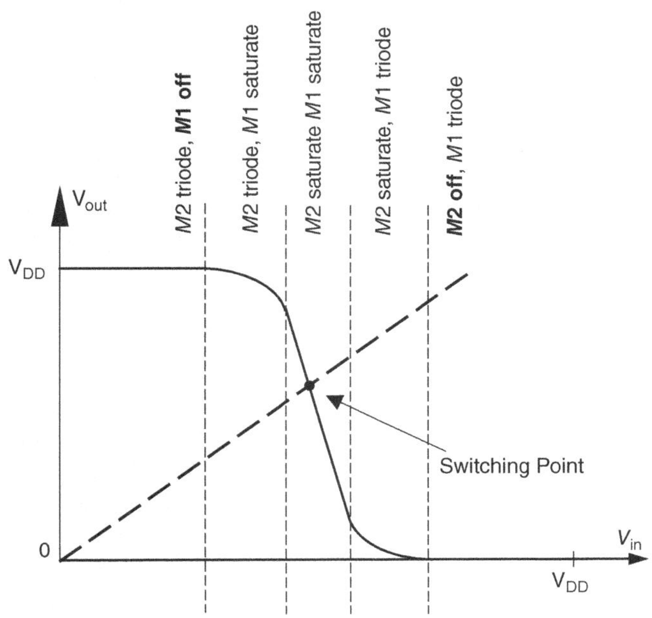

Figure A4.7f

As illustrated in Figure A4.7f, there exists an input voltage range over which both M1 and M2 are in saturation. The inverter essentially works like an amplifier in this region, and there is a DC current flowing through M1 and M2 during this transition period (usually short). The *switching point* is defined as the input voltage at which the output voltage equals that of the input. As both transistors are in saturation at the switching point, we have the following equation:

$$I_{D1} = \frac{K_{pn} \cdot W_n}{2 \cdot L} \cdot (V_{SW} - V_{TH})^2 = I_{D2} = \frac{K_{pp} \cdot W_p}{2 \cdot L} \cdot (\text{VDD} - V_{SW} - V_{TH})^2$$

This yields:

$$V_{SW} = \frac{\sqrt{\dfrac{K_{pp} \cdot W_p}{K_{pn} \cdot W_n}} \cdot \text{VDD} + \left(1 - \sqrt{\dfrac{K_{pp} \cdot W_p}{K_{pn} \cdot W_n}}\right) \cdot V_{TH}}{1 + \sqrt{\dfrac{K_{pp} \cdot W_p}{K_{pn} \cdot W_n}}}$$

where V_{SW} is the switching point of the inverter. For simplicity, the threshold voltages of both NMOS and PMOS devices are assumed to be equal in magnitude and are represented by V_{TH}.

Usually we would prefer to set the switching point at the middle of the voltage swing, i.e., VDD/2. This results in the following relationship:

$$\frac{W_n}{W_p} = \frac{K_{pp}}{K_{pn}}$$

It is sometimes referred to as a "β-matched" condition (β is the device transconductance parameter, which is defined as $K_p \cdot W/L$). For example, if $K_{pn} = 3 \cdot K_{pp}$, then making the width of the PMOS three times that of the NMOS, i.e., $W_p = 3 \cdot W_n$, would keep the switching point at VDD/2.

Note that a "β-matched" inverter would also have equal (symmetrical) rise and fall times if we ignore the difference between the threshold voltage of NMOS and PMOS transistors.

Due to the high gain of the inverter around its switching point, a slightly higher or lower input voltage around this point will drive the inverter output to logic low or logic high. In practice, due to large inverters offset and circuit noise, a margin is added to the switching point when defining the acceptable logic low and logic high voltages.

Example 3: CMOS inverter input and output capacitance

In a digital circuit system, most logic gates, such as inverters, drive the input of subsequent logic gates. The parasitic capacitance of transistors often dominates the total load capacitance.

As shown in Figure A4.7g, the major parasitic capacitances of the inverter in Figure A4.7a include the gate capacitances and the drain depletion capacitances of both NMOS and PMOS transistors (see *Appendix 1, Section 1, MOSFET (long-channel) equations*). Let us evaluate the equivalent input capacitance first.

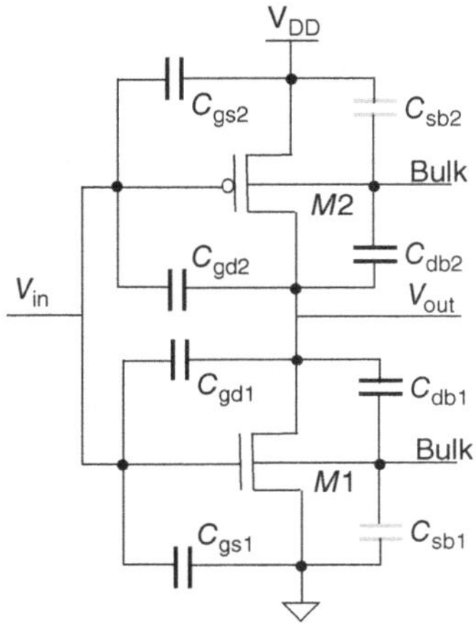

Figure A4.7g

The input capacitance of the inverter includes gate-to-source capacitance, C_{gs}, and gate-to-drain capacitance, C_{gd}, of both M1 and M2. Because C_{gd} is connected between the input and output of the inverter, Miller effect takes place (see *Appendix 2, Section 4, Miller's theorem*). Assuming the gain of the inverter is A_v, the total input capacitance of the inverter can be written as:

$$C_{in} = C_{gs1} + C_{gs2} + \left(C_{gd1} + C_{gd2}\right) \cdot (1 - A_v)$$

Unlike that of a linear amplifier, voltage transition of an inverter is a large signal nonlinear process: both the gain and gate capacitance change during the transition. This makes estimating the input capacitance a difficult task, especially the effective Miller capacitance. Simply put, when the input changes from 0V to VDD, the output of the inverter changes from VDD to 0V. From a large signal point of view, this results in a gate-to-drain capacitance that is twice its actual value. But C_{gd} of both M1 and M2 changes value during the transition when M1 and M2 experience different operating regions (cutoff, saturation, and triode). For example, C_{gd} could be as small as the gate-to-drain overlap capacitance while the transistor is in saturation region, but could be as large as half of the total gate capacitance in triode region (see *Appendix 1, Section 1, MOSFET (long-channel) equations*). Because of this reason, sometimes we use the sum of the total gate capacitances of both the NMOS and PMOS devices as a quick, rough estimation of the input capacitance and rely on circuit simulations for more accurate results:

$$C_{in} \approx C_{ox1} + C_{ox2}$$

where C_{ox1} and C_{ox2} are the total gate capacitances of M1 and M2, respectively.

Similarly, the output capacitance of the inverter can be written as:

$$C_{out} \approx C_{db1} + C_{db2} + C_{gd1} + C_{gd2}$$

where C_{db1} and C_{db2} are the drain-to-bulk (depletion) capacitances of M1 and M2.

And again, both C_{gd} and C_{db} change during the transitions, making an accurate estimation of the output capacitance difficult. One common practice is to ignore the depletion capacitances (C_{db1} and C_{db2}) and compensate for the capacitance loss by using the total gate capacitance (larger than the sum of C_{gd1} and C_{gd2}) as a starting point for manual estimations:

$$C_{out} \approx C_{ox1} + C_{ox2}$$

More studies and discussions on CMOS inverters can be found through Question 37, Inverter-based ring oscillators, Question 38, Inverter optimization of propagation delay and duty cycle distortion, Question 39, Designing buffers for driving long *RC* lines, and Question 40, Inverter switching current and power consumption.

General References

1 Allen PE and Holberg DR. *CMOS Analog Circuit Design*. 3rd ed. New York (NY): Oxford University Press; 2012.
2 Baker RJ. *CMOS: Circuit Design, Layout, and Simulation*. 3rd ed. Hoboken (NJ): Wiley/IEEE Press; 2010.
3 Boylestad RL. *Introductory Circuit Analysis*. 13th ed. [place unknown]: Pearson; 2015.
4 Gray PR, Hurst PJ, Lewis SH and Meyer RG. *Analysis and Design of Analog Integrated Circuits*. 6th ed. Hoboken (NJ): Wiley; 2024.
5 Nise NS. *Control Systems Engineering*. 8th ed. [place unknown]: Wiley; 2020.

Index

Wiley The manufacturer's authorized representative according to the EU General Product Safety Regulation is Wiley-VCH GmbH, Boschstr. 12, 69469 Weinheim, Germany, e-mail: Product_Safety@wiley.com.

Printed and bound by CPI Group (UK) Ltd, Croydon, CR0 4YY
24/08/2026
02201835-0001